Futures

Futures: Imagining Socioecological Transformation brings together leading scholars to explore how we might know, enact and struggle for, the conjoined social and ecological transformations we need to achieve just and sustainable futures. The question of transformation, and how it might be achieved, is explored across a variety of topics and geographical sites, and through heterodox analytical and theoretical approaches, in a collective effort to move beyond a form of critique that hands down judgements, to one that brings new ideas and new possibilities to life. Chapters are lively and original engagements with concrete situations that sparkle with creativity. Together, they add up to an impressive study of how to live, and what to struggle for, in the complex socioecological landscapes of the Anthropocene. This book was previously published as a special issue of the *Annals of the Association of American Geographers*.

Bruce Braun is a Professor of Geography at the University of Minnesota, USA, and a specialist in environmental thought and politics. Current research includes geosocial formations, green urbanism, and apparatuses of government in an age of climate change. His books include *The Intemperate Rainforest: Nature, Culture and Power on Canada's West Coast* (2002), *Political Matter: Technoscience, Democracy and Public Life* (with Sarah Whatmore, 2010) and *Remaking Reality: Nature at the Millennium* (with Noel Castree, 1998).

Futures

Imagining socioecological transformation

Edited by
Bruce Braun

Routledge
Taylor & Francis Group
LONDON AND NEW YORK

First published 2016
by Routledge
2 Park Square, Milton Park, Abingdon, Oxon, OX14 4RN, UK

and by Routledge
711 Third Avenue, New York, NY 10017, USA

Routledge is an imprint of the Taylor & Francis Group, an informa business

British Library Cataloguing in Publication Data
A catalogue record for this book is available from the British Library

ISBN 13: 978-1-138-66869-0

Typeset in Goudy
by diacriTech, Chennai

Publisher's Note
The publisher accepts responsibility for any inconsistencies that may have arisen during the conversion of this book from journal articles to book chapters, namely the possible inclusion of journal terminology.

Disclaimer
Every effort has been made to contact copyright holders for their permission to reprint material in this book. The publishers would be grateful to hear from any copyright holder who is not here acknowledged and will undertake to rectify any errors or omissions in future editions of this book.

Contents

Citation Information

The chapters in this book were originally published in the *Annals of the Association of American Geographers*, volume 105, issue 2 (March 2015). When citing this material, please use the original page numbering for each article, as follows:

Chapter 1
Futures: Imagining Socioecological Transformation—An Introduction
Bruce Braun
Annals of the Association of American Geographers, volume 105, issue 2 (March 2015) pp. 239–243

Chapter 2
The Future of Environmental Expertise
Rebecca Lave
Annals of the Association of American Geographers, volume 105, issue 2 (March 2015) pp. 244–252

Chapter 3
Knowing Climate Change, Embodying Climate Praxis: Experiential Knowledge in Southern Appalachia
Jennifer L. Rice, Brian J. Burke, and Nik Heynen
Annals of the Association of American Geographers, volume 105, issue 2 (March 2015) pp. 253–262

Chapter 4
Temporalities in Adaptation to Sea-Level Rise
Ruth Fincher, Jon Barnett, and Sonia Graham
Annals of the Association of American Geographers, volume 105, issue 2 (March 2015) pp. 263–273

Chapter 5
Translating Climate Change: Adaptation, Resilience, and Climate Politics in Nunavut, Canada
Emilie Cameron, Rebecca Mearns, and Janet Tamalik McGrath
Annals of the Association of American Geographers, volume 105, issue 2 (March 2015) pp. 274–283

Chapter 6
Environmental Politics After Nature: Conflicting Socioecological Futures
Becky Mansfield, Christine Biermann, Kendra McSweeney, Justine Law, Caleb Gallemore, Leslie Horner, and Darla K. Munroe
Annals of the Association of American Geographers, volume 105, issue 2 (March 2015) pp. 284–293

Chapter 7
The Place and Time of the Political in Urban Political Ecology: Contested Imaginations of a River's Future
Ryan Holifield and Nick Schuelke
Annals of the Association of American Geographers, volume 105, issue 2 (March 2015) pp. 294–303

Chapter 8
Toward an Interim Politics of Resourcefulness for the Anthropocene
Kate Driscoll Derickson and Danny MacKinnon
Annals of the Association of American Geographers, volume 105, issue 2 (March 2015) pp. 304–312

Chapter 9
Climate Change and the Adaptation of the Political
Joel Wainwright and Geoff Mann
Annals of the Association of American Geographers, volume 105, issue 2 (March 2015) pp. 313–321

Chapter 10
A Manifesto for Abundant Futures
Rosemary-Claire Collard, Jessica Dempsey, and Juanita Sundberg
Annals of the Association of American Geographers, volume 105, issue 2 (March 2015) pp. 322–330

Chapter 11
The Art of Socioecological Transformation
Harriet Hawkins, Sallie A. Marston, Mrill Ingram, and Elizabeth Straughan
Annals of the Association of American Geographers, volume 105, issue 2 (March 2015) pp. 331–341

Chapter 12
These Overheating Worlds
Kendra Strauss
Annals of the Association of American Geographers, volume 105, issue 2 (March 2015) pp. 342–350

Chapter 13
When Horses Won't Eat: Apocalypse and the Anthropocene
Franklin Ginn
Annals of the Association of American Geographers, volume 105, issue 2 (March 2015) pp. 351–359

Chapter 14
Imaginaries of Hope: The Utopianism of Degrowth
Giorgos Kallis and Hug March
Annals of the Association of American Geographers, volume 105, issue 2 (March 2015) pp. 360–368

Chapter 15
On the Possibilities of a Charming Anthropocene
Holly Jean Buck
Annals of the Association of American Geographers, volume 105, issue 2 (March 2015) pp. 369–377

Chapter 16
Banking Spatially on the Future: Capital Switching, Infrastructure, and the Ecological Fix
Noel Castree and Brett Christophers
Annals of the Association of American Geographers, volume 105, issue 2 (March 2015) pp. 378–386

Chapter 17
Biomimetic Futures: Life, Death, and the Enclosure of a More-Than-Human Intellect
Elizabeth R. Johnson and Jesse Goldstein
Annals of the Association of American Geographers, volume 105, issue 2 (March 2015) pp. 387–396

Chapter 18
Agro-Ecology and Food Sovereignty Movements in Chile: Sociospatial Practices for Alternative Peasant Futures
Beatriz Cid Aguayo and Alex Latta
Annals of the Association of American Geographers, volume 105, issue 2 (March 2015) pp. 397–406

Chapter 19
School Gardens as Sites for Forging Progressive Socioecological Futures
Sarah A. Moore, Jeffrey Wilson, Sarah Kelly-Richards, and Sallie A. Marston
Annals of the Association of American Geographers, volume 105, issue 2 (March 2015) pp. 407–415

Chapter 20
From Incremental Change to Radical Disjuncture: Rethinking Everyday Household Sustainability Practices as Survival Skills
Chris Gibson, Lesley Head, and Chantel Carr
Annals of the Association of American Geographers, volume 105, issue 2 (March 2015) pp. 416–424

Chapter 21
Transforming Household Consumption: From Backcasting to HomeLabs Experiments
Anna R. Davies and Ruth Doyle
Annals of the Association of American Geographers, volume 105, issue 2 (March 2015) pp. 425–436

Notes on Contributors

Beatriz Cid Aguayo is a researcher in the Department of Sociology and Anthropology at the University of Concepción, Chile.

Jon Barnett is a Professor in the School of Geography at the University of Melbourne, Australia.

Christine Biermann is an Assistant Professor in the Department of Geography at the University of Washington, USA.

Bruce Braun is a Professor in the Department of Geography, Environment, and Society at the University of Minnesota, USA.

Holly Jean Buck is a PhD candidate in the Department of Development Sociology at Cornell University, USA.

Brian J. Burke is an Assistant Professor in the Goodnight Family Sustainable Development Department at Appalachian State University, USA.

Emilie Cameron is an Assistant Professor in the Department of Geography and Environmental Studies at Carleton University, Canada.

Chantel Carr is a researcher in the Australian Centre for Cultural Environmental Research at the University of Wollongong, Australia.

Noel Castree is a Professor in the Department of Geography & Sustainable Communities at the University of Wollongong, Australia.

Brett Christophers is a Senior Lecturer in the Institute for Housing and Urban Research at Uppsala University, Sweden.

Rosemary-Claire Collard is an Assistant Professor in the Department of Geography, Planning and Environment at Concordia University, Canada.

Anna R. Davies is a Professor in the Department of Geography at Trinity College Dublin, Ireland.

Jessica Dempsey is an Assistant Professor in the Department of Geography at the University of British Columbia, Canada.

Kate Driscoll Derickson is an Assistant Professor in the Department of Geography, Environment, and Society at the University of Minnesota, USA.

Ruth Doyle is a PhD candidate in the Department of Geography at Trinity College Dublin, Ireland.

Ruth Fincher is a Professor in the School of Geography at the University of Melbourne, Australia.

Caleb Gallemore is a Lecturer in the Department of Geography and Environmental Studies at Northeastern Illinois University, USA.

Chris Gibson is a Professor in the Australian Centre for Cultural Environmental Research at the University of Wollongong, Australia.

Franklin Ginn is a Lecturer in the Institute of Geography, School of Geosciences at the University of Edinburgh, UK.

Jesse Goldstein is an Assistant Professor in the Department of Sociology at Virginia Commonwealth University, USA.

Sonia Graham is a Lecturer in the School of Social Sciences at the University of New South Wales, Australia.

Harriet Hawkins is a reader in the Department of Geography at Royal Holloway University of London, UK.

Lesley Head is a Professor in the School of Geography at the University of Melbourne, Australia.

Nik Heynen is a Professor in the Department of Geography at the University of Georgia, USA.

Ryan Holifield is an Associate Professor in the Department of Geography at the University of Wisconsin–Milwaukee, USA.

Leslie Horner is a researcher in the Department of Geography at the Ohio State University, USA.

Mrill Ingram is a researcher in the School of Geography and Development at the University of Arizona, USA.

Elizabeth R. Johnson is a research fellow in the Department of Geography at the University of Exeter, UK.

Giorgos Kallis is a Professor in the Department of Geography at Universitat Autónoma de Barcelona, Spain.

Sarah Kelly-Richards is a PhD candidate in the School of Geography and Development at the University of Arizona, USA.

Alex Latta is an Associate Professor in the Department of Global Studies at Wilfrid Laurier University, Canada.

Rebecca Lave is an Associate Professor in the Department of Geography at Indiana University, USA.

Justine Law is an Assistant Professor of Environmental Studies at Denison University, USA.

Geoff Mann is a Professor in the Department of Geography, Centre for Global Political Economy at Simon Fraser University, Canada.

Becky Mansfield is a Professor in the Department of Geography at the Ohio State University, USA.

Hug March is a research fellow in the Internet Interdisciplinary Institute at Universitat Oberta de Catalunya, Spain.

Sallie A. Marston is a Professor in the School of Geography and Development at the University of Arizona, USA.

Janet Tamalik McGrath is an independent scholar in Ottowa, Canada.

Danny MacKinnon is a Professor in the Centre for Urban and Regional Development Studies at Newcastle University, UK.

Kendra McSweeney is a Professor in the Department of Geography at the Ohio State University, USA.

Rebecca Mearns is a research student at Nunavut Sivuniksavut, Canada.

Sarah A. Moore is an Assistant Professor in the Department of Geography at the University of Wisconsin, USA.

Darla K. Munroe is a Professor in the Department of Geography at the Ohio State University, USA.

Jennifer L. Rice is an Assistant Professor in the Department of Geography at the University of Georgia, USA.

Nick Schuelke is a doctoral student in the Department of Geography at the University of Wisconsin–Milwaukee, USA.

Elizabeth Straughan is Honorary Research Associate in the School of Geographical and Earth Science at the University of Glasgow, UK.

Kendra Strauss is an Assistant Professor in the Department of Sociology & Anthropology at Simon Fraser University, Canada.

Juanita Sundberg is an Associate Professor in the Department of Geography at the University of British Columbia, Canada.

Joel Wainwright is an Associate Professor in the Department of Geography at the Ohio State University, USA.

Jeffrey Wilson is a Lecturer in the School of Geography and Development at the University of Arizona, USA.

Futures: Imagining Socioecological Transformation—An Introduction

Bruce Braun

Department of Geography, Environment, and Society, University of Minnesota

> Action cannot be delayed because time does not flow from the present to the future—as if we had to choose between scenarios, hoping for the best—but as if time flowed from what is coming ("l'avenir" as we say in French to differentiate it from "le future") to the present. Which is another way to consider the times in which we should live as "apocalyptic." Not in the sense of the catastrophic (although it might be that also), but in the sense of the revelation of things that are coming toward us.
>
> —Latour (2013, 12)

Paul Klee's famous painting *Angelus Novus* (1920) has over the years been subject to numerous interpretations, perhaps most famously by Benjamin ([1940] 1968) in his *Theses on the Philosophy of History*. In Walter Benjamin's reading, the "Angel of History" faces the past and is propelled into a future to which its back is turned, caught up in a storm of progress that piles wreckage upon wreckage at the angel's feet. Although the angel would like to make the past whole, the violence of the storm renders him powerless to do so. More recently, Bruno Latour (2013) has provided a different reading: It is precisely our modern tendency to face backward while continuously attempting to flee the horror and destruction of the past, he suggests, that renders us unable to see or to face the looming catastrophe that we have inadvertently allowed to pile up behind our backs. It is only now, with the advent of the Anthropocene, that we have begun to see the full horror not of the past but of the shape of things to come.[1]

According to Latour's provocative reading, we have entered a period in which our experience of time has changed. If in modernism time was seen to flow from the present to the future, today we increasingly experience time coming toward us, from the future to the present. Across a diverse interdisciplinary literature we find two related themes: (1) that socioecological transformations are coming, although the form and shape of what is to come is not easily predicted; and (2) that socioecological changes are also necessary, if we are to avoid the catastrophic futures that appear to be coming toward us. The language of "planetary boundaries" and "tipping points" has become commonplace, as has the recognition that it is human activity—or the activities of certain humans—that threatens to push earth systems into unpredictable and potentially turbulent states. Without social, political, and economic change, the future of our species and many others is seen to be at risk.

For geographers this presents a host of new challenges. Although a robust critical literature has done much to help us understand how we have arrived at this juncture and has highlighted the deeply uneven geographies of socioecological change, it has been far less successful at imagining and engendering just and sustainable alternatives to existing political, economic, and ecological practices. In part, this reflects a particular critical stance that remains wedded to a linear conception of time, namely, that by understanding the past we might be able to anticipate and shape the future. Certainly the past still matters when we face the future (see Collard, Dempsey, and Sundberg, this issue), but in the Anthropocene the shape of things to come is increasingly seen to be nonanalogous with what existed in the past (indeed, this assumption informs many political technologies of security, from preparedness planning to resilience). We might also locate the discipline's inability to imagine alternatives in its widespread and principled rejection of prescriptive or normative approaches to political or ecological change and its widely held suspicion of utopian thought. Whether these stances remain viable today is a matter for debate.

The theme for this special issue ("Futures: Imagining Socioecological Transformation") strongly reflects the current historical juncture and emerged from discussion with members of the *Annals* editorial board during the summer of 2012. It was proposed to the Association of American Geographers (AAG) publications committee in fall of that year and in the following months, 220 scholars or teams of scholars submitted abstracts that were read, evaluated, and ranked by reviewers drawn from the journal's editorial board. Thirty-one contributors were invited to submit full manuscripts for review by referees, from which the twenty articles appearing in this issue were selected.

Many more promising abstracts were submitted than could be accepted, an encouraging sign that the discipline, and nature–society studies more generally, has begun to pivot toward the future and to consider how socioecological transformation might be imagined, anticipated, or enacted.

Thematic relevance and a clear contribution to geographical scholarship were emphasized in the selection process, but authors were not constrained by genre or style. The result is articles ranging from traditional case studies to more experimental forms like manifestos. Due to tight word limits, trade-offs between empirical detail and conceptual or theoretical development were often necessary, adding further to the diversity of the articles included. But although the special issue is topically diverse, it is less geographically diverse than we had hoped. The vast majority of abstracts were submitted by institutionally situated scholars in North America, Europe, or Australia and often focused on sites in those regions. Regrettably, several invited submissions that focused on other regions failed to materialize. In part this geographical distribution reflects global disparities that demand our continued attention, both in terms of where and by whom academic scholarship is produced and disseminated but also in terms of where socioecological innovation is seen to be occurring and which regions have the resources to anticipate and adapt to future changes. One need look no further than the frenzied activity of architects, planners, artists, and city officials busily designing experimental buildings, infrastructures, and environments in New York City, which during the Bloomberg era was increasingly promoted as a laboratory for resilient urbanism (Wakefield and Braun 2014). This does not mean that communities in the Global South are less resourceful or less creative. Indeed, in the face of looming ecological change, many important social and political innovations are emerging there and diffusing widely. This will likely accelerate in coming years as cities in the Global South become the global cities of the future. Yet with a few notable exceptions like the innovative transportation systems developed in the Colombian cities of Bogotá and Medellín, these experiments often proceed with far fewer resources, are less grandiose in scale and flashy in style, and struggle to gain the attention of global media and North American or European scholars. Further, with the shift of focus in climate change scholarship from global mitigation to local adaptation, researchers have often turned to those communities in which they are institutionally situated. There is much that is salutary in the publicly engaged and participatory research that has resulted, but in the aggregate it runs the risk of narrowing the geographical focus of research. At the very least, the uneven representation of world regions in this issue suggests the need to reflect more broadly on the geography of scholarship in the Anthropocene, as well as where we locate or seek important socioecological changes.

Although the articles in the issue are diverse, they nevertheless address a number of common themes. Many of the articles are concerned with marginalized populations, issues of social and environmental justice, and the question of who is empowered and authorized to imagine and define socioecological futures. For Derickson and MacKinnon, these are critical questions in the context of historically marginalized black communities in Atlanta and take us to the heart of how and by whom climate justice is defined and struggled for. For Cameron, Mearns, and McGrath, the imagination of climate futures in Nunavut must grapple with the practice and politics of translation, as notions of resilience, adaptation, and climate change can have very different meanings in English and Inuktitut. To not attend to these differences in the context of a broader global shift from mitigation toward adaptation risks blindly—or perhaps deliberately—perpetuating a colonial present. Likewise, Collard, Dempsey, and Sundberg, in their hard-hitting yet hopeful "A Manifesto for Abundant Futures," argue that "postnatural" environmentalism cannot simply turn its back on the past in the name of composing socioecological futures but must also reckon with the "ruins" of existing colonial and capitalist relations, confront onto-epistemological difference, and recognize the autonomy of nonhuman others. Although in the Anthropocene time might come toward us from the future, each of these articles insists that the past continues to haunt the present and that ignoring this leaves us poorly equipped to address crucial social differences in how we face the future.

Other articles take up questions that are at once epistemological and political. Increasingly we inhabit a world in which models and scenarios generate policy and inform politics, nowhere better illustrated than in the multiple reports of the Intergovernmental Panel on Climate Change (IPCC). With every model and scenario, and with every socioecological proposal that takes wing from them, pressing questions arise about what counts as expertise and how and by whom knowledge is constructed, stabilized, trusted, or contested. Although the "climategate" affair is now seen as an unwarranted attack on climate change scientists by climate change deniers, there is a growing interest in the diversity of ways that socioecological change is known and experienced. In her contribution, Lave charts the changing landscape of scientific expertise in and beyond the university and asks searching questions about how, where, and by whom knowledge is likely to be generated in coming years. For their part, Rice, Burke, and Heynen consider the role of experiential knowledge in how Appalachian residents respond to climate change, and Fincher, Barnett, and Graham contrast different understandings of time and temporality between Australian climate scientists and state planners on the one hand and residents of several ocean-side communities who have incommensurate understandings of temporality on the other. What emerges in these and other articles is a growing recognition of multiple ways of knowing socioecological change and the need to navigate among them.

Perhaps one of the most notable developments in recent years has been the growing importance of literature, film, and art for how individuals and groups figure, imagine, or anticipate what is to come. Indeed, as future scenarios have taken greater significance in public life, the line separating science from fiction has become increasingly blurred, reflected in the emerging genre of "cli-fi," the proliferation of apocalyptic novels and film, and the reemergence and redeployment of utopian and dystopian fiction. For Strauss, the recent works of Canadian novelist Margaret Atwood and American writer Barbara Kingsolver open opportunities for exploring political imaginaries of climate change, blending utopian and dystopian imaginations of socioecological transformation while developing explicitly feminist themes. For Ginn, the Hungarian filmmaker Béla Tarr's *The Turin Horse* takes us to the very limits of human life and reveals an underplayed political dimension to apocalyptic visions insofar as they force viewers to feel the conditions of the Anthropocene. Kallis and March, on the other hand, mobilize the science fiction of Ursula Le Guin to provide critical perspective on the growing "degrowth" movement, and Hawkins and her coauthors explore how art projects provoke new ways of understanding environmental change. Although the intertwining of aesthetics and politics has often been viewed with suspicion, each of these articles takes up in different ways Yusoff's (2010) suggestion that in the face of climate change aesthetics can function as a valuable space "that configures the realm of what is possible" in any politics.

One of the appeals of art and aesthetics in the Anthropocene is that they are seen to open much needed space for creativity and experimentation and to work on an affective and emotional level often seen as lacking in the more technocratic language of policy and planning. This is consistent with the wider turn to experimentation and invention in recent nature–society scholarship and the view that experimentation is necessary to engender new forms of knowing and dwelling in and with human and nonhuman others (see, e.g., Gibson-Graham and Roelvink 2010; Whatmore 2013; Lorimer and Driessen 2014). Experimentation here is understood less in terms of the hypothesis testing of positivist science than in terms of the staging of encounters through which new possibilities for politics might emerge along with new political subjectivities. The conceit here is that by working with materials and material others we can potentially open ourselves to events that surprise and disturb, allowing the world to force thought (cf. Deleuze and Guattari 1996; Stengers 2005). Understood thus, the turn to experimentation is often attached to a decidedly optimistic politics that stands in contrast to gloomy prognostications of catastrophic climate change and earth systems pushed past tipping points. Contributors to this special issue advance our understanding of the relation between experimentation and socioecological transformation across a wide range of sites and practices. Gardening projects in struggling schools are examined for their ability to disrupt standardized educational practices and present opportunities to deepen relationality and solidarity in spaces of education (Moore et al.), agroecological and architectural experiments are explored for their ability to reawaken a sense of wonder and an ethic of care (Buck), and experimental backcasting methods are explored as a means by which future environmental goals can be brought forward into the present to inform how households are managed (Davies and Doyle). Other articles identify novel sites and practices that might provide key resources for socioecological transformation. Approaching household practices somewhat differently than Davies and Doyle, Australian geographers Gibson, Head, and Carr investigate everyday household sustainability practices as the potential bases for survival skills that might be needed to endure catastrophic change. Others locate political and ecological alternatives in sites and practices as diverse as biomimicry (Johnson and Goldstein), peasant movements in Chile (Aguayo and Latta), and even financial markets (Castree and Christophers), with the latter arguing that finance capital might provide important pragmatic means to switch credit money into green infrastructures.

Throughout virtually all of the articles, we find the recognition that with the advent of the Anthropocene the separation of nature and society, and the human and nonhuman, is no longer tenable, if it ever was. This recognition does not mean that humans fully control earth processes, although for some so-called eco-pragmatists such control is deemed both possible and desirable (i.e., Brand 2010). Nor should we too quickly assume symmetry between human actions and earth processes, because despite the immense impact of human activity, we remain subject to, rather than the authors of, many large-scale events (Clark 2010). Challenging nature–society dualisms, though, means that environmentalism today faces the challenge of reimagining environmental politics after nature, a position that has gained traction among mainstream environmental nongovernmental organizations like the Breakthrough Institute and the Nature Conservancy, as well as environmental geographers and historians (i.e., Cronon 1995; Braun 2002). Whereas Collard, Dempsey, and Sundberg trace the convergence of postnatural environmentalism and neoliberal environmental governance, Mansfield and colleagues derive a somewhat different lesson from the same developments; namely, that today, environmentalism focused on external nature must be seen as only one of many competing environmentalisms organized around different socioecological projects. For Mansfield and her coauthors, socioecological futures will be shaped through struggles not over what is natural but over what natures are to be produced, by whom, and for whose benefit. For their part, Wainwright and Mann suggest the need to return to notions of natural history that include humanity, such as found in the writings of Karl Marx, as part of the dual task of challenging existing and emerging forms of planetary management and engendering a climate politics today that can create a just and livable planet in the future.

For Wainwright and Mann, the urgency of their intervention is heightened by the character of contemporary political life in the United States, Europe, and elsewhere, which they argue prevents rather than encourages the radical responses required by climate change. This concern is articulated most forcefully by Swyngedouw (2007), who argues that the proliferation of carefully managed means of participation in political institutions and processes is symptomatic of the absence of the political, insofar as the political is defined by the recognition of and antagonism between radically different political demands. Swyngedouw's claim that much environmental management today is "postpolitical" has generated considerable debate and is taken up by numerous authors in this issue. Holifield and Schuelke, for example, turn to science studies, pragmatist philosophy, and aesthetics to trace the political trajectories of environmental matters of concern in Milwaukee, Wisconsin. In doing so, they suggest that the political cannot be reduced only to those exceptional moments in which prevailing orders are disrupted by egalitarian challenges but must include the diverse processes by which knowledge is produced, desires constituted, and claims made. Derickson and MacKinnon and Wainwright and Mann also seek to extend, contest, or complicate Swyngedouw's thesis, whereas other articles, like those of Collard, Dempsey, and Sundberg and Cameron, Mearns, and McGrath, can be seen to bring into view precisely what Swyngedouw thinks is missing: the recognition of, and struggle for, radically different possible socioecological futures. Noting this might be an appropriate way to close this introduction, not only because the dangers of technocratic management and its democratic deficits are very real but also because in the pages of this issue there are significant and meaningful disagreements about how the antagonisms that constitute "the political" in the Anthropocene should be understood, and from where, by whom, and in what ways transformations toward a just and livable planet should be generated. We invite you to read this special issue with this sense of disagreement in mind, with the hope that the convergences, tensions, and gaps in the issue are themselves productive of new directions for thought, practice, and politics in the Anthropocene.

Acknowledgments

Many thanks to the countless reviewers who devoted precious hours to reading and reviewing submissions and to members of the editorial board for their help shaping the issue. Jessica Lehman and Sara Nelson provided much needed editorial and administrative assistance. Thanks also to Morgan Adamson and Kate Driscoll Derickson for thoughtful comments and suggestions.

Note

1. The term *Anthropocene* was suggested by atmospheric chemist Paul Crutzen in 2002 as the name

for the present geological epoch in which humanity has become a significant geological agent. It is often accompanied by the more controversial concept of planetary boundaries, within which a safe operating space for humanity can be defined (Rockström et al. 2009).

References

Benjamin, W. [1940] 1968. Thesis on the philosophy of history. In *Illuminations*, trans. H. Zohn, 253–64. New York: Schocken Books.

Brand, S. 2010. *Whole earth discipline: Why dense cities, nuclear power, transgenic crops, restored wildlands, and geoengineering are necessary*. New York: Penguin.

Braun, B. 2002. *The intemperate rainforest: Nature, culture and power on Canada's west coast*. Minneapolis: University of Minnesota Press.

Clark, N. 2010. *Inhuman nature: Sociable life on a dynamic planet*. London: Sage.

Cronon, W. 1995. The trouble with wilderness; or, getting back to the wrong nature. In *Uncommon ground: Rethinking the human place in nature*, ed. W. Conon, 69–90. New York: Norton.

Deleuze, G., and F. Guattari. 1996. *What is philosophy?* New York: Columbia University Press.

Gibson-Graham, J. K., and G. Roelvink. 2010. An economic ethics for the Anthropocene. *Antipode* 41:320–46.

Latour, B. 2013. Telling friends from foes at the time of the Anthropocene. Paper written for the EHESS-Center Koyré-Sciences Po symposium "Thinking the Anthropocene," Paris.

Lorimer, J., and C. Driessen. 2014. Wild experiments at the Oostvaardersplassen: Rethinking environmentalism in the Anthropocene. *Transactions of the Institute of British Geographers* 39:169–81.

Rockström, J., W. Steffen, K. Noone, A. Persson, F. Chapin, E. Lambin, T. Lenton, et al. 2009. A safe operating space for humanity. *Nature* 461:472–75.

Stengers, I. 2005. Deleuze and Guattari's last enigmatic message. *Angelaki* 10 (2): 151–67.

Swyngedouw, E. 2007. Impossible "sustainability" and the postpolitical condition. In *The sustainable development paradox: Urban political ecology in the United States and Europe*, ed. R. Krueger and D. Gibbs. New York: Guilford.

Wakefield, S., and B. Braun. 2014. Governing the resilient city. *Environment and Planning D: Society and Space* 32 (1): 4–11.

Whatmore, S. 2013. Earthly powers and affective environments: An ontological politics of flood risk. *Theory, Culture and Society* 30 (7–8): 33–50.

Yusoff, K. 2010. Biopolitical economies and the political aesthetics of climate change. *Theory, Culture and Society* 27 (2): 73–99.

The Future of Environmental Expertise

Rebecca Lave

Department of Geography, Indiana University

Many have observed the decline of scientific authority over the last three decades, for reasons ranging from the toxic legacies of Cold War science (Beck 1992), to the current commercialization and privatization of knowledge production (Mirowski 2011), to the success of social constructivist critique (Latour 2004). Whatever the cause(s), it seems clear that the relationship among academia, the military, and state and economic elites is shifting once again. A new regime of knowledge production is emerging (Pestre 2003) in which academia carries significantly less clout than it has over the previous half-century, and broadly legitimate knowledge claims are increasingly developed outside of the academy. These changes carry obvious implications for the future of academic legitimacy and institutions. The implications for environmental and social justice are less obvious, although perhaps even more important, as the ways in which knowledge is vetted and the questions investigated (or ignored) shift. In this article, I use exploration of the changing relationship between academic and extramural knowledge producers to lay out potential futures for the production of environmental knowledge. I argue that although academics have been notably unsuccessful in challenging private-sector, commercialized environmental knowledge claims, we are increasingly successful in leveraging our remaining authority to enable the democratization of knowledge production to intellectually and politically progressive ends.

过去三十年来，由于冷战科学的毒害遗绪（Beck 1992）、当前知识生产的商业化与私有化（Mirowski 2011），以至社会建构论的成功批判（Latour 2004），诸多人已观察到，科学权威因而衰落。不论肇因为何，学术、军事和国家暨经济菁英之间的关係，似乎的确再度有所转变。新的知识生产体制正在浮现（Pestre 2003），其中学术较前半世纪而言，明显拥有更加无足轻重的影响力，而广泛具有正当性的知识主张，亦逐渐在学术圈之外构成。这些转变，为学术正当性及机构的未来，带来了显着的意涵。当知识调查的方式与探问（或忽略）的问题转变之时，该转变对环境与社会公义的意涵，儘管或许是更为重要的，但却较不明显。我将在本文中，运用学术和外界知识生产者之间改变中的关係之探讨，藉此展示环境知识生产的潜在未来。我主张，儘管学术界在挑战私部门、商业化的环境知识主张上相当不成功，但我们仍在採取自身仅有的权威、使知识生产得以民主化以追求知识和政治上的激进目标方面，逐渐取得成功。

Son muchas las personas que han observado la declinación de la autoridad científica durante las últimas tres décadas, debido a razones que incluyen desde los tóxicos legados de la ciencia de la Guerra Fría (Beck 1992), la actual comercialización y privatización de la producción de conocimiento (Mirowski 2011), hasta el éxito de la crítica social constructivista (Latour 2004). Sean cuales fueren las causas, parece claro que la relación entre academia, lo militar y el estado y las élites económicas está de nuevo cambiando. Un nuevo régimen de producción de conocimiento está tomando forma (Pestre 2003), en el que la academia comanda significativamente menor influencia de la que ejercía en la anterior media centuria, y ampliamente legítimos reclamos sobre el conocimiento son cada vez más desarrollados fuera de la academia. Estos cambios llevan consigo obvias implicaciones para el futuro de la legitimidad académica y para sus instituciones. Las implicaciones para la justicia ambiental y social son menos obvias, aunque quizás sean mucho más importantes, al propio tiempo que cambian el modo como se evalúa el conocimiento y las preguntas investigadas (o ignoradas). En este artículo utilizo la exploración de la cambiante relación entre los generadores de conocimiento, académicos y extramurales, para entrever futuros potenciales sobre producción de conocimiento ambiental. Argumento que aunque los académicos han tenido poco éxito al retar las reivindicaciones de conocimiento ambiental del sector privado comercializado, sí tenemos cada vez más éxito en hacer valer nuestra restante autoridad para habilitar la democratización de la producción de conocimiento hacia fines intelectual y políticamente progresistas.

The lament that scientific authority is losing currency has become increasingly common, as authors with quite different epistemological and political commitments offer pessimistic narratives of decline. To Beck (1992) that decline is a product of public consciousness of the disasters enabled by environmental science during the twentieth century. To Latour (2004) the fault lies not in academics' failures but in our success, as the dissemination of social constructivist critiques fatally undermined science's claim to epistemological superiority. Mirowski (2011), by contrast, lays the blame squarely on structural causes: the neoliberalization of science policy and practice.

Regardless of how science's reduced authority is explained, we are clearly experiencing a deep shift in the character and organization of knowledge production. In the phrasing of historian of science Pestre (2003), we are witnessing the birth of a new *science regime*: a distinctive reconfiguration of the relations among scientists and state, military, and economic elites. Pestre's research demonstrates that the historical and geographic particularities of science regimes profoundly affect not just scientists' conditions of production but also the content they produce and how that content circulates.

The key to the science regime concept is the acknowledgment that, contrary to the post–World War II ideal of an independent academia protected from external pressures, scientific practice has always been shaped by external forces. Pestre (2005) points out that "for at least the last five centuries, what we now call scientific knowledge—be it characterized as pure or applied . . . has been of crucial interest to the political and economic powers since knowledge led to the material and social techniques of control" (30). Thus in any given place, historical moment, and institutional context, scientists have distinctive working conditions and relations with military, state, and economic elites that lead to the production of characteristic kinds of knowledge. As Pestre (2005) puts it, "The fact that Galileo successively worked in a university, then for the Republic of Venice, and finally at the court of the Grand Duke of Tuscany is of direct relevance to the kind of knowledge he produced" (30).

What particular configuration of scientific institutions, markets, and state power characterizes the new science regime, whose arrival Beck, Latour, and Mirowski call out?[1] We have some indications from current practices (Lave 2012b), but at the very least it seems clear that academia will carry significantly less clout than it has over the previous half-century and that broadly legitimate knowledge claims will increasingly be developed outside of the academy.

The stakes in this shift are clear for academics: Our intellectual authority and our jobs are increasingly imperiled. What the new science regime might mean for society more broadly is not so obvious, though. The evolving horizontality of knowledge production is already shifting how knowledge is vetted, which questions are investigated, and which are ignored. Given the well-documented failures of the peer review system and the legacies of deep discrimination in universities and funding agencies, some of these changes could be very welcome. Other characteristics of the new science regime seem to promise far less progressive results, however. The key point is that the new science regime carries potentially very different consequences for the environment and social justice: the domination of commercially interested corporate knowledge claims or the democratization of knowledge production to more socially and environmentally just ends.

To illustrate both the risk of this current moment of transition and its emancipatory potential, I focus here on the relationship between environmental knowledge claims produced inside and outside the university. I first present a typology of forms of knowledge production outside the academy and then describe in more detail two increasingly prominent forms of extramural knowledge: free-range science and citizen science. These comparative case studies, although brief, illustrate the democratizing potential that forms the Janus face of the loss of authority that Beck, Latour, and Mirowski bemoan. These cases also demonstrate that although academics have been strikingly unable to debunk commercially driven extramural science, we have been notably successful in lending our authority, however reduced, to the production of knowledge for progressive, justice-focused ends. I thus argue that we might achieve more of our political and intellectual goals by embracing the progressive aspects of our reduced authority than by fighting its erosion.

Producing "Amateur" Environmental Knowledge

Today, it is broadly assumed that legitimate knowledge claims about the environment can be produced only by professional scientists who have undergone lengthy academic training and work with expensive equipment in specially designed settings. People who do not have or do those things yet claim to produce environmental knowledge are commonly viewed as dabblers, or even crackpots. Yet this taken-for-granted, commonsense privileging of professional knowledge producers is relatively recent. Until the late nineteenth century, full-time environmental knowledge producers and those we would today regard as amateurs were closely associated. This is not to downplay very real differences of class, race, and gender[2] between those able to dedicate their working hours to the pursuit of knowledge and those who pursued such passions in the interstices of their lives. But historians of science consistently demonstrate that what today appear to us as totally different animals used to regard each other as part of the same species. In the United States and Europe they worked with the same texts, pooled both knowledge and collections under the auspices of local scientific societies, and disseminated their results in the same publications (Allen 1976; Keeney 1992; Secord 1994; Knell 2000; Kohler 2006). Part- and full-time environmental knowledge producers depended on each other. The latter required part-time scientists to collect specimens, and some of the former received crucial financial support and up-to-date information from full-time scientists (Reingold 1976; Keeney 1992).

Beginning in the middle of the nineteenth century, however, full-time environmental scientists began to shut out their former colleagues through a new emphasis on certification via university degrees and on membership in professional societies to which amateurs could not belong. This process of professionalization included the foundation of journals open only to professional authors, and a critical switch from field to lab science, which exponentially increased entry costs to scientific discussion: Without expensive equipment and lab space, one could no longer actively participate in the shaping of legitimate knowledge claims (Reingold 1976; Keeney 1992; Livingstone 2003). The relocation of scientific practice to places open only to the elite was a particularly effective tactic of exclusion; as historian of science Secord (1994) and others have noted, a substantial amount of scientific practice in the United States and Europe had previously been located in venues that were open across classes, such as coffee shops or the pubs that were the center of amateur botany in England.[3]

Professionalization was accompanied by appropriation. The scientific canon at that point was a product of collective work, built on the contributions of full- and part-time knowledge producers. By the end of the nineteenth century, it had been repackaged as the product of white, male, Western professionals. Not only did full-time knowledge producers no longer cite the broader community whose labor and research enabled their findings, but they also reduced public access to the still expanding collections of specimens that that public had been instrumental in collecting (Secord 1994). By the end of the nineteenth century, only professional knowledge producers were considered legitimate members of the scientific community (Reingold 1976).

Part-time knowledge producers did not disappear in a puff of professionalizing smoke; they continued their scientific work. They were excluded from academic scientists' consideration until roughly 1980, however, when there was a striking resurgence in attention to nonacademic environmental knowledge production within academic, policy, and corporate spheres. This attention ranged from Canada's decision that environmental management policies must incorporate First Nation knowledge (Nadasdy 1999), to big pharma reemphasizing bioprospecting and biopiracy to prop up the sagging drug pipeline (Brush and Stabinsky 1996; Shiva 2001; Hayden 2003). Environmental justice activism took off across the United States (Brown and Mikkelsen 1990; Ottinger 2013), and the Convention on Biological Diversity asserted that preserving biodiversity required preserving indigenous communities and their ecological knowledge. During this same time period, academic interest in nonacademic environmental knowledge production exploded; a basic Web of Knowledge search shows that 98.8 percent of all articles on what we might term *extramural knowledge* production have been published since 1980.[4] Thus across the board in academia, in corporations, and in policy there has been a dramatic increase in interest in environmental knowledge claims produced outside the academy (Moore et al. 2011).

The academic literature on extramural knowledge describes a wide variety of knowledge producers who I group into seven different categories (see Lave [2012b] for more detail). *Amateur scientists* are by definition not professional scientists, but they have a recognized role and, in some cases, a high level of legitimacy within academic communities, particularly in the fields of archaeology, astronomy, and paleontology. *Indigenous knowledge* is typically used to describe the agro-ecological knowledge of marginalized peoples in the developing world; *local knowledge* is most frequently used to describe similar agro-ecological knowledge when held by marginalized white people in developed countries. *Crowdsourcing* describes the much newer practice of distributed communal problem solving, which ranges from high-profile mathematics (e.g., the contest to improve Netflix's recommendation algorithm) to the drudgery of electronic piecework on Mechanical Turk. For physical scientists, *citizen science* describes the legions of unpaid volunteers who now form the backbone of many large studies, particularly in ecology and astronomy, whereas social scientists use the term to describe activist knowledge production in the environmental justice

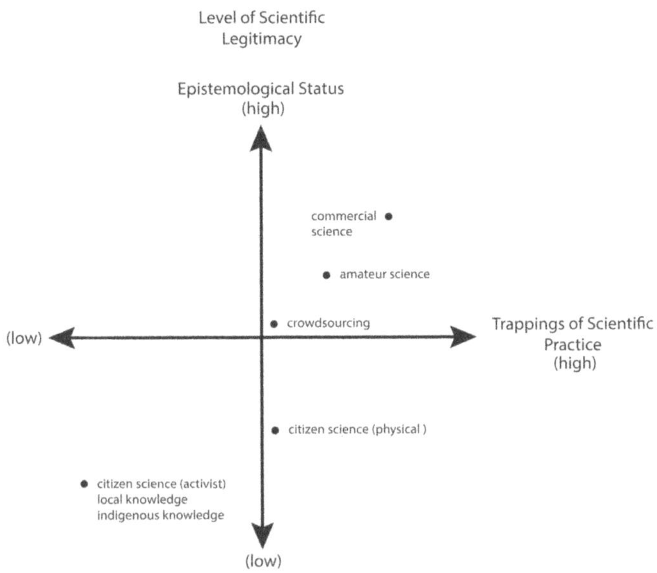

Figure 1. Relative status of extramural knowledge forms.

movement (e.g., popular epidemiology or street science). Finally, *commercial science* describes the practice of knowledge production in the private sector, from the famously academic atmosphere of Bell Labs to contract research organizations to the free-range scientists I discuss in the following section.[5]

These forms of extramural knowledge production are assigned notably different epistemological statuses by academics (Figure 1). Whereas amateur scientists are usually given credit for their discoveries, indigenous knowledge is typically viewed as static and as held, rather than as developed over time through painstaking research.[6] The stature of extramural knowledge producers does seem to be growing, however, as academic legitimacy declines. In the following section I compare two increasingly visible types of extramural science: free-range science, a variant of commercial science, and citizen science.

Free-Range Science Versus Citizen Science

To explore what the growing horizontality of knowledge production can tell us about the future of environmental expertise, it is useful to compare what initially appear to be two extremes of extramural knowledge production: commercial science in stream restoration and range ecology that has catalyzed a shift in focus for academic researchers and citizen science (in physical scientists' use of that term), which appears to have no influence on academic practice beyond the availability of unpaid labor.

Free-Range Science

Not all commercial science has the budget and prestige of Bell Labs. Some looks more like local knowledge for a profit: low-budget, informal, strongly regional, and without the trappings of professionalized laboratories and tools. I refer to this as *free-range science*, a form of extramural environmental knowledge production in which scientific authority stems from market take-up rather than prestigious academic degrees, funding, or employment. The distinctive character of free-range science stems not from the educational qualifications (or lack thereof) of its practitioners but from the commercial roots of their scientific legitimacy.

The strong currents of anti-intellectualism in U.S. culture and the autonomy of U.S. state and federal agencies in determining accepted qualifications for public contracts clearly play a role in the rise of free-range science. Yet I think it would be a mistake to assume that the phenomenon is therefore restricted geographically. As Lawless and Williams (2010) note, a similar dynamic is unfolding in forensic science in England. Further, the increasing emphasis on contract research and public–private partnerships for state-based research within the European Union and in many other developed countries seems likely to expand the range of free-range scientists beyond their current U.S. base. For the purposes of this article, however, I focus on two free-range scientists in the United States—Dave Rosgen and Allan Savory—who have not only become widely recognized scientific experts but have shifted the course of their respective fields of environmental science: fluvial geomorphology and range science. I describe each of them in turn and then lay out the notable commonalities between them.

In 1985, Dave Rosgen left the U.S. Forest Service and embarked on a career as an independent stream restoration consultant, researcher, and teacher. The basic principle of his natural channel design (NCD) approach is to work with nature to create stable channels rather than to use concrete and riprap to force channels into place. With his cowboy hat, boots, enormous belt buckle, and unapologetically rural turns of phrase, Rosgen appears far more like a rancher than a scientist. Yet despite his Western self-presentation and his relative lack of formal scientific training,[7] he has been extremely successful, becoming the most widely legitimate scientific expert on river restoration in the United States. He has taught more than 15,000 people how to restore streams, and his classification system and NCD method are institutionalized in many federal agencies (including the U.S. Environmental Protection Agency, the U.S. Forest Service, and the Natural Resource Conservation Service), as well as in dozens of states (Lave 2012a).

Rosgen has disseminated his work via self-published textbooks, conference presentations, and a series of short courses. State and federal agency support has been central to his success, as many project managers require use of NCD to win bids for publicly funded stream restoration work. In an unprecedented show of support, the U.S. Forest Service purchased a copy of Rosgen's first book for every hydrologist at the agency. When the Natural Resource Conservation Service (NRCS) published the volume of the National Engineering Handbook devoted to stream restoration design, Rosgen's work was given pride of place (NRCS 2007). Perhaps most instrumentally, the U.S. Environmental Protection Agency has funded thousands of its own and state agencies' employees to attend Rosgen's short courses (Lave 2012a).

In response to Rosgen's initial success, university- and federal research agency-based scientists embarked on a disorganized but passionate campaign to debunk him and the NCD approach, denouncing both in scathing terms beginning in the mid-1990s. But Rosgen's continuing influence also pushed river scientists into applied work on restoration in attempts to counter his work. As Gary Parker, a sediment transport researcher at the University of Illinois, put it:

> [U]ntil Rosgen started actually doing things, most academics had not the slightest intention of getting involved in an applied project and saying how things ought to be done. They spent most of their time telling people what they couldn't do. . . . Rosgen has had the effect of moving the entire field of river geomorphology more in the direction of thinking about how to solve practical problems. (interview with author, 26 June 2006)

Allan Savory's career is strikingly parallel to Rosgen's, although his influence extends much farther. Savory immigrated to the United States from Rhodesia in the early 1970s and began promoting his approach to grazing in arid and semiarid landscapes, now referred to as holistic resource management (HRM; Hadley 2000). The basic principle of HRM is that forage plants and grazing animals evolved together, so that there is no inherent reason why grazing should cause ecological problems. What we need to do, Savory argues, is not decrease stocking levels in response to desertification and ecological degradation but increase them in combination with intensive management to mimic the ways livestock moved across the landscape before humans arrived (Savory 1983). HRM's evolutionary rationale made intuitive sense to many, and Savory's call for increased stocking rates was highly appealing to ranchers and others invested in the ranching industry, as higher stocking rates can mean higher profits. Despite Savory's relative lack of formal scientific training (he holds a BS in ecology), he quickly became one of the most broadly legitimate international experts on range science and has remained so for decades.

Savory disseminates the HRM approach via short courses, self-published textbooks, and a series of nonprofit and profit organizations, from his initial nonprofit Holistic Management International to his current for-profit Savory Institute. In the United States and abroad, resource agencies and nonprofits have promoted the use of Savory's work, in part by tying improvement grants and other forms of aid to ranchers to their use of HRM. University- and research agency–based scientists have strongly critiqued Savory's work, but it has been so widely adopted that they have been unable to ignore it. Range scientists have changed their research programs to address HRM, conducting repeated field trials in attempts to debunk it. More recently, some range scientists have turned to social science approaches to explain the persistence of HRM in the face of long-term critique (Briske et al. 2011). Debates over Savory's work continue to be a prominent feature of both academic and applied range science conferences.

Savory has continued to spark controversy, arguing in a March 2013 TED talk that radically increasing stocking rates and employing his HRM approach could catalyze sufficient carbon take up to halt global warming. The talk (Savory 2013) has garnered more than 2.5 million views. Its conclusions have been touted by luminaries including Michael Pollan and rebutted in popular press outlets such as *Slate* (McWilliams 2013). Clearly, Savory has powerfully influenced the visibility and content of range science.

There are notable parallels in these cases. Both Rosgen and Savory are emphatically low-brow commercial knowledge producers. Yet they are making the claims that are most broadly viewed as legitimate, have developed the most widely applied techniques, and are the primary source of training for practitioners in their respective fields. In both river and range science, academics' boundary work (Gieryn 1999)[8] was unsuccessful, which is remarkable given the intensity of their opposition and the extent to which Rosgen and Savory eschew the typical trappings of academic science. Perhaps most notably, in both cases the content and focus of academic research shifted in response to extramural environmental knowledge claims. Fluvial geomorphology now tackles far more applied research topics, and range scientists have redirected their research first to test Savory's approach; and, more recently to consider social explanations of ecological outcomes of HRM.

What accounts for Rosgen's and Savory's success? Charisma clearly plays a role but is a far from sufficient answer: The world is littered with charismatic failures. Instead, I would point to three key commonalities. First, there are strong structural factors pushing both range management and stream restoration: Rising property values in the U.S. West (Sayre 2002) mean that the economic pressure to increase stocking rates is intense, and the U.S. Clean Water Act mandates stream restoration as a permit requirement for projects that negatively affect fluvial systems. Thus in both cases, environmental knowledge production is being driven by economic and regulatory practice. Second, both rivers and rangelands are highly complex systems, and the sciences that study them have concomitantly high levels of uncertainty. This makes it extremely difficult to evaluate the success or failure of either NCD or HRM, so their scientific opponents have been strikingly unable to dispatch them. Finally, as described earlier, the monopoly of peer review and other academic methods on vetting new knowledge claims is fading. Other forms of validation are rising, perhaps most notably market take-up.[9] Thus, one of the most influential arguments made by supporters of Savory and Rosgen is baldly economic: They must be correct or they would not be so financially successful. Academics get paid even if they are wrong; consultants do not.

What sort of new science regime does the success of free-range knowledge producers portend? Academics have been notably ineffective at delegitimizing these knowledge claims, in the end changing their own practices rather than Rosgen's or Savory's. This suggests a future of increasing marginalization for academics, sidelined by the claims of commercially interested, private-sector knowledge producers. The environmental consequences of this shift are not yet clear, however; if NCD and HRM turn out to restore natural function to streams and rangelands, then the increased legitimacy of at least some subset of commercial knowledge producers might be worth celebrating.

The Democratic Possibilities of Citizen Science

Citizen science appears to lie at the other end of the spectrum of epistemological influence from Rosgen and Savory. No one describes the multitude of unpaid volunteers as scientists. The primary controversy is not whether citizen scientists deserve academic recognition for their work but whether the data they gather have any scientific utility at all; to many academics, the work of citizen scientists is irredeemably compromised by its origin with untrained eyes and hands (e.g., Cohn 2008).

Citizen scientists are at work on a striking range of projects, from identifying galaxy types for Galaxy Zoo to parsing out hard-to-decipher field notes and records for museums of natural history via Notes from Nature; even the Kinsey Institute has gotten into the citizen science game, launching the Kinsey Reporter app for reporting sexual behavior in 2013. The Cornell Laboratory of Ornithology has been one of the strongest promoters of citizen science through programs such as eBird and Project FeederWatch; many of their volunteers not only give their time but actually pay to be involved (Dickinson, Zuckerberg, and Bonter 2010, 151).

All traces of the increasing horizontality of knowledge production so evident in the cases of Rosgen and Savory seem to vanish in citizen science, which most commonly reinforces the current hierarchy of knowledge producers: academic scientists firmly on top, with their volunteers earning a pat on the head and a virtual merit badge.[10] And yet, there are several trends that suggest that these highly skewed power relations might be shifting.

Academic environmental researchers' rapidly increasing dependence on citizen scientists to carry out large-scale studies in an era of shrinking scientific funding is worth considering. It is not clear that large ecological studies are feasible any longer without citizen scientists, and with dependence could come greater status. Also, to fortify volunteer-collected data against their colleagues' critiques and to increase the utility of those data, academic scientists are giving their volunteers more training, developing new research protocols around them, and thus building their research capacities (Kinchy, Jalbert, and Lyons 2014). That, too, could lead to more substantive participation over the long term.

Further, there are already a handful of cases where citizen scientists are setting the intellectual agenda and academics are following their lead. For example, volunteer water quality monitoring has proliferated rapidly in response to concerns over hydraulic fracturing in the Marcellus Shale play. As Kinchy and Perry (2012) note, academics at Penn State have received

National Science Foundation funding to work with those volunteer-collected data, jumping in on a project defined by volunteers in a reversal of the typical power dynamics of citizen science. Other academics are attempting to democratize knowledge production by bringing lay people into substantive roles in setting research agendas, collecting data, and interpreting its results via approaches such as participatory action research (Pain et al. 2012), participatory statistics (http://developmentbookshop.com/whocounts, last accessed 20 May 2014), and Extreme Citizen Science (http://www.ucl.ac.uk/excites, last accessed 20 May 2014). Thus, even the strongly hierarchical current practices of citizen science could be becoming more horizontal. The key factor in these democratizing trends is academics' willingness to endorse and promote the legitimacy of extramural knowledge production.

The Public Laboratory for Open Technology and Science (Public Lab) serves as both example and inspiration for how academics can substantively contribute to democratizing environmental knowledge production, blurring what has been until recently the stark distinction between activist- and academic-driven forms of citizen science.[11] Public Lab was created by a consortium of activists and academics in 2010 in large part as a response to the Deep Water Horizon oil spill. The community of critical makers associated with Public Lab develops inexpensive, carefully tested, open source environmental monitoring hardware ranging from spectrometers to aerial mapping kits, as well as the software to process the data. In Public Lab "chapters" and individually, at sites across the United States and now in many other countries, activist groups and laypeople pick research problems and investigate them with tools they assemble themselves, producing results that have been used by, among others, the U.S. Federal Emergency Management Agency in its response to Hurricane Sandy and the U.S. Environmental Protection Agency in its Superfund cleanup of the Gowanus Canal in Brooklyn, New York. Public Lab's D.I.Y. R&D platform places scientific tools for monitoring and analyzing environmental conditions at the service of community agendas, democratizing the production of environmental knowledge. It is not clear whether the potential for substantively influential citizen science embodied in organizations such as Public Lab will spill over into other currently low-status forms of extramural knowledge production (Figure 1), but it does strongly suggest that academics can play a critical role in enabling emancipatory knowledge production.

In this future, too, academia's current corner on the market for scientific legitimacy erodes but by academics' own hands. Further, the environmental and justice implications are more straightforwardly positive than in the case of free-range science.

Conclusion

I began this article with scholars' laments over the decline of scientific authority and our precarious academic future. Although I very much agree that the future of the university is precarious, I would argue that what we are seeing is not the decline of scientific authority but its redistribution, as a far wider range of knowledge producers are accorded intellectual legitimacy. It is tempting to regard any moves toward the democratization of knowledge production as progressive, but the implications of this increasingly horizontal knowledge production are not yet clear. The same trends that allow the Environmental Working Group and Public Lab to transcend accusations of amateurism also elevate Merck and Monsanto above concerns over commercial conflict of interest.

Will the new science regime devolve to corporate control of knowledge production, extending the dystopian conditions described by Sunder-Rajan (2006) and Mirowski (2011), or can we direct the changing dynamics of scientific legitimacy toward more just ends? Can academics influence the future of environmental knowledge production given the declining authority of the university?

The contrast between free-range and citizen science is instructive on this point. Academics have been strikingly ineffective at stopping the spread of commercial knowledge claims; market take-up is an increasingly powerful source of legitimacy and one that we seem ill equipped to counter (Lave 2012a). Where we seem to be far more effective is in using our legitimacy, however reduced, to enable noncommercial forms of extramural knowledge claims to gain influence. We can and should strive to shape the soul of the new science regime and to promote the progressive potential of increasingly horizontal knowledge production. We face new constraints, but we are far from helpless.

Acknowledgments

This article has benefited greatly from the thoughtful comments of four anonymous reviewers, carefully considered and prioritized by Bruce Braun. I am also grateful to Gwen Ottinger, Tom Gieryn, and the Indiana University STS Working Group for insightful comments on earlier versions of the text.

Notes

1. It might be argued that the science regime concept is implicitly too restrictive to describe the current diversification of knowledge production. Pestre's definition, however, seems to me sufficiently open-ended to describe any configuration of power relations between elites and knowledge producers, regardless of where the latter are located.
2. The history I offer here is predominantly Western and thus has obvious limitations. As Adas (1989) and others have documented, there were typically far sharper status distinctions between indigenous and Western knowledge producers under colonial governments, although Dove (2011) has noted a more benign aspect to that separation, as mutual ignorance and imagination facilitated trade relations.
3. For very helpful reviews of science & technology studies (STS) and geographic scholarship on the emplacedness of science see Henke and Gieryn (2008) and Powell (2007).
4. Search terms: amateur science, citizen science, crowdsourcing, indigenous knowledge, indigenous ecological knowledge, indigenous technical knowledge, and local knowledge.
5. I have not included regulatory science in this typology because it is publicly funded and thus not typically subject to commercial conflicts of interest (cases of agency capture aside) and because the knowledge claims produced by regulatory science are typically incorporated into the academic literature without reservations. For example, in academic writing on stream restoration, publications by researchers at the U.S. Geological Survey, the Agricultural Research Service of the U.S. Department of Agriculture, and the U.S. Army Corps of Engineers are cited as authoritative.
6. There is a substantial critical literature in anthropology and geography that challenges the epistemological assumptions shown in Figure 1. Authors such as Blaikie (1985), Braun (2002), Brosius (1997), Fairhead and Leach (1996), Mitchell (2002), and Robertson (2006) address the politics of science and the question of whose knowledge claims are awarded social legitimacy. In STS, authors such as Irwin (1995), Wynne (1996), and Collins and Evans (2007) have mounted an epistemological defense of extramural knowledge, arguing that nonacademic experts must have a place at the decision-making table to increase the intellectual robustness of scientific decision making.
7. Although Rosgen earned a PhD from the University of East Anglia in 2004, there was no course work involved in that degree, and for most of his career he had only a BS in forestry from California State University, Humboldt.
8. Gieryn's concept of boundary work describes efforts to redraw the maps of what constitutes legitimate science to define particular research subjects or approaches as unscientific and therefore illegitimate forms of knowledge production.
9. The rise of non-peer-reviewed forms for communicating knowledge claims, such as blogs, and of crowdfunded research seem likely to accelerate these trends.
10. See, for example, http://www.notesfromnature.org (last accessed 20 May 2014).
11. Data in this paragraph are drawn from personal communications with Sarah Wylie, one of Public Lab's founders, participant observation at a Public Lab workshop, and publiclab.org. There are some interesting parallels between Public Lab and Science Shops; see Leydesdorff (2005) for a useful overview of the latter.

References

Adas, M. 1989. *Machines as the measure of men: Science, technology, and ideologies of Western dominance*. Ithaca, NY: Cornell University Press.
Allen, D. E. 1976. *The naturalist in Britain*. London: Penguin.
Beck, U. 1992. *Risk society: Towards a new modernity*. Thousand Oaks, CA: Sage.
Blaikie, P. 1985. *The political economy of soil erosion in developing countries*. New York: Wiley.
Braun, B. 2002. *The intemperate rainforest: Nature, culture and power on Canada's West Coast*. Minneapolis: University of Minnesota Press.
Briske, D. D., N. Sayre, L. Huntsinger, M. Fernandez-Gimenez, B. Budd, and J. D. Derner. 2011. Origin, persistence, and resolution of the rotational grazing debate: Integrating human dimensions into rangeland research. *Rangeland Ecology & Management* 64 (4): 325–34.
Brosius, J. P. 1997. Endangered forest, endangered people: Environmentalist representations of indigneous knowledge. *Human Ecology* 25 (1): 47–69.
Brown, P., and E. Mikkelsen. 1990. *No safe place: Toxic waste, leukemia, and community action*. Berkeley: University of California Press.
Brush, S. B., and D. Stabinsky, eds. 1996. *Valuing local knowledge: Indigenous people and intellectual property rights*. Washington, DC: Island Press.
Cohn, J. P. 2008. Citizen science: Can volunteers do real research? *Bioscience* 58 (3): 192–97.
Collins, H. M., and R. Evans. 2007. *Rethinking expertise*. Chicago: University of Chicago Press.
Dickinson, J. L., B. Zuckerberg, and D. N. Bonter. 2010. Citizen science as an ecological research tool: Challenges and benefits. *Annual Review of Ecology, Evolution, and Systematics* 41:149–72.
Dove, M. 2011. *The banana tree at the gate: A history of marginal peoples and global markets in Borneo*. New Haven, CT: Yale University Press.
Fairhead, J., and M. Leach. 1996. *Misreading the African landscape: Society and ecology in a forest-savana mosaic*. Cambridge, UK: Cambridge University Press.
Gieryn, T. F. 1999. *Cultural boundaries of science: Credibility on the line*. Chicago: University of Chicago Press.
Hadley, C. J. 2000. The wild life of Allan Savory. *Rangelands* 21 (5): 6–10.

Hayden, C. 2003. *When nature goes public: The making and unmaking of bioprospecting in Mexico*. Princeton, NJ: Princeton University Press.

Henke, C., and T. F. Gieryn. 2008. Sites of scientific practice: The enduring importance of place. In *The handbook of science & technology studies*, ed. E. J. Hackett, O. Amsterdamska, M. E. Lynch, and J. Wajcman, 353–76. Cambridge, MA: MIT.

Irwin, A. 1995. *Citizen science: A study of people, expertise and sustainable development*. London and New York: Routledge.

Keeney, E. B. 1992. *The botanizers: Amateur scientists in nineteenth century America*. Chapel Hill: University of North Carolina Press.

Kinchy, A., K. Jalbert, and J. Lyons. 2014. What is volunteer water monitoring good for? Fracking and the plural logics of participatory science. *Political Power and Social Theory* 27:259–89.

Kinchy, A., and S. L. Perry. 2012. Can volunteers pick up the slack? Efforts to remedy knowledge gaps about the watershed impacts of Marcellus Shale gas development. *Duke Environmental Law & Policy Forum* 22 (1): 303–39.

Knell, S. J. 2000. *The culture of English geology, 1815–1851: A science is revealed through its collecting*. Burlington, VT: Ashgate.

Kohler, R. E. 2006. *All creatures: Naturalists, collectors, and biodiversity, 1850–1950*. Princeton, NJ: Princeton University Press.

Latour, B. 2004. Why has critique run out of steam? From matters of fact to matters of concern. *Critical Inquiry* 30:225–48.

Lave, R. 2012a. Bridging political ecology and STS: A field analysis of the Rosgen wars. *Annals of the Association of American Geographers* 102 (2): 366–82.

———. 2012b. Neoliberalism and the production of environmental knowledge. *Environment & Society: Advances in Research* 3 (1): 19–38.

Lawless, C., and R. Williams. 2010. Helping with inquiries, or helping with profits? The trials and tribulations of a technology of forensic reasoning. *Social Studies of Science* 40 (5): 731–55.

Leydesdorff, L. 2005. Science shops: A kaleidoscope of science–society collaborations in Europe. *Public Understanding of Science* 14 (4): 353–72.

Livingstone, D. N. 2003. *Putting science in its place: Geographies of scientific knowledge*. Chicago: University of Chicago Press.

McWilliams, J. 2013. All sizzle and no steak: Why Allan Savory's TED talk about how cattle can reverse global warming is dead wrong. *Slate*. http://www.slate.com/articles/life/food/2013/04/allan_savory_s_ted_talk_is_wrong_and_the_benefits_of_holistic_grazing_have.single.html (last accessed 20 May 2014).

Mirowski, P. 2011. *Science-Mart: Privatizing American science*. Cambridge, MA: Harvard University Press.

Mitchell, T. 2002. *Rule of experts: Egypt, techno-politics, modernity*. Berkeley: University of California Press.

Moore, K., D. L. Kleinman, D. J. Hess, and S. Frickel. 2011. Science and neoliberal globalization: A political sociological approach. *Theory and Society* 40 (5): 505–32.

Nadasdy, P. 1999. The politics of TEK: Power and the "integration" of knowledge. *Arctic Anthropology* 36 (1–2): 1–18.

National Resources Conservation Service. 2007. *Stream restoration design. National Engineering Handbook Part 654*. Des Moines, IA: U.S. Department of Agriculture National Resources Conservation Service.

Ottinger, G. 2013. *Refining expertise: How responsible engineers subvert environmental justice challenges*. New York: New York University Press.

Pain, R., G. Whitman, D. Milledge, and Lune Rivers Trust. 2012. *Participatory action research toolkit: An introduction to using PAR as an approach to learning, research and action*. Durham, UK: Durham University.

Pestre, D. 2003. Regimes of knowledge production in society: Towards a more political and social reading. *Minerva* 41 (3): 245–61.

———. 2005. The technosciences between markets, social worries and the political: How to imagine a better future? In *The public nature of science under assault: Politics, markets, science and the law*, ed. H. Nowotny, D. Pestre, E. Schmidt-Assman, H. Schultze-Fielitz, and H. Trute, 29–52. Berlin: Springer.

Powell, R. 2007. Geographies of science: Histories, localities, practices, futures. *Progress in Human Geography* 31 (3): 309–29.

Reingold, N. 1976. Definitions and speculations: The professionalization of science in America in the nineteenth century. In *The pursuit of knowledge in the early American republic*, ed. A. Oleson and S. C. Brown, 33–70. Baltimore: Johns Hopkins University Press.

Robertson. M. 2006. The nature that capital can see: Science, state and market in the commodification of ecosystem services. *Environment and Planning D: Society and Space* 24 (3): 367–87.

Savory, A. 1983. The Savory grazing method or holistic resource management. *Rangelands* 5 (4): 155–59.

———. 2013. How to fight desertification and reverse climate change. http://www.ted.com/talks/allan_savory_how_to_green_the_world_s_deserts_and_reverse_climate_change.html (last accessed 20 May 2014).

Sayre, N. F. 2002. *Ranching, endangered species, and urbanization in the Southwest: Species of capital*. Tucson: University of Arizona Press.

Secord, A. 1994. Science in the pub: Artisan botanists in early 19th century Lancashire. *History of Science* 32 (3): 269–315.

Shiva, V. 2001. *Protect or plunder? Understanding intellectual property rights*. London: Zed Books.

Sunder-Rajan, K. 2006. *Biocapital: The constitution of postgenomic life*. Durham, NC: Duke University Press.

Wynne, B. 1996. Misunderstood misunderstandings: Social identities and public uptake of science. In *Misunderstanding science? The public reconstruction of science and technology*, ed. A. Irwin and B. Wynne, 19–46. Cambridge, UK: Cambridge University Press.

Knowing Climate Change, Embodying Climate Praxis: Experiential Knowledge in Southern Appalachia

Jennifer L. Rice,* Brian J. Burke,† and Nik Heynen*

*Department of Geography, University of Georgia
†Goodnight Family Sustainable Development Department, Appalachian State University

Whether used to support or impede action, scientific knowledge is now, more than ever, the primary framework for political discourse on climate change. As a consequence, science has become a hegemonic way of knowing climate change by mainstream climate politics, which not only limits the actors and actions deemed legitimate in climate politics but also silences vulnerable communities and reinforces historical patterns of cultural and political marginalization. To combat this "post-political" condition, we seek to democratize climate knowledge and imagine the possibilities of climate praxis through an engagement with Gramscian political ecology and feminist science studies. This framework emphasizes how antihierarchical and experiential forms of knowledge can work to destabilize technocratic modes of governing. We illustrate the potential of our approach through ethnographic research with people in southern Appalachia whose knowledge of climate change is based in the perceptible effects of weather, landscape change due to exurbanization, and the potential impacts of new migrants they call "climate refugees." Valuing this knowledge builds more diverse communities of action, resists the extraction of climate change from its complex society–nature entanglements, and reveals the intimate connections between climate justice and distinct cultural lifeways. We argue that only by opening up these new forms of climate praxis, which allow people to take action using the knowledge they already have, can more just socioecological transformations be brought into being.

科学知识在今日，无论是用来支持或阻碍行动，皆较过往更作为气候变迁政治论述的首要架构。因此，科学成为主流气候政治用来理解气候变迁的霸权方式，而这不仅限制了在气候政治中被认可的行动者与行动，亦同时使得脆弱的社群维持沉默，并强化了文化与政治边缘化的历史模式。为了对抗此般"后政治"境况，我们透过涉入葛兰西主义政治生态学与女性主义科学研究，寻求民主化气候知识，并想像气候实践的可能性。此一框架，强调反阶层与经验性的知识形式，如何可能颠覆技术专家的治理模式。我们对阿帕拉契南部的人们进行民族志研究，以此描绘上述取径的潜能；该地人们的气候变迁知识，是根据可感知的天气效应、由超城市化所导致的地景改变，以及其所称之为"气候难民"的新移民所带来的潜在冲击。重视这种知识，可建构更为多元的行动社群、抵抗将气候变迁抽离自其复杂的社会—自然之交错，并揭露气候正义与特殊文化生活方式之间的亲密连结。我们主张，唯有透过开启此般让人们得以运用自身已拥有的知识进行行动的崭新气候实践形式，更多的公义社会生态变迁才能应运而生。

Bien que se utilice para apoyar la acción o para impedirla, más que nunca el conocimiento científico es ahora el marco primario para el discurso político sobre cambio climático. En consecuencia, la ciencia se ha convertido en una vía hegemónica para conocer el cambio climático a través de la corriente principal de la política climática, que no solo limita a los actores y acciones consideradas legítimas en política climática sino que también silencia comunidades vulnerables y refuerza patrones históricos de marginamiento cultural y político. Para combatir esta condición "pospolítica," buscamos democratizar el conocimiento climático e imaginamos las posibilidades de la práctica climática haciendo un compromiso con la ecología política gramsciana y con los estudios de ciencia feminista. Tal marco enfatiza la manera como pueden operar las formas de conocimiento antijerárquicas y experienciales para desestabilizar los modos tecnocráticos de gobernar. Ilustramos el potencial de tal enfoque por medio de investigación etnográfica con pobladores de los Apalaches meridionales, cuyo conocimiento del cambio climático se basa en los efectos percibidos del tiempo atmosférico, cambios del paisaje ocasionados por la exurbanización y los impactos potenciales de nuevos migrantes a los que ellos denominan "refugiados climáticos." Dándole valor a este conocimiento se construyen comunidades activas más diversas, se resiste la extracción del cambio climático desde su complejo entrelazamiento sociedad–naturaleza, y se revelan las íntimas conexiones entre la justicia climática y los distintos estilos culturales de vida. Sostenemos que solo abriendo estas nuevas formas de praxis climática, que le permiten a la gente actuar con el conocimiento que ya tienen, se pueden producir transformaciones socioecológicas más justas.

On 3 July 2012, North Carolina lawmakers approved House Bill 819 (HB 819), prohibiting the state's Department of Environment and Natural Resources from projecting sea level rise until 2016, thereby stalling regulatory action. Democrats attacked the bill and then-Governor Beverly Perdue declined to sign it, stating, "North Carolina should not ignore science when making public policy" (Gannon 2012). The media characterized the event as yet another antiscience tactic by climate skeptics (Glass and Pilkey 2013), and editorial pages erupted with letters written by or quoting scientific experts, explaining the physical processes that could produce up to three feet of sea level rise on North Carolina's coast by 2100 (Soucheray 2014). Supporters of HB 819 provided their own scientific interpretation, using historical analyses to predict sea level rise of only 8 inches by 2100 (Zucchino 2012), and Republican Senator David Rouzer "argued that House Bill 819 doesn't ignore the science, but rather requires that the state look at all available studies and data on the issue when it develops policies regarding sea-rise" (Gannon 2012).

Calls by those working against climate change policy to consider "all available studies" illustrate the extent to which climate politics has, like most environmental politics, become a politics of "expertise and counter-expertise" (Eden 1996, 193) and, as such, has become largely depoliticized. As Demeritt (2006, 468) argues,

> The case of climate change shows how a technically framed and expert-led politics of sound science can be debilitating. . . . The instrumental role of science in legitimating policy invites interest groups to contest political decisions by questioning science (and scientists), rather than debating the reasons for the policy itself.

This ethos means that much of mainstream climate politics seeks to influence decision making through science education and communication, based on the presumption that more relevant or "usable" science will change attitudes and behaviors (Moser 2010; McNie 2007). Swyngedouw (2013) argues that this is symptomatic of a wider "post-political" condition, which is characterized by a systematic failure to engage in debate about the fundamental moral, ethical, and economic foundations of climate change (see also Ranciere 2004; Mouffe 2005). Democracy, disagreement, direct action, and dissent have been replaced by a mode of politics where "scientific expertise [is] the foundation and guarantee for properly constituted politics/policies" (Swyngedouw 2010, 217). This mode of technocratic governing is "science-driven and expert-oriented" in its pursuit of technological (over social) solutions to climate change (Bäckstrand 2004, 696), and it becomes exclusionary through its dismissal of alternative ways of knowing.

We argue that mainstream attempts to respond to climate change are fundamentally limited because they do not directly confront the hegemonic conditions under which science dominates climate change politics. We are imagining a notion of "hegemony" here to be closely aligned with Gramsci (1971) but also Williams (1977) and Hall (1986). Swyngedouw (2010, 228–29) argues that a truly political politics "requires foregrounding and naming different socio-environmental futures and recognizing conflict, difference and struggle over the naming and trajectories of these futures." Climate politics urgently needs to be repoliticized to include more democratic debate and argument based in a wider discussion of values, norms, and experiences. This requires, among other things, a discussion of the politics of knowledge underpinning our current political condition.

To accomplish this, we use Gramscian political ecology and feminist science studies to show how nonhierarchical knowledge production can foster a more inclusive and egalitarian climate praxis. Empirically, we draw on our research in southwestern North Carolina, a region at the edge of politically conservative southern Appalachia, to solicit a more diverse set of epistemologies on climate change that are gained through experiential, placed-based, and nonscientific knowledge. Valuing people's everyday experiences of climate change and diverse ways of knowing climate (even when they might be scientifically imprecise) provides the possibility for people and communities to act on climate change through the knowledge and experience they already have. We show how in southern Appalachia, nonscientific ways of knowing include intimate experiences and family histories of changing weather, concerns for landscape changes associated with rapid exurban development, and threats to culturally valued, historical ways of life by an influx of climate migrants. Recognizing experiential ways of knowing has three advantages for climate praxis: It enables and legitimates more diverse communities of action, it resists the extraction of climate change from its complex socionatural entanglements that have place-based meaning, and it provides culturally specific understandings of what is at stake with climate justice.

Repoliticization Through Antihierarchical Knowledge Production

As scientific practice has become the hegemonic way of knowing climate, the processes through which knowledge hierarchies are produced deserve careful examination (Ekers, Loftus, and Mann 2009). This is especially the case because the individuals and communities most often marginalized through expert-only politics are often more likely to experience negative consequences from socioecological changes (Smith 2006). Confronting scientific hegemony is, therefore, a problem of both theory (understanding the causes and ramifications of hegemony) and practice (providing a pathway for those most vulnerable to participate). For these reasons, we claim that climate praxis, rooted in identifying and valuing alternative epistemologies, can counter post-political dynamics and help us imagine more just socioecological transformations. By contrast, socioecological transformations that retrench expert authority have the potential to reproduce the status quo.

Scientific and technical expertise construct knowledge about climate change as a globally scaled, mathematically modeled, carbon-centric problem (Demeritt 2001). Because science understands climate through earth system models, physical proxies, and technological observations, "we have universalized the idea of climate change, detached it from its cultural setting" (Hulme 2008, 9). This reification of technoscientific expertise often marginalizes nonscientific ways of knowing climate change that are meaningful to nonexperts (Geoghegan and Leyson 2012; Leyshon and Geoghegan 2012). The result is, according to Hulme (2008, 5), a "politics that is seemingly powerful, yet fundamentally fragile" because it depends on expanding the hegemony of positivist science and Western (neo)liberalism. Even where scientists have created more place-based, socially relevant information, science often provides the standard by which knowledge is judged valuable or legitimate.

The normatively, theoretically, and politically problematic hegemony of climate science can be countered, in part, through Gramscian political ecology, which points scholars toward praxis-oriented research "that would both make sense of the world and *help change* the situation under the microscope" (Ekers, Loftus, and Mann 2009, 288; see also Mann 2009). This involves, first, adopting Gramsci's approach to science, which "situates scientific practices on the same plane as all other acts involving knowledge production" (Wainwright and Mercer 2009, 247), such that "neither nature (the so-called real world) nor science (qua objective view of Nature) may be treated as unquestionable sources of truth" (352). Second, Gramscian political ecology aims for collective analysis of socionatural transformation and imaginations through counterhegemonic projects that respect "the struggles and conceptions of . . . 'subalterns'" (Loftus 2012, xxiv) and support ways of knowing that enable alternative forms of consciousness and action. Here, the Gramscian distinction between traditional intellectuals, whose knowledge is grounded in formal expertise and supportive of elite interests, and organic intellectuals, whose knowledge is grounded in the everyday experiences and interests of working-class life, is particularly useful (Gramsci 1971; see also Hall 1992). In the case of climate, organic intellectuals could articulate the knowledge of ordinary people and subalterns in place-based, culturally attuned ways that spark more inclusive and just climate actions, thus replacing traditional *intellegentsia* with a more egalitarian politics of knowledge.

Identifying climate science as a hegemonic way of knowing is not a critique of the practice of science itself, or of scientists as knowledge producers. The practice of climate science is diverse; for example, many scientists are providing place-based science at smaller scales and others are working to communicate science in accessible ways. We are concerned, however, that the way in which scientific thought comes to dominate political discourse, and the corresponding community of technical experts that is called into importance, provides a narrow pathway of understanding and action that is not sufficient to produce change because of its exclusionary politics. Marginalization is a complex process, furthermore, and we are concerned with one specific aspect of that process here—the ways in which mainstream climate politics marginalizes nonscientific ways of knowing that are often held by nonelites.

Gramscian attention to organic intellectuals is bolstered by feminist science studies, which opens up what we "honor as knowledge" (Harding 1986, 24) by recognizing that all knowledge is culturally and geographically situated (Haraway 1988; Merrifield 1995; Rose 1997; Nightingale 2003). Applying this to climate politics, Israel and Sachs (2013) suggest that "rethinking (but certainly not *dismissing*) the climate science that grounds concerns about global climate creates space for new ways of thinking about climate change and for new forms of activism" (34, emphasis in the original). This does not mean, however, that all claims about climate change are equally true. Instead, feminists urge us to take seriously the experiential, embodied, and even contradictory ways that people understand the world and to ask "whose interests are served by the knowledge projects that are overlooked or ignored" (Tuana 2013, 18). Social scientists must engage, therefore, in new forms of praxis-oriented research that consider situated knowledge through a full accounting of the epistemologies and experiences through which people come to know climate change. In other words, it is necessary "to reclaim climate from the natural sciences and to treat it unambiguously as a manifestation of both Nature and Culture, to assert that the idea of climate can only be understood when its physical dimensions are allowed to be interpreted by their cultural meanings" (Hulme 2008, 6).

Brace and Geoghegan (2011) show that place-based knowledge of climate change "enables us to ask how a variety of publics make sense of climate change, as witnessed and responded to in ordinary, everyday-life scenarios, such as walking, gardening, fishing, sailing, and working on land" (289). Although these specific forms of knowledge are central, this also requires a broader political engagement with climate praxis as a method of challenging expert-oriented hierarchies of knowledge. To be clear, challenging knowledge hierarchies does not value experiential or place-based knowledge more than scientific knowledge. Instead, it requires that we be attuned to the ways that scientific and nonscientific ways of knowing develop political significance through their interactions. Ultimately, we must work harder to understand "how lay knowledges might ignore, resist or remain indifferent to science but still motivate people to act on climate change" (Brace and Geoghegan 2011, 296). To illustrate this, we turn to our engagement with experiential knowledge in southwestern North Carolina and the politics this knowledge enables.

Climate Change Knowledge in Southern Appalachia

Our research is part of the Coweeta Long Term Ecological Research (LTER) Program, a research network studying ecological processes in southern Appalachian forests. Southern Appalachia, and southwest North Carolina in particular, is experiencing dramatic changes due to exurbanization, driven by second home owners and retirees from Florida and Georgia (Kirk, Bolstad, and Manson 2011; S. Gustafson et al. 2014). Southern Appalachia is a valuable site for research on climate praxis and knowledge politics because it hosts a confluence of new residents from other regions, an upsurge in libertarian politics, and skepticism of formal government and scientific expertise informed by the historical political and economic marginalization of the region (Fisher and Smith 2012; Newfont 2012). On a more local level in southwestern North Carolina, uneven development is exemplified by the lower income levels of permanent residents of the region, when compared with higher income seasonal residents who are responsible for the newest and most ecologically disruptive forms of development (Pollard 2003; J. S. Gustafson 2014).

The Coweeta Listening Project (CLP) exists within the Coweeta LTER as an action-research collective that seeks to listen to residents of Southern Appalachia, integrate social and ecological science through the coproduction and democratization of knowledge, and build useful and meaningful connections between scientists and the public. The data for this article are based on CLP research with two informal environmental groups in southwestern North Carolina, conducted from March to October 2013, and broader ethnographic work in the region since 2010. The first group formed when we co-organized a series of four community dialogues about environmental stewardship and climate change at a historic general store. Each of the conversations included seven to ten people, mostly white, ranging in age from their early forties to late seventies, spanning lower to upper middle-class incomes. Some had lived in the region their entire lives, others moved to the area as adults (primarily from the southeastern United States, but also from Peru), and others were second home owners. The second group was initiated when a local artist assembled her environmentalist contacts to brainstorm strategies for climate resilience and adaptation in the region. The participants in these meetings were also mostly white, although slightly younger in age (ranging from their twenties to early fifties), low to middle income, with varied lengths of residency in the region. We acted as participant-observers during the first two meetings of this group, helped facilitate a third meeting to develop a mission and action plan, cosponsored two other events about local and scientific perspectives on weather and climate, and conducted interviews with some group members. Drawing on southern Appalachia's long history of popular education (Freire 1970; Horton 2003), we organized each meeting around a participatory research activity, such as discussing "What environmental changes have you seen in the region?" and then mapping their causal connections or asking "What can we do to address climate change?" and developing an action plan. Although "climate change" denotes a complex, hybrid, and varied set of social and ecological processes operating at a variety of scales, we chose to use the term in our research because it is a widely recognizable discourse about the existence of those changes. All conversations and interviews were recorded, transcribed, and coded for instances of climate-related knowledge, weather and climate experiences, and community or government responses to climate change.

Here, we highlight three features of experiential ways of knowing climate change that are particularly salient for addressing science hegemony and climate praxis provided by our research participants. We do not want to suggest that these narratives represent all southern Appalachians or that the people we spoke with are not aware of climate science. What distinguishes these forms of knowledge, however, is that they challenge the globally scaled, carbon-centered discourses of climate change so often referred to by scientists, politicians, and activists and instead focus on embodied experiences and place-based accounts of socionatural change.

Experiences and Memories of Weather Through the Years

Many people we talked with concluded that the climate is changing based on their own long-term observations and stories passed down from ancestors, drawn from embodied, emplaced, and emotionally resonant experiences of weather through the years. Extreme weather events like floods, blizzards, and tropical storms leave particularly strong impressions, but people also perceive, remember, and even record subtle trends through a range of proxies: Clothes worn on the first days of school indicate changes in summer and fall weather. The depth at which one buries water pipes, the (in)ability to bury dead bodies, the fate of overwintering insects, and the extent to which snow remains on the ground all indicate the depth of hard frosts. School days missed, snowmen built, and other measures of the quantity and quality of snow suggest changes in winter moisture and temperature. Observations about the emergence of spring wildflowers, the patterns of migratory birds, the invasion of foreign plants and animals, and the behavior of salamanders signal environmental change.

Consider, for example, a weather memory shared by a recently retired woman, Ella, whose childhood fascination with weather led her to record old-timers' weather stories and to volunteer for more than twenty years as a certified weather observer for the National Oceanic and Atmospheric Administration.

> I've heard my mom, my dad, aunts and uncles talk about . . . snow storms. It was very rare in the winter that there were not big snowstorms, and there would be three feet and over. They said that there would be drifts that were higher than a six plus foot man. We lived out on Brush Creek. . . . The neighborhood boys congregated at our home on . . . Sundays. . . . Well we had a big snow, and how those boys got to our house that Sunday I don't know, because that snow was every bit two, two and a half feet. Well, they decided they were gonna build me a snowman . . . and back then snow would pack . . . you could actually roll it and [build] your snowman. . . . Well, it got to where that snowball for the base was higher than them . . . and it got momentum . . . and that snowball was rollin' away from them. . . . But those boys was so determined that I was gonna have a snowman that they all got down there and they rolled that ball of snow back up that bank into our yard and I, that was the prettiest snowman I've ever seen in my life. I've never seen one like that again. But that's just one of the many stories that I know and that I've heard family tell about the snow and everything.

Ella's story does more than claim that snowfall has decreased from the days of her childhood. What is so important is how she tells her story. Like many multigenerational residents of the region, Ella fills her story with extensive detail about the local environment and people that she considers important for making sense of the climate changes she has seen. She cannot tell the story of snow without embedding it within a dense web of genealogical and historical connections, emotionally rich accounts of community, and the daily rhythms and changing nature of rural life. For many we have talked to, weather memories and other environmental narratives are also memories of family, of connectedness to land and people, and of "where we come from" and "who we are." These place-based perceptions, experiences, and memories of weather permit people's acceptance that climate is changing. Many residents of southern Appalachia understand, and ultimately respond to, climate change through embodied experiences of dwelling in local places—places that are simultaneously social and natural, that are culturally and historically meaningful, and that are seen as embedded in interconnected, regional socioecological systems.

Exurban Development and Landscape Change as Drivers of Climate Change

During an activity to map the causes of climate change, participants hardly mentioned carbon emissions and climate models. Instead, personal experiences of exurbanization were identified as key links to climate change, and altered mountain landscapes served as disturbing, visual evidence that climate change is happening. Participants discussed a variety of aspects of local development—road building, construction of the first Wal-Mart, loss of forested lands, new residential subdivisions built on mountainsides rather than valley bottoms—as contributors to climate change. The following exchange captures the ways that exurban development, spoken of here as the spread of "suburbia," symbolizes climate and landscape change:

> *Jerry:* I think suburbia is leading to climate change as well because you have, you know, man has gone from living in caves to moving into cities and now moving out of cities into suburbia and then that destroys the natural landscape when you do that. And that leads to development and so on.
>
> *Anna:* Suburbia people want to bring some of the city with them and that's the deal . . .
>
> *Richard:* They want the nice little yard, so on and so forth.
>
> *Anna:* They get the suburbia out there, but say, "Man, it's just too far to drive all the way into the city to go to Wal-Mart, we gotta have one out here and now."

Whereas this group continued discussing indirect impacts of development, such as land cover change and altered hydrologic regimes, Ella made the causal connection between development and climate change even more directly in a different meeting:

> [The mountains were] a fortress. And whatever weather hit those mountains . . . had to be strong enough to make it over. . . . When I went to school here in the 50s and 60s we'd be doin' good if we got four days of school in in January. . . . But then, in the 70s, we had . . . Highway 74. . . . You got Interstate 40, and look at the gorge that it cut in the mountains of Haywood County. And then over near Asheville you got I-26, and it cut a big gorge up through Madison County. . . . I believe that when those routes were cut in these mountains it broke down our solid fortress of weather . . . and that's when I started seein' our weather and our climate changing. It affected temperatures, it affected precipitation, it affected wind flow, and everything. . . . Construction and development . . . I think that's the big thing as far as our climate and weather.

These narratives show that residents are keenly aware of their region's connections to the outside, yet they emphasize the regional landscape changes and development processes that contribute to climate change and its impacts far more than distant and global processes. Disentangling climate change from social processes of exurbanization is not possible (or productive) for these individuals.

From Weather Migrants to "Climate Refugees": Cultural Impacts of Climate Change

Many residents we spoke with made predictions of climate impacts by linking environmental knowledge with knowledge of historical demographic and economic changes. From the international trade of deerskins, to the rise of mining, forestry, and agriculture, and now their replacement by the service economy, residents of the region have learned that economic and demographic shifts reshape local landscapes, often to the benefit of wealthy outsiders. The most significant changes today arise from what has been called "amenity migrants" (Gosnell and Abrams 2011), whose mark on the landscape has made development and its ecological impacts a well-known and controversial issue.

One of the most prominent conversations about the interactions between exurbanization and climate change came during highly contested public debates about how steep-slope development—a relatively new form of development driven primarily by amenity migrants—is made riskier by the region's heavy rainfall events. Perhaps most interesting, however, is how historical knowledge of regional connections and exurbanization have raised predictions that weather-based amenity migrants will increasingly transition into "climate refugees"—a term that originated from the people we spoke to, not our own research team. Keenly aware of the interactions between exurbanization and climate change, some residents are concerned that environmental hazards like heat waves, sea level rise, and deteriorating water quality and quantity will drive people from coastal regions toward southern Appalachia. Mark, a participant in our climate adaptation meetings, identified this as one of his major concerns when he said:

> Um, climate refugees. People who will move to the mountains when sea level rises more and severe weather patterns are hitting the coast, and I'm concerned about the carrying capacity of the mountains. How much more development can we handle without losing our water table and, uh, taxing the water?

Maria also said extreme weather and water shortages are likely to drive more people to the region's mountains. She knows several people who came to southwestern North Carolina after Hurricane Katrina and have permanently relocated to the region. She and others worry that "climate refugees" will be a major element of future demographic and economic changes that negatively impact "old-timers" in the region through cultural disruptions. As another gentleman said, "Climate refugees from the coast is going to be an important thing. You know, head to the hills, run to the mountains."

Climate Praxis Through Engagements with Nonscientific Ways of Knowing

To illustrate how nonscientific knowledge can foster climate praxis, we turn to Adam, a local resident who is the director of a regional environmental organization. When we asked Adam whether the experiential knowledge of residents could generate effective climate action during an interview with him, he was skeptical. Instead, he emphasized the need for more education about the scientific causes of climate change to correct misunderstandings by people like Ella who blame road cuts in the mountainside over atmospheric concentrations of carbon dioxide. Although Adam's work is consistent with the dominant mode of politics, his approach overlooks how situated, embodied, and historical experiential knowledge are precisely what make climate an issue of concern to many people. When mainstream, expert-driven politics suppresses everyday climate knowledge in this way, it becomes oppressive, diminishing people's power to make decisions and pursue their own actions. This is especially troubling when the communities whose knowledge is marginalized are often the most vulnerable to climate change.

So, how can we imagine climate praxis becoming embodied through experiential knowledge? Consider, first, the importance of weather memories to people in the region. Although climate scientists are careful not to conflate short-term weather variability with long-term climate change, an emphasis on weather might be exactly what is needed. Discussing the ways that weather has changed can create a robust community of concern among farmers, gardeners, and other outdoor enthusiasts with intimate and everyday experiences of weather (Geoghegan and Leyson 2012). In fact, people in southwestern North Carolina are assembling in groups to discuss and validate local experiences and knowledge of weather, which is galvanizing broader networks that can empower people to act. During the time of this study, this included a meeting in Macon County called "Record Rainfall, but Is It Climate Change?" hosted by a local watershed association director, which focused primarily on people's experiences of changing weather. Acknowledging the importance of weather to individuals, and being careful not to dismiss their experiential knowledge through complex scientific explanations, can help facilitate action when people are not expected to fall in line with overly scientific ways of explaining the problem.

Second, the connections between exurbanization and climate change emphasized by residents also show the complex socioecological assemblages that make climate change culturally meaningful. Politicians and activists might leverage this complex entanglement of environmental, social, and economic processes of exurbanization in ways that galvanize communities to protect their natural environmental and social heritage through controls on development. This is not an uncontroversial topic in the region; a steep-slope building ordinance has been unsuccessfully pushed in Macon County based on technical and scientific aspects of landslide hazards (J. S. Gustafson 2014). Yet, reframing the issue to consider local concern for the changing social and ecological character of the region might spur wider concern, where trying to disentangle climate change out of this rich and complex socioecological context can be counterproductive (Leyshon and Geoghegan 2012). Although not directly about climate change, confronting the issue of exurban development as one with social, cultural, and community effects that are important to people in the region could generate support for local planning efforts that emphasize sustainable and compact growth and other climate-relevant actions.

Finally, the forms of climate praxis enabled by the use of the term "climate refugee" require careful consideration. Voluntary migration to rural mountain communities by upper class individuals certainly does not fit the conditions of refugees, but it facilitates nontraditional engagement with climate politics through a consideration of the historical class politics of the region. As privileged individuals are able to cope with climate change through voluntary migration, the most vulnerable people already living in southwestern North Carolina have little political capacity to resist the accompanying socionatural transformations. As is the case in southern Appalachia, more broadly, government intervention to mitigate the associated socioecological impacts is not historically present, exacerbating uneven development of the region and increasing social stratification. This brings the issue of climate justice into view in new and nontraditional ways—many of the most vulnerable populations in southwestern North Carolina are rural, white, and older individuals (Pollard 2003). Discussing the accelerated migration to the region that could take place under a changing climate with the receiving communities who are concerned about it negatively impacting their local culture might make the importance of climate change more evident and incite more ethical and political (not scientific) debate on the issue.

Conclusion

Socioecological transformations have never been neutral or undeliberately orchestrated; they have always been first imagined and then enacted. Our analysis draws forth the experiential, place-based knowledges of climate change held by people in southern Appalachia as an example of culturally salient epistemologies of climate change and ways to imagine more egalitarian socioecological transformation moving forward. Soliciting and valuing marginalized knowledge of silenced communities helps us transcend the narrow politics exemplified by the North Carolina legislature and the climate debate more generally. Considering "all available studies," as North Carolina legislators have requested, will never mobilize broad constituencies toward unambiguous solutions, and it will only deepen historical patterns of exclusion and marginalization reflected in hierarchies of knowledge. By contrast, our approach seeks to enhance people's power to make decisions by destabilizing the dominance of scientific knowledge only to create space for more pluralistic knowledge of the problem, its effects, and possible solutions. In this way we are working toward new forms of climate praxis.

Engaging in this cultural politics of climate change presents several new opportunities for climate praxis that move us beyond debilitating and depoliticizing debates about science and technology. This approach enables more diverse communities of action when people are not expected to fully understand or accept scientific ways of explaining the problem. Experiential manifestations of climate change are important to people, and they are especially important when they allow concerned individuals to work within their own culturally specific socionatural entanglements to produce change. This approach also demands a reallocation of efforts and resources away from science education and toward facilitating democratic debate that involves dissent and disagreement about what the problem is and what its solution might be, privileging moral and ethical considerations, not techno-scientific ones.

Perhaps most important, expert-only politics runs the risk of excluding the knowledge of individuals who do not prioritize scientific explanations, who in some cases might also be the most vulnerable. Insisting on "climate literacy" might actually be a way of working *on* these communities rather than working *with* them. Any truly revolutionary politics that has the ability to confront our post-political condition and produce egalitarian socioecological futures must grapple not only with the physical drivers of climate change but also with the dynamics of political marginalization and silencing produced by techno-scientific hegemony. Moving away from universal ways of knowing climate change, such as those provided by scientific analysis, to more differentiated and embodied ways of knowing climate change helps combat the post-political condition through an explicit engagement with marginalized framings of the problem and its solutions.

Acknowledgments

The authors would like to thank four anonymous reviewers and Bruce Braun for their helpful comments, which greatly improved this article. We would also like to thank our research participants, who generously shared their time, knowledge, and experiences with us.

Funding

This research is supported by funding from the Coweeta Long Term Ecological Research Site (National Science Foundation DEB-0823293).

References

Bäckstrand, K. 2004. Scientisation vs. civic expertise in environmental governance: Eco-feminist, eco-modern and post-modern responses. *Environmental Politics* 13 (4): 695–714.

Brace, C., and H. Geoghegan. 2011. Human geographies of climate change: Landscape, temporality, and lay knowledges. *Progress in Human Geography* 35 (3): 284–302.

Demeritt, D. 2001. The construction of global warming and the politics of science. *Annals of the Association of American Geographers* 91 (2): 307–37.

———. 2006. Science studies, climate change, and the prospects for constructivist critique. *Economy and Society* 35 (3): 453–79.

Eden, S. 1996. Public participation in environmental policy: Considering scientific, counter-scientific and non-scientific contributions. *Public Understanding of Science* 5 (3): 183–204.

Ekers, M., A. Loftus, and G. Mann. 2009. Gramsci lives! *Geoforum* 40 (3): 287–91.

Fisher, S. L., and B. E. Smith, eds. 2012. *Transforming places: Lessons from Appalachia.* Urbana: University of Illinois Press.

Freire, P. 1970. *Pedagogy of the oppressed*, trans. M. Bergman Ramos. New York: Continuum.

Gannon, P. 2012. Sea-level rise bill becomes law. *Star News Online.* http://www.starnewsonline.com/article/20120801/ARTICLES/120809970 (last accessed 12 May 2014).

Geoghegan, H., and C. Leyson. 2012. On climate change and cultural geography: Farming on the Lizard Peninsula, Cornwall, UK. *Climatic Change* 113 (1): 55–66.

Glass, A., and O. Pilkey. 2013. Denying sea-level rise: How 100 centimeters divided the state of North Carolina. *EARTH Magazine.* http://www.earthmagazine.org/article/denying-sea-level-rise-how-100-centimeters-divided-state-north-carolina (last accessed 9 May 2014).

Gosnell, H., and J. Abrams. 2011. Amenity migration: Diverse conceptualizations of drivers, socioeconomic dimensions, and emerging challenges. *GeoJournal* 76 (4): 303–22.

Gramsci, A. 1971. *Selections from the prison notebooks.* New York: International Publishers.

Gustafson, J. S. 2014. Urban political ecology and exurban environmental knowledge in post-2008 southern Appalachia. Doctoral dissertation, University of Georgia, Department of Geography, Athens.

Gustafson, S., N. Heynen, J. L. Rice, T. Gragson, J. M. Shepherd, and C. Strother. 2014. Megapolitan political ecology and urban metabolism in southern Appalachia. *The Professional Geographer* 66 (4): 664–75.

Hall, S. 1986. Gramsci's relevance for the study of race and ethnicity. *Journal of Communication Inquiry* 10 (2): 5–27.

———. 1992. Cultural studies and its theoretical legacies. In *Cultural studies*, ed. L. Grossberg, C. Nelson, and P. Treichler, 277–94. London and New York: Routledge.

Haraway, D. 1988. Situated knowledges: The science question in feminism and the privilege of partial perspective. *Feminist Studies* 14 (3): 575–99.

Harding, S. 1986. *The science question in feminism.* Ithaca, NY: Cornell University Press.

Horton, M. 2003. *The Myles Horton reader: Education for social change*, ed. D. Jacobs. Knoxville: University of Tennessee Press.

Hulme, M. 2008. Geographical work at the boundaries of climate change. *Transactions of the Institute of British Geographers* 33 (1): 5–11.

Israel, A. L., and C. Sachs. 2013. A climate for feminist intervention: Feminist science studies and climate change. In *Research, action and policy: Addressing the gendered impacts of climate change*, ed. M. Alston and K. Whittenbury, 33–51. Amsterdam: Springer.

Kirk, R. W., P. V. Bolstad, and S. M. Manson. 2011. Spatio-temporal trend analysis of long-term development patterns (1900–2030) in a southern Appalachian county. *Landscape and Urban Planning* 104 (1): 47–58.

Leyshon, C., and H. Geoghegan. 2012. Anticipatory objects and uncertain imminence: Cattle grids, landscape and the presencing of climate change on the Lizard Peninsula, UK. *Area* 44 (2): 237–44.

Loftus, A. 2012. *Everyday environmentalism: Creating an urban political ecology.* Minneapolis: University of Minnesota Press.

Mann, G. 2009. Should political ecology be Marxist? A case for Gramsci's historical materialism. *Geoforum* 40:335–44.

McNie, E. C. 2007. Reconciling the supply of scientific information with user demands: An analysis of the problem and review of the literature. *Environmental Science & Policy* 10 (1): 17–38.

Merrifield, A. 1995. Situated knowledge through exploration: Reflections on Bunge's "geographical expeditions." *Antipode* 27:49–70.

Moser, S. C. 2010. Communicating climate change: History, challenges, process and future directions. *Wiley Interdisciplinary Reviews: Climate Change* 1 (1): 31–53.

Mouffe, C. 2005. *On the political.* London and New York: Routledge.

Newfont, K. 2012. *Blue Ridge commons: Environmental activism and forest history in western North Carolina.* Athens: University of Georgia Press.

Nightingale, A. J. 2003. A feminist in the forest: Situated knowledges and mixing methods in natural resource management. *ACME: An International E-Journal for Critical Geographies* 2 (1): 77–90.

Pollard, K. M. 2003. *Appalachia at the millennium: An overview of results from Census 2000.* Washington, DC: Population Reference Bureau.

Ranciere, J. 2004. Introducing disagreement. *Angelaki: Journal of the Theoretical Humanities* 9 (3): 3–9.

Rose, G. 1997. Situating knowledges: Positionality, reflexivities and other tactics. *Progress in Human Geography* 21 (3): 305–20.

Smith, N. 2006. There's no such thing as a natural disaster. In *Understanding Katrina: Perspectives from the social sciences.* Brooklyn, NY: Social Science Research Council Forum. http://understandingkatrina.ssrc.org/Smith/ (last accessed 14 May 2014).

Soucheray, S. 2014. Sea level rise threatens public health infrastructure. *North Carolina Health News.* http://www.northcarolinahealthnews.org/2014/04/14/sea-level-rise-threatens-public-health-infrastructure/ (last accessed 10 May 2014).

Swyngedouw, E. 2010. Apocalypse forever? Post-political populism and the spectre of climate change. *Theory, Culture & Society* 27 (2–3): 213–32.

———. 2013. The non-political politics of climate change. *ACME: E-Journal of Critical Geography* 12 (1): 1–8.

Tuana, N. 2013. Gendering climate knowledge for justice: Catalyzing a new research agenda. In *Research, action and policy: Addressing the gendered impacts of climate change*, ed. M. Alston and K. Whittenbury, 17–31. Amsterdam: Springer.

Wainwright, J., and K. Mercer. 2009. The dilemma of decontamination: A Gramscian analysis of the Mexican transgenic maize dispute. *Geoforum* 40 (3): 345–54.

Williams, R. 1977. *Marxism and literature*. Oxford, UK: Oxford University Press.

Zucchino, D. 2012. N.C. to sea level forecasters: Ignore climate change data for now. *Los Angeles Times*. http://articles.latimes.com/2012/jul/03/nation/la-na-nn-north-carolina-climate-change-predictions-20120703 (last accessed 12 May 2014).

Temporalities in Adaptation to Sea-Level Rise

Ruth Fincher, Jon Barnett, and Sonia Graham

School of Geography, University of Melbourne

Local residents, businesspeople, and policymakers engaged in climate change adaptation often think differently of the time available for action. Their understandings of time, and their practices that invoke time, form the complex and sometimes conflicting temporalities of adaptation to environmental change. They link the conditions of the past to those of the present and the future in a variety of ways, and their contemporary practices rest on such linking explicitly or implicitly. Yet the temporal connections between the present and distant future of places are undertheorized and poorly considered in the science and policy of adaptation to environmental change. In this article we address this theoretical and practical challenge by weaving together arguments from social and environmental geography with evidence from small coastal communities in southeastern Australia. We show that the past conditions residents' imagined futures and that these local, imagined futures are incongruent with scientific, popular, and policy accounts of the future. Thus we argue that the temporalities of adaptation include incommensurate and unacknowledged ways of knowing and that these affect adaptation practices. We propose that strategies devised by governments for adapting to environmental change need to make visible—and calibrate policies with—the diverse temporalities of adaptation. On this basis, the times between the present and the long-term future can be better navigated as a series of short and negotiated policy steps.

涉入气候变迁调适的在地居民、商人和政策制定者，在思考可行动的时间方面，经常有所不同。他们对于时间的理解，及其诉诸时间的实践，构成了调适环境变迁的复杂且时而冲突的时间性。他们以不同的方式，将过去的境况连结至现在与未来的境况，而他们当下的实践，则明确地或暗自地根据上述的连结。但现在与遥远的未来之地之间的时间性连结，却未能充分进行理论化，且亦未能被调适环境变迁的科学与政策善加考量。我们在本文中，透过将社会与环境地理学的主张，和来自澳大利亚东南部的小型海岸社区的证据相互整合，处理此一理论与实践的挑战。我们显示，过往决定了居民所想像的未来，而这些在地的、想像的未来，与科学、流行和政策对未来的描述不尽相同。我们因而主张，调适的时间性，包含了不相称与不被认知的理解方式，并影响了调适的实践。我们主张，政府为适应环境变迁之需求所规划的策略，必须使不同的调适时间性被看见——并随之调整政策。在此一基础上，便能够更佳地航行于当下与长期的未来之间的时间，作为一系列短期与协商的政策步骤。

Los residentes locales, gente de negocios y hacedores de políticas relacionadas con adaptación al cambio climático a menudo piensan diferentemente sobre el tiempo disponible para la acción. Su entendimiento del tiempo, y sus prácticas que implican tiempo, forman las complejas y a veces conflictivas temporalidades de adaptación al cambio ambiental. Ellos conectan las condiciones del pasado con las del presente y el futuro en una variedad de maneras, y sus prácticas actuales descansan explícita o implícitamente en tales conexiones. Sin embargo, las conexiones temporales entre el presente de los lugares y su futuro distante tienen poca teorización y merecen poca atención en la ciencia y la política de adaptación al cambio ambiental. En este artículo abocamos este reto teórico y práctico entrelazando argumentos de la geografía social y ambiental con la evidencia registrada en pequeñas comunidades costeras del sudeste de Australia. Mostramos que el pasado condiciona los futuros imaginados de los residentes y que estos futuros imaginados locales son incongruentes con las representaciones científicas, populares y políticas del futuro. Por eso argüimos que las temporalidades de adaptación incluyen modos de conocer inconmensurables y no reconocidos y que estos afectan las prácticas de adaptación. Proponemos que las estrategias diseñadas por los gobiernos para adaptarse al cambio ambiental necesitan hacer visibles—y con ello calibrar las políticas—las diversas temporalidades de adaptación. A partir de esa base, los tiempos entre el presente y el futuro a largo término pueden navegarse mejor como una serie de etapas de políticas, cortas y negociadas.

Climate change is expected to alter ways of life in places, and planning for these futures is a troubling prospect for scientists adept at conceptualizing sociospatial processes in the present (Hulme 2010; Swyngedouw 2010; Nielsen and Sejersen 2012). The prospect of changed environments in the future challenges researchers to provide theories of the relationship between the present and the future that can inform adaptation decisions. In this article we address this challenge, drawing on geographic literature about time in places and on evidence collected in small coastal communities in southeastern Australia. We argue that there are diverse temporalities associated with adapting to climate change and that revealing and reconciling these is necessary to manage risks as they unfold over time. We show how the temporalities relevant to places imply different sets of expectations of and practices for managing the future and that there can be different, even conflicting, temporalities coexisting within the thinking and practices associated with a place. Our findings are in keeping with recent geographical research that shows that expert and local knowledge might have different temporalities, requiring innovative responses to managing change (Brace and Geoghegan 2011; Lane et al. 2011; Lane et al. 2013).

Sea-level rise is a major theme in the scientific and popular representations of the dangers of climate change (Farbotko 2010b). It poses risks to the lived values of 600 million people and two thirds of the world's major cities that are located in low-lying coastal areas (McGranahan, Balk, and Anderson 2007), yet in most places these risks are unlikely to materialize for decades. Under most scenarios sea levels are expected to rise incrementally for the next twenty-five years or so and then rise more rapidly to between 26 and 82 cm by the end of the century (Intergovernmental Panel on Climate Change 2013). There are few easy options for adapting to these projected rises in sea level (Moser, Williams, and Boesch 2012).

Local governments in the Gippsland East region on the southeast coast of Australia are being required by their state government to plan for sea-level rise of between 0.2 and 0.8 m by the years 2040 and 2100, respectively (Victorian Government 2013). In a three-year-long study in five small communities in this region (Figure 1) in partnership with local and state government organizations, we have been working to identify socially equitable ways of adapting to future sea-level rise. Here, most homes and infrastructure are located within 2 m of current mean sea level. Coastal flooding, which inundates roads, jetties, homes, and businesses, is experienced when king tides and strong winds coincide (Department of Climate Change [DCC] 2009). The median age of the population is considerably older than the state and Australian medians, and median weekly household incomes are considerably lower than the Australian median (Table 1). In this article, we focus on the temporalities associated with these places and the importance of revealing them as pivotal for different participants in decisions about the future there.

Between November 2010 and April 2013 multiple methods were used to understand how local residents, second home owners, and policy actors view the past, present, and future of the five communities. A detailed explanation of the full suite of methods is provided in Barnett et al. (2014). The data presented here were generated using four methods from this suite.

1. Semistructured interviews were conducted with forty-two residents and second home owners. Questions were asked about the history of their connection to their town, their ongoing links to the place and people, if and how they would like the place to be different, and their experiences of coastal flooding. The questions were designed with reference to past research on community responses to environmental change (specifically, Stolp et al. 2002; Measham 2003; Brown, Lloyd, and Murray 2006; Waters et al. 2010).
2. Eight focus groups were conducted with forty-nine residents who had previously participated in the research, to gauge local reactions to alternative policy frameworks for responding to sea-level rise.
3. Semistructured interviews with thirty policy actors, working at national, state, regional, and local levels of government, were conducted to ascertain how their organizations are involved in decision making about climate change adaptation in Gippsland East and their perspectives on past, current, and future policy options.
4. A day-long workshop with researchers and regional and local policymakers was used to explore the possibilities for timing future decisions about adapting to climate change, particularly sea-level rise.

Data collected from the use of these four methods were audio recorded, transcribed, and thematically analyzed. Using these data, we offer grounded

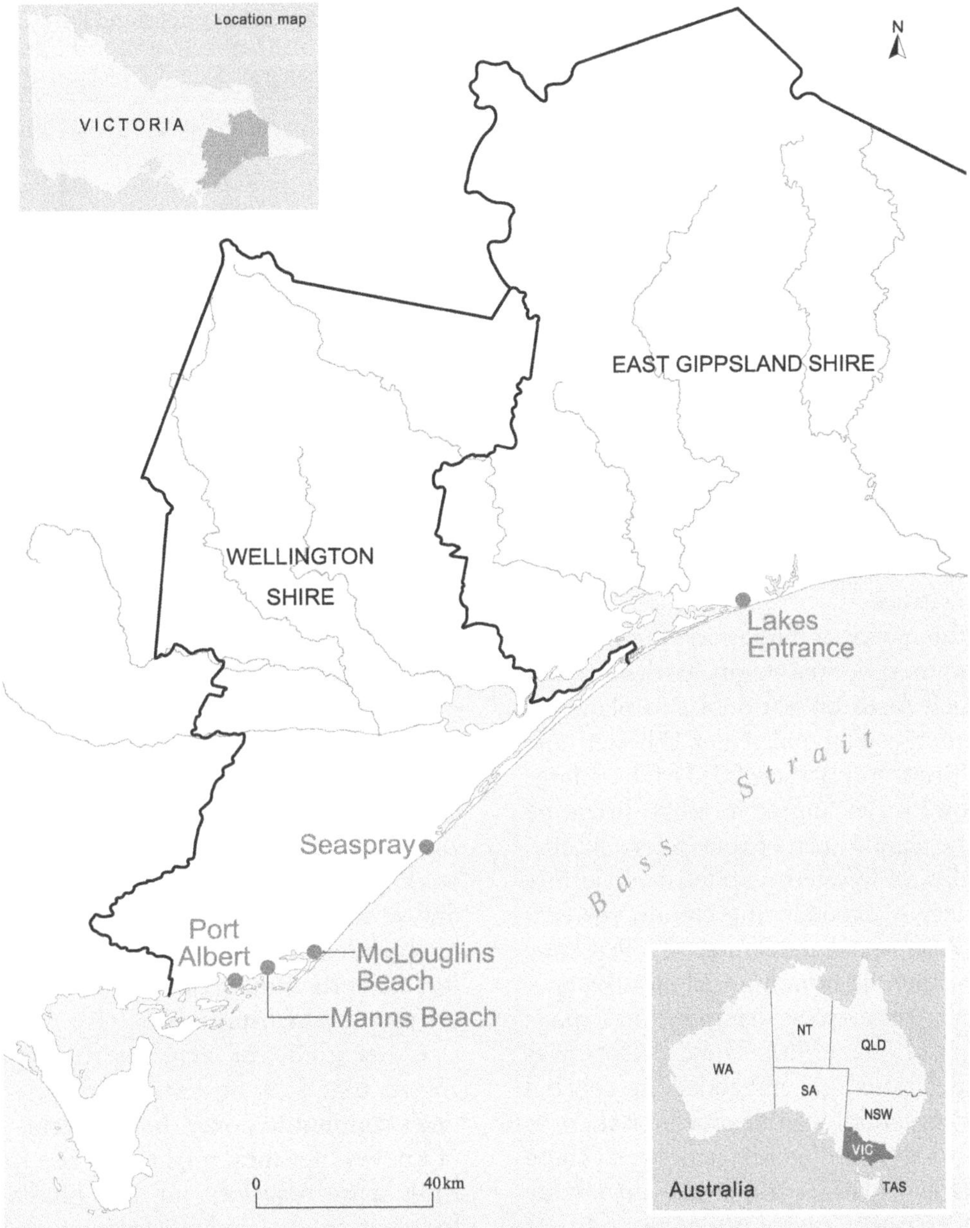

Figure 1. Five communities in East Gippsland, Victoria, Australia. (Color figure available online.)

theorizations of the various temporalities associated with adaptive responses to an anticipated future of sea-level rise in this region.

The Many Times Involved in Responding to Climate Change

The numerous temporalities associated with the coastal places of Gippsland East, drawn from interviewees' comments about those places, can be understood as interplays between the past, the present, and the future. The spatial scale of our analysis, its point of focus, is the locality, where there are many times operating in the local practice and interpretation of the everyday. In observing the everyday, where "there is no single time but a variety of times" (Urry 2009, 180), we do not assume that people bend their own local knowledge to take in expert scientific opinion formed at larger scales. Rather, we recognize, as Lane and colleagues (Lane et al. 2011; Lane et al. 2013) and Robin (2012) show, that local knowledges

Table 1. Sociodemographic information for the five study sites in Gippsland East

Towns	Population	Median household income (weekly)	Median age		Number of people per household	
			Population	Interview sample	Population	Interview sample
Lakes Entrance	5,020	$694	49	42.0	2.2	2.8
Port Albert	247	$608	57	49.5	1.9	2.3
Seaspray	316	$933	48	62.0	2.4	2.4
McLoughlins Beach*	255	$773	51	60.0	2.2	2.0
Manns Beach*		$635	53	62.0	2.1	3.0
Australia	23,156,138	$1,234	37		2.6	

Note: Population data from Australian Bureau of Statistics (2011).
*McLoughlins Beach and Manns Beach are contained within larger collection districts and therefore the numbers presented here might not accurately reflect the populations of these small communities.

are valuable, diverse, and have different temporalities and other features that might cause them to conflict with "expert" knowledge.

We note that the spatial and temporal scales of our analysis differ from most contemporary analyses of climate change, which focus on national and planetary scales and on immense temporal scales (Nielsen and Sejersen 2012; O'Brien and Barnett 2013). These large spatial and temporal scales appear in much literature conceptualizing the significance of time for social life. For example, studies of Western capitalist modernity indicate that society is experiencing certain patterns of acceleration (Rosa and Scheuerman 2009; Crang 2010). Analysis of such metatiming includes the economic and social "time–space compression" made familiar by Harvey (1990, 1996). These perspectives indicate that populations are embedded in epochal changes of major structural significance, yet there is little attention paid to any disconnect between these immense temporal and spatial scales and everyday life or indeed to how we might enable our analyses to see any such variation within the broad patterns delineated. Demonstrating the scaling up of local times, or the scaling down of the temporalities of epochs, is not our task in this article. We note that it is an important task, however, even as we focus our attentions on the varied temporalities of the local and the everyday.

How Pasts Influence Presents

Geographic exploration of time–space has long explained how the when of a local place is as important as the where of it (Hagerstrand 1970; Whitehead 2005; Taylor 2009). This research has taken different forms. Each of those forms of examining the local has made visible new links between pasts and presents. So, Hagerstrand's (1970) pioneering work on the time–space prisms through which individuals' lives are enacted—highlighting the role of schedule in the activities of a day or week—showed how the accreted past in the form of temporal and spatial constraints shapes life in daily practice in a place. Viewing time differently, feminist analysis has revealed the temporality of caring. It has recognized how the relentless clock time of the workplace is generally less defining in caring work (Tronto 2003). Caring is an ever-present task, always there and often invisible because its timing is unregulated and therefore it is unnoticed. This characteristic of the timing of caring is as familiar in the present as it was in distant pasts.

Other studies of local environmental cultures have shown that looking back to the past experiences of one's community, over the intergenerational timescale of known lifetimes, might give rise to continuity with those past practices, in the present. For example, focusing on the dryland farming communities of the Mallee region in southeastern Australia, D. Anderson (2008, 68) finds that drought, a persistent feature, defines the way of life of rural communities and "can be viewed as a cultural term whose primary connotations are less related to rainfall than to an overarching, mythic narrative of endurance." The imagining of present-day lives in places, by those living there and familiar with the places, often draws on the past.

Philosopher Elizabeth Grosz (2004, 251) explains the interactions between past, present, and future thus: "Every present is driven by memory . . . equally, the present always spreads itself out to the imminent future, that future a moment ahead for which the present prepares itself by reactivating the past in its most immediate and active forms, as habit, recognition, understanding." Futures, then, are framed by past-influenced presents. In considering long-term futures in places, both dwellers in those places and policymakers bring to their thinking their linking of past and present. Because the pasts and presents of all those involved in framing futures will not be the same, the temporalities they see and enact will not necessarily cohere.

In our empirical research in Gippsland East, we found (like D. Anderson 2008) that residents of small settlements deemed vulnerable to sea-level rise drew parallels between their present-day experiences of flooding due to tidal and riverine events and their families' past experiences of flooding. They built their environmental knowledge on the basis of known family and community histories and made clear their understanding of the similarities between these experiences in the past, present, and future. They drew on the past to explain their own presents and the continuities they foresaw for their futures. For example:

> [Flooding is] something that's been happening here—families have been around here for years and it was happening when our grandfathers were here. The same, when the king tides and the rivers got a lot of rain, it was happening way back then.

In a focus group, when one person agreed that flooding affected her home quite frequently, saying, "Yeah we can't get out, certainly not with the motorbike, so we're blocked in for a few days," the significance of her view was countered by another who said that this situation has long been happening (and therefore should be seen as something that is coped with, as needed).

> But that's been going on ... we used to come as kids and I mean there were lots of times when we'd go oh we've got to leave before the tide comes in because we won't be able to get around it. I think you don't move down here without expecting it.

Well-functioning drains are seen as the way to combat inundation in these small, low-lying settlements and as the frontline for dealing with flooding of greater frequency if that should occur. "As long as the drains are maintained and things are maintained, it [the flood water] just runs away. You know, it's been doing that for millennia," was a comment in another focus group.

To be clear, our point here is not to romanticize or valorize these statements of knowledge as records of environmental change (on this issue see, e.g., Griffiths and Robin 1997). Rather, our point is to show how futures are anticipated in terms of pasts, in this case as involving continued and successful coping with flooding. The temporal scale of the lifetime and the coping of families and ancestors with environmental vagaries exceed in significance the occasional everyday difficulties of flooding, even when occasional flooding is expected to increase in frequency. And local residents want the future to be like their past and their present. Said one interviewee, part of a family who has been in the area for a long time:

> We're actually falling into the same kind of routine that his parents did. ... Knowing the place so well ... knowing that there's a certain expectation it's going to be the same. ... So there's that nonchanging thing about it that's quite nice and reassuring ... our children really like it, too.

We emphasize that one important aspect of these local perspectives, which hold that flooding is to be expected and can be anticipated, is to view flooding as a short-term inconvenience that can be managed. It is temporary; one can work out when it is going to occur and make preparations.

> You've got a fair indication by looking out at the tides with the wind and everything. ... So you say, well the weather's not going to be too good, we could have a high tide. If you've got to go into town, you go into town, get what you have to get and get home.

> [And] We were well warned that it was going to happen and when it floods. ... So I stayed on this side of the town and it was just over in a few nights in the end. ... We had to wait for the water to go down.

In this context, popular messages of permanent, climate-driven catastrophe are incommensurable with local time-spaces, as is explained by one local leader:

> When they can look back on one hundred years of rainfall figures from their property passed down from their grandfather and their father and now on to them, and they hear a prime minister in a suit talking about catastrophic climate change, they're not inclined to believe them. They can just look at their own figures and say ... it's always changing, things are different, we'll adapt, we'll manage.

In Gippsland East, as in many places, the past strongly conditions attitudes to environmental change (Whitmarsh 2008; Hulme et al. 2009; Myers et al. 2013). The urgency that many decision makers express in making plans now for a future of higher sea levels and more frequent flooding is not shared by local residents. Perhaps these are communities living in what Urry

(2009) has termed *glacial time*, which exists together with (or perhaps in reaction to) "speeded up" time and that slows living down to "nature's speed." Indeed, there are many residents in these settlements who have retired from full-time work or second home owners who are on vacation when in the region, and many do indeed regard their lives in these places as leisurely lives, describing their speed of living (in one interviewee's words) as "relaxed ... pretty cruisey ... there's not too much going on." But also, for many people among the aging population of this area, slowness is not merely a valued characteristic of their everyday lives but a necessary one given aging and illness. For example, a man caring for his ill wife describes their everyday lives as being about "resting and recuperating" through slow routines. For another woman, aging means routines slow down: "I took Sid to the beach on Saturday ... [it takes] about half an hour to an hour, now, because we're both getting old—the dog and me." Numerous interviewees, even those who were not themselves elderly or ill, mentioned the special difficulties that environmental emergencies posed for much older residents:

> The drains, when we did have that flood, were so full of water that there was no footpath. ... The oldies would have to walk down the street.

> It could be devastating for the older people ... we have quite a number of older people ... they don't like to impose on other people and that would have a devastating effect on them because they wouldn't even really let us know how they were going.

We might conclude that a slower speed of living is at risk from rapid responses to anticipated sea-level rise in this place, as well as being at risk from sea-level rise itself, acknowledging that what is at risk includes values, life course characteristics, and psychosocial needs (Graham et al. 2013). An older population in a place, with all that advanced age implies in terms of limited mobility and greater experience of daily constraints, is particularly likely to wish for stability and to value a slower time for mutual caring.

So, in the accounts of residents of the small settlements of Gippsland East, places whose populations are of older median age than the national or state populations (Table 1), the temporalities of major relevance to coping with environmental change are the intergenerational (we will put up with this, as our families always have done; cf. D. Anderson 2008), and the daily (we will cope with flooding because we can manage on our own over periods of several days; cf. Hagerstrand 1970). Both temporalities are threaded through with the slower practices of caring, especially among the elderly population. Importantly, many residents make sense of their present-day lives with reference to their families' pasts in the area. Pasts are influencing presents, and they in turn influence interpretations of futures.

How Presents Frame Futures

Recent scholarship is showing how the idea of future changes influences contemporary meanings and experiences of places. This is shown, for example, in the way projections of obesity and associated anticipatory policies make possible futures felt in the present (Evans 2009). Similar arguments about temporality have been made with respect to climate change, where practices aimed at managing future risks might preempt undesirable outcomes (N. Anderson 2010; Brace and Geoghegan 2011; Marino and Ribot 2012). In the case of climate change and small islands, social scientists have shown how the idea of "disappearing" islands undermines arguments for sustainable development (Barnett and Adger 2003), transforms popular representations of islands and islanders (Farbotko 2010a), and can manifest itself in local narratives about culture and the future (Rudiak-Gould 2012).

Somewhat like small islands, places along the Gippsland Coast, and in particular Lakes Entrance, are becoming icons for the problem of sea-level rise in Australia. To be sure, these are places that are low-lying, adjacent to a sandy coast, and on a lakes system that is well known for its environmental fluctuations, but they are hardly unique in this respect. What is unique is the intensity of research on climate impacts in the region. Since 2004 there have been six major studies investigating the effects of sea-level rise, changes in wind and waves, and subsidence on sea levels. Framing the future in sets of fixed, risk-laden points in time (2030 and 2070), these studies conclude that environmental change will cause an increase of the 100-year flood level at Lakes Entrance by between 2 and 20 cm in 2030 relative to present levels, and by between 4 and 59 cm by 2070.

Such expert accounts of a distant and discontinuous future contrast strongly with the continuity anticipated by residents across their remembered pasts, lived presents, and imagined futures. Lane et al. (2011; Lane et al. 2013) observe a similar contrast between expert and local knowledge in their work in a flood-prone British locality and community. This incongruity is enhanced by the way it abets a long-standing stigmatization of settlements in the region as economically and environmentally unsustainable places. The residents of Lakes Entrance and staff working in local government offices feel this stigma. Two respondents in our study describe the state-wide reporting by media of a recent flood event:

> I was watching the news and they were talking about flooding in Lakes Entrance ... they didn't mention that it was only a couple of puddles and they were doing the interview in Bairnsdale in front of the hotel there, the one right on the Mitchell River that floods every time.

> Well, down at the motel the bloke was kneeling in the water taking the photos so ... instead of standing up, he was down low and it gave the impression the water was a lot higher than what it was. That was just irresponsible.

This led another respondent to conclude that the problem of flooding was not the fact of water in the town but rather "the perceptions of the flooding, in the wider community and especially in the corridors of power."

Popularly imagined futures for places can influence the way policy narratives frame those places. For example, in the words of the Australian Government's landmark report on sea-level rise, Gippsland East is "one of the most vulnerable coastal areas in Australia" (DCC 2009, 93). This view is not without material effects, for five of the six landmark decisions about planning for sea-level rise in the State of Victoria have related to the Gippsland East area (Macintosh 2012). In perhaps the most important of these decisions, one of the members of the Victorian Civil and Administrative Appeals Tribunal (VCAT) noted that "rising sea-levels are likely to and will have an influence on the future shape of the Victorian coastline."

Scientific, popular, and policy-based imaginaries of the future simultaneously constitute the present meaning of places. When places are seen as being imperiled, by actors in markets and policy communities, and their thinking informs decisions that affect the present, then the present itself becomes vulnerable to ideas of the future.

Acting on anticipated futures can reify future dangers in lived presents (N. Anderson 2010). In Lakes Entrance, these ideas of the future materialize in the present in various ways. For local businesspeople, futures are unambiguously seen through the lens of the viability of property and the possibilities of "development." The aforementioned VCAT decision, which led to a series of controls on commercial expansion (Macintosh 2012), is widely perceived by locals to have stifled development.

> [It] ruined all that investment. So unfortunately Lakes Entrance is becoming, along the Esplanade here, an area of dumps [deteriorated buildings]. People can't invest in it.

In housing markets, too, there appears to have been reduced activity. Local people see these effects manifested in various ways, including through falling property prices:

> People that I talk to in real estate ... that have been holding onto property for development, they're all very disappointed, they're saying the property to developers was worth X amount, now it's probably worth a third of that.

Stigmatizing reporting by the media of flooding in Lakes Entrance affects local businesses:

> Like, on the news, don't come to Lakes, it's flooding. ... It was close to one of the school holidays and there was a lot of accommodation cancelled because of this "flooding."

Locals also believe that people in the low-lying parts of Lakes Entrance pay higher insurance premiums now due to the popular idea of Lakes Entrance as a flood-prone place:

> I had a vehicle crashed into the front of my shop. I had to put in an insurance claim for my signage and the window has to be replaced. All the paperwork keeps coming back: water damage, flood damage ... we keep saying, you've got it wrong, it wasn't a flood. ... I think it's just in people's minds that if there's a damage claim in Lakes Entrance, it must be floods.

Finally, local residents also suggest that local governments are now reluctant to invest in the infrastructure necessary for the orderly functioning of their communities for fear of losses arising from flooding and inundation.

> Local government and federal government and state governments, how they use the tool [of climate change predictions is] to really destroy community infrastructure. That's what's happening at the moment ... [your local community] will not get anything now because in 2100 there's going to be .8 of a meter of sea-level rise.

Indeed, this view held by local residents is not without substance, as a policy actor suggests:

> The discussion about climate change affecting these communities means ... in decision-making about infrastructure replacement there's always a two-minute

> discussion about climate change and whether it's good value [to maintain infrastructure].

Future dangers are reified in material presents, and as Grosz (2004) makes clear, they are also inseparable from the interpretations of pasts that make their way into presents. But what is evident in this place is that the shaping of the future by governmental and media outsiders generates images of the future of this area that are unsympathetic to (or unaware of) the emphasis on continuities and of coping that many residents espouse. The absence of perspectives about continuities and coping from the publicized view of the future of their place makes some locals angry.

As the future unfolds in a distant time, the local residents who spoke to us in interviews and focus groups will make their futures in these places, using the habits and routines of their everyday living—in caring, in responding to constraints on their daily lives, in making sense of things with reference to their familiar pasts (in the manner conceptualized by Hagerstrand 1970; Tronto 2003; D. Anderson 2008; and many others). The official future makers, at present, do not seem to be hearing the stories that local people are telling about what those futures could and should include. The conditions in which the producers of expert knowledge might slow down to listen (cf. Whatmore and Landstrom 2011) are evidently not present. This is not to suggest that local people should be the sole arbiter of knowledge and overly privileged in decision making, but it is to say that effective planning of adaptation requires recognition of diverse temporalities and the inclusion of the varied perspectives of local people.

The Temporalities of Adaptive Responses to Sea-Level Rise

There are varied temporalities associated with places such as Gippsland East, and these "temporal dimensions of a place are at least as important as the spatial ones" (Robin 2012, 75). This in part explains why measures to adapt to sea-level rise have been far from successful (Moser, Williams, and Boesch 2012), for they have by and large proposed singular spatial fixes to manage long-term changes that are unpredictable and discontinuous from the temporal reality of the everyday. These spatial fixes do not fit with local understandings of time–space, for they seek to effect solutions rapidly and are seen to transfer many of the costs of adaptation onto present generations, in the form of devalued assets and constraints on local economic development. Institutions fail to acknowledge that, given uncertainties about impacts, some of these costs might be better distributed into the future when risks are more evident. Fitting institutions to align appropriately with varied spatial scales (Young 2002), then, requires fitting institutions to align with temporal scales. Those temporal scales would be a varied set, recognizing that many temporalities coexist in places (cf. Urry 2009). They would include institutional attempts to characterize the specificity of connections to the past in places and to understand the particular constraints on and hopes for the future that the everyday lives of local residents exhibit (cf. Hagerstrand 1970).

Getting the institutions of risk management to take into account the temporalities of local places and the groups within them will itself take time. Thus far, however, adaptation institutions seem to be in a hurry, working to political and bureaucratic schedules and not local times. For example, in our study areas some policies anticipating sea-level rise and adaptation to it have, in the words of one respondent, interpreted "planning for" sea-level rise as "plan now, rather than giving people, organizations, applicants, or the industry, the opportunity to actually establish strategies." Some locally situated policy actors are frustrated because institutional priorities expect "that all the answers are required now." This situation has prompted new kinds of thinking about adaptation to long-term environmental change, with some decision makers among our interviewees suggesting that adaptation needs to "slow down and have a process that we trust." In the words of one policy actor:

> We've now got the data, we've now got the tools and we have to explain these very well over the next three to five years to the populace and industry sectors as well so they can understand when we start to consolidate and start to nail policy positions.

The sense of urgency in climate change mitigation policy—captured in Australia in talk of the "critical decade"—should not pervade discussions about adaptation: there is time to plan for adaptation, although this planning needs to begin now and the form it will take requires discussion. We accept that this urgency to establish rules now for the distant future might be a characteristic of the governance of this jurisdiction in particular, and that other local parts of the world or indeed the nation might envy the attention being paid to designing adaptive responses to likely sea-level rise in this place. The point we are making, though, is not that there should be no planning for environmental change in the future but, rather, that the planning should not be so hasty as to create unintended damage. It should be "well-planned" planning, and this requires a little time to be taken to know its context.

As part of their slowing down, and perhaps as the reason for their slowing down, policies of adaptation to long-term environmental change require reconciliation with local characteristics and temporalities (cf. Hulme et al. 2009; Lane et al. 2011; Whatmore and Landstrom 2011; Lane et al. 2013). Adaptation in this context might usefully proceed as a series of short steps in the long and unending (and certainly not linear) process of adjusting to sea-level rise. The time for adaptation to sea-level rise is not an empty, temporal space between the present and some distant future point, decades away. Rather, the time between now and the distant future is an eventful period that can be calibrated and constructed as having particular, material characteristics for decision makers to respond to as events unfold (Brace and Geoghegan 2011). Recognizing this requires accepting a level of indeterminacy and abandoning the possibility of a one-off, spatio-temporal fix to manage the risks of sea-level rise.

Short-term timing of responses to long-term environmental futures offers a means to build in iteration and adjustment as environmental changes unfold. Sea-level rise means that the future will be unlike the present; we cannot predict precisely how the future will appear when seen from the viewpoint of the everyday. Waiting and acting in short steps, each step being triggered by a change in environmental (or social) conditions, offers a fairer and more practicable approach. There are a variety of times in action, simultaneously, as Urry (2009) has said, and this approach is more likely to capture them and the benefits of acknowledging them than is the current overarching and simple policy strategy. The approach need not entail a uniform set of choices for all people and places, for choices within each step in the pathway can be tailored for different groups and circumstances.

The principle that adaptation to projected environmental futures is a process to be managed in steps along a pathway is slowly being recognized (Bloetscher, Heimlich, and Meerof 2011; Haasnoot et al. 2013; Ranger, Reeder, and Lowe 2013; Abunnasr, Hamin, and Brabec 2015). Such an approach creates time and space for meaningful local engagement and for further information about local time–spaces to be collected; it better calibrates the responsibility for decision making across generations; it enables iterative responses that take account of new knowledges, technologies, the resolution of uncertainties, and changed values and political and economic conditions; it helps overcome the opposition of climate skeptics (as most costly actions can be avoided until there is more evidence of change); it distributes the costs of adaptation more efficiently across generations; it enables no-regrets actions to commence in the short term; and it creates (temporal) spaces in which people can be given more choices about how to adapt. Such approaches do not propose to determine all actions now, for the future, but rather craft a set of early actions and a flexible process that aims to achieve shared vision of the future for places. Effective adaptation responses to sea-level rise therefore require multiple actions in time, acknowledging the many temporalities at play, rather than single fixes in space.

Conclusion

From the perspective of social geographies of time–space, contemporary geographical research into climate change privileges the spatial and engages insufficiently with the temporal. We have suggested how the past and the future are present in the contemporary meaning and experience of local places and that there are many temporalities in adaptation to sea-level rise. In Gippsland East, where the past circulates in residents' accounts, the future arrives through expert knowledge, popular imaginaries, and the effects of decisions.

Imagining an ongoing planning framework that includes local residents and business people along with researchers and governmental decision makers in managing pathways to the future is a difficult matter. It is our conclusion that developing a grounded view of the future of everyday lives in places requires that the "long term" is viewed in a nonlinear way as a series of "short terms," of different lengths, kinds of overlap, and characteristics. Understanding the long term as made up of a series of varying short terms is conceptually a more astute starting point for institutions concerned with environmental futures than one in which the present is seen as having certain features and the future as having others, with the time between them remaining unexamined. Bringing the relevant contributors to the table to devise a view of these overlapping short terms and the trigger points between them is a key step for further research and for policy development.

Acknowledgments

We acknowledge the work of colleagues Dr. Anna Hurlimann and Colette Mortreux on the broader project of which this article is part.

Funding

This research was funded through Australian Research Council Linkage Grant LP100100586, with support from the East Gippsland Shire Council, the Gippsland Coastal Board, the Victorian Department of Planning and Community Development, the Victorian Department of Sustainability and Environment, and Wellington Shire Council.

References

Abunnasr, Y., E. Hamin, and E. Brabec. 2015. Windows of opportunity: Addressing climate uncertainty through adaptation plan implementation. *Journal of Environmental Planning and Management* 58 (1): 135–55.

Anderson, D. 2008. Drought, endurance and "the way things were": The lived experience of climate and climate change in the Mallee. *Australian Humanities Review* 45:67–81.

Anderson, N. 2010. Preemption, precaution, preparedness: Anticipatory action and future geographies. *Progress in Human Geography* 34 (6): 777–98.

Australian Bureau of Statistics. 2011. *2011 census quickstats*. Canberra: Australian Bureau of Statistics.

Barnett, J., and N. Adger. 2003. Climate change and atoll countries. *Climatic Change* 61 (3): 321–37.

Barnett, J., R. Fincher, A. Hurlimann, S. Graham, and C. Mortreux. 2014. *Equitable local outcomes in adaptation to sea-level rise: Year 1 project report 2011*. Melbourne, Australia: University of Melbourne.

Bloetscher, F., B. Heimlich, and D. Meerof. 2011. Development of an adaptation toolbox to protect southeast Florida water supplies from climate change. *Environmental Reviews* 19:397–417.

Brace, C., and H. Geoghegan. 2011. Human geographies of climate change: Landscape, temporality, and lay knowledges. *Progress in Human Geography* 35 (3): 284–302.

Brown, C. S., S. Lloyd, and S. A. Murray. 2006. Using consecutive rapid participatory appraisal studies to assess, facilitate and evaluate health and social change in community settings. *BMC Public Health* 6:68–75.

Crang, M. 2010. The calculus of speed: Accelerated worlds, worlds of acceleration. *Time & Society* 19 (3): 404–16.

Department of Climate Change. (DCC). 2009. *Climate change risks to Australia's coasts: A first pass national assessment*. Canberra: Commonwealth of Australia.

Evans, B. 2009. Anticipating fatness: Childhood, affect and the pre-emptive "war on obesity." *Transactions of the Institute of British Geographers* 35:21–38.

Farbotko, C. 2010a. "The global warming clock is ticking so see these places while you can": Voyeuristic tourism and model environmental citizens on Tuvalu's disappearing islands. *Singapore Journal of Tropical Geography* 31 (2): 224–38.

———. 2010b. Wishful sinking: Disappearing islands, climate refugees and cosmopolitan experimentation. *Asia Pacific Viewpoint* 51 (1): 47–60.

Graham, S., J. Barnett, R. Fincher, A. Hurlimann, C. Mortreux, and E. Waters. 2013. The social values at risk from sea-level rise. *Environmental Impact Assessment Review* 41:45–52.

Griffiths, T., and L. Robin, eds. 1997. *Ecology and empire: Environmental history of settler societies*. Edinburgh, UK: Keele University Press.

Grosz, E. 2004. *The nick of time*. Sydney, Australia: Allen & Unwin.

Haasnoot, M., J. Kwakkel, W. Walker, and J. ter Maat. 2013. Dynamic adaptive policy pathways: A method for crafting robust decisions for a deeply uncertain world. *Global Environmental Change* 23 (2): 485–98.

Hagerstrand, T. 1970. What about people in regional science? *Papers of the Regional Science Association* 24:7–21.

Harvey, D. 1990. *The condition of postmodernity*. Oxford, UK: Blackwell.

———. 1996. *Justice, nature and the geography of difference*. Oxford, UK: Blackwell.

Hulme, M. 2010. Problems with making and governing global kinds of knowledge. *Global Environmental Change* 20:558–64.

Hulme, M., S. Dessai, I. Lorenzoni, and D. Nelson. 2009. Unstable climates: Exploring the statistical and social constructions of "normal" climate. *Geoforum* 40:197–206.

Intergovernmental Panel on Climate Change. 2013. Climate change 2013: The physical science basis. Working Group I contribution to the IPCC Fifth Assessment Report. http://www.climatechange2013.org/report/ (last accessed 19 August 2014).

Lane, S., V. November, C. Landstrom, and S. Whatmore. 2013. Explaining rapid transitions in the practice of flood risk management. *Annals of the Association of American Geographers* 103:330–42.

Lane, S., N. Odoni, C. Landstrom, S. Whatmore, N. Ward, and S. Bradley. 2011. Doing flood risk science differently: An experiment in radical scientific method. *Transactions of the Institute of British Geographers* NS 36:15–36.

Macintosh, A. 2012. Coastal climate hazards and urban planning: How planning responses can lead to maladaptation. *Mitigation and Adaptation Strategies for Global Change* 18:1035–55.

Marino, E., and J. Ribot. 2012. Adding insult to injury: Climate change and the inequities of climate intervention. *Global Environmental Change* 22:323–28.

McGranahan, G., D. Balk, and B. Anderson. 2007. The rising tide: Assessing the risks of climate change and human settlements in low elevation coastal zones. *Environment and Urbanization* 19:17–37.

Measham, T. G. 2003. Learning and change in rural regions: Understanding influences on sense of place. Unpublished PhD thesis, Australian National University, Canberra, Australia.

Moser, S., S. Williams, and D. Boesch. 2012. Wicked challenges at land's end: Managing coastal vulnerability in climate change. *Annual Review of Environment and Resources* 37:51–78.

Myers, T., E. Maibach, C. Roser-Renouf, K. Akerlof, and A. Leiserowitz. 2013. The relationship between personal experience and belief in the reality of global warming. *Nature Climate Change* 3:343–47.

Nielsen, J. Ø., and F. Sejersen. 2012. Earth system science, the IPCC and the problem of downward causation in human geographies of global climate change. *Geografisk Tidsskrift* 112:194–202.

O'Brien, K., and J. Barnett. 2013. Global environmental change and human security. *Annual Review of Environment and Resources* 38:373–91.

Ranger, N., T. Reeder, and J. Lowe. 2013. Addressing "deep" uncertainty over long-term climate in major infrastructure projects: Four innovations of the Thames Estuary 2100 Project. *EURO Journal on Decision Processes* 1 (3–4): 233–62.

Robin, L. 2012. Global ideas in local places: The humanities in environmental management. *Environmental Humanities* 1:69–84.

Rosa, H., and W. Scheuerman, eds. 2009. *High-speed society: Social acceleration, power and modernity*. University Park: Pennsylvania State University Press.

Rudiak-Gould, P. 2012. Promiscuous corroboration and climate change translation: A case study from the Marshall Islands. *Global Environmental Change* 22 (1): 46–54

Stolp, A., W. Groen, J. van Vliet, and F. Vanclay. 2002. Citizen values assessment: Incorporating citizens' value judgements in environmental impact assessment. *Impact Assessment and Project Appraisal* 20 (1): 11–23.

Swyngedouw, E. 2010. Apocalypse forever? Post-political populism and the spectre of climate change. *Theory, Culture & Society* 27 (2–3): 213–32.

Taylor, P. 2009. Time: From hegemonic change to everyday life. In *Key concepts in geography*. 2nd ed., ed. N. Clifford, S. Holloway, S. Rice, and G. Valentine, 140–52. Los Angeles: Sage.

Tronto, J. 2003. Time's space. *Feminist Theory* 4 (2): 119–38.

Urry, J. 2009. Speeding up and slowing down. In *High-speed society: Social acceleration, power and modernity*, ed. H. Rosa and W. Scheuerman, 179–98. University Park: Pennsylvania State University Press.

Victorian Government. 2013. *Victorian climate change adaptation plan*. Melbourne, Australia: Victorian Government.

Waters, E., F. McKenzie, C. McCarthy, and S. Pendergast. 2010. *The drying lake: Lake Boga's experience of change and uncertainty*. Melbourne, Australia: Victoria Department of Planning and Community Development.

Whatmore, S., and C. Landstrom. 2011. Flood apprentices: An exercise in making things public. *Economy and Society* 40:582–610.

Whitehead, M. 2005. Between the marvellous and the mundane: Everyday life in the socialist city and the politics of the environment. *Environment and Planning D* 23:273–94.

Whitmarsh, L. 2008. Are flood victims more concerned about climate change than other people? The role of direct experience in risk perception and behavioural response. *Journal of Risk Research* 11 (3): 351–74.

Young, O. 2002. *The institutional dimensions of environmental change: Fit, interplay, and scale*. Cambridge, MA: MIT Press.

Translating Climate Change: Adaptation, Resilience, and Climate Politics in Nunavut, Canada

Emilie Cameron,* Rebecca Mearns,† and Janet Tamalik McGrath‡

*Department of Geography and Environmental Studies, Carleton University, Ottawa, Canada
†Nunavut Sivuniksavut, Ottawa, Canada
‡Independent Scholar, Ottawa, Canada

This article examines the translation of key terms about climate change from English into Inuktitut, considering not only their literal translation but also the broader context within which words make sense. We argue that notions of resilience, adaptation, and climate change itself mean something fundamentally different in Inuktitut than English and that this has implications for climate policy and politics. To the extent that climate change is translated into Inuktitut as a wholly environmental phenomenon over which humans have no control, both adaptation and resilience come to be seen as appropriate and distinctly Inuit modes of relating to shifting climatic conditions, calling on practices of patience, observation, creativity, forbearance, and discretion. If translated as a matter of unethical harm of *sila*, however, Inuit frameworks of justice, relationality, and healing would be activated. In the context of a broader global shift away from mitigation and toward enhancing the adaptive capacities and resilience of particular populations, current modes of translating climate change, we argue, are deeply political.

本文检视将气候变迁的主要概念从英文转成伊努伊特语的翻译，不仅考量字面的翻译，更探讨赋予文字意义的更广阔脉络。我们主张，恢復力、调适与气候变迁本身的概念，在伊努伊特语中，与在英语中有着根本上不同的意义，并对气候政策及政治具有意涵。气候变迁在伊努伊特语中被翻译成人类无法控制的全然环境现象之程度，使调适和恢復力皆被视为适切的，并且是将自身连结至改变中的气候环境的特殊伊努伊特模式， 更号召耐心、 观测、 创造力、宽容与谨慎斟酌。但若翻译成*sila*的不道德伤害，那麼伊努伊特的公平、相关性与康復之框架便会被激活。我们主张，在从缓和转向强化特定族群的调适能力与恢復力的更广泛的全球脉络中，当前气候变迁的翻译模式，便具有深刻的政治性。

Este artículo examina la traducción de términos claves acerca del cambio climático del inglés al inuktitut, considerando no solo la traducción literal sino también el contexto más amplio dentro del cual tengan sentido las palabras. Sostenemos que las nociones de resiliencia, adaptación y el propio cambio climático significan en inuktitut algo fundamentalmente diferente al significado que tienen en inglés, lo cual tiene implicaciones en políticas climáticas y en la política. En la medida en que cambio climático es traducido al inuktitut como un fenómeno completamente ambiental sobre el cual los humanos carecen de control, adaptación y resiliencia son vistas como modos apropiada e indistintamente Inuit de relacionarse con cambiantes condiciones climáticas, que demandan prácticas de paciencia, observación, creatividad, autocontrol y discreción. Sin embargo, si fuesen traducidas como algo moralmente dañino para la *sila*, el marco de la justicia Inuit, la relacionalidad y la curación serían activados. En el contexto de un desplazamiento global más amplio de la mitigación y en pro del fortalecimiento de las capacidades adaptativas y resiliencia de poblaciones particulares, los actuales modos de traducción del cambio climático, argüimos, son profundamente políticos.

Climate change has been a component of Inuit political mobilization for decades.[1] Initially, Inuit climate politics emphasized mitigation and framed climate change as a geopolitical, cultural, and economic phenomenon with disproportionate impacts on Inuit lives and lands (Inuit Circumpolar Council [ICC] 2002; Government of Nunavut [GN] 2003; Watt-Cloutier 2004). Climate change, it was argued, was a problem with an identifiable cause (greenhouse gas emissions) and a set of actors associated with that cause (industrial political economies). These causes could and should be slowed or stopped, and those responsible should be held accountable for the ways in which excessive greenhouse gas emissions undermine Inuit lands and livelihoods, as well as a broader "right to be cold" (ICC 2005; Watt-Cloutier 2009). The ICC's (2002) Kuujjuaq Declaration, for example, called on governments "to enact domestic legislation and promote and implement multi-lateral agreements to reduce and/or eliminate harmful environmental damage," mandated regional ICC offices to "lobby their respective governments to immediately ratify" the Kyoto Protocol, and instructed ICC to "work in partnership with Arctic and other governments and appropriate NGOs to develop global initiatives to combat climate change" (ICC 2002). Climate change, Inuit emphasized, directly compromised Inuit rights, and in 2005 the ICC brought a petition to the Inter-American Commission on Human Rights aiming to hold the United States responsible for global warming (ICC 2005).

Inuit leaders continue to emphasize their right to predictable and stable ice regimes, viable coastal communities, land-based livelihoods, and cultural continuity, all of which are threatened by climatic change (Simon 2009; Watt-Cloutier 2009; ICC 2010a; Lynge 2012). Over the last decade, however, Arctic climate research and policy have shifted away from matters of mitigation, responsibility, justice, and rights and toward the identification of strategies and resources that might promote Inuit adaptation and resilience (Ford et al. 2010). As the most recent GN climate strategy makes clear, "Addressing the causes of climate change" remains among "high priorities for Nunavut," but the current policy priority is to "enabl[e] Nunavummiut [people of Nunavut] to better adapt to current and future changes brought on by climate change"(GN 2011, 4–5). Climate change, it is suggested, is now inevitable, and the strategy is thus focused on identifying climate impacts and promoting adaptation and resilience among Nunavummiut, not on efforts to slow or stop climatic change. Consistent with global shifts toward adaptation and resilience, ICC called in its 2010 Nuuk Declaration for ongoing pursuit of "all available avenues to combat human-induced climate change" but also for the development of "ways to adapt to the new Arctic reality including insisting on the inclusion of Arctic Inuit communities in the proposed 20 billion dollar international climate change adaptation fund" (ICC 2010b). Although this shift from mitigation to adaption has been highly contentious, and many continue to advocate for immediate globalized action to slow the production and emission of greenhouse gases, adaptation and resilience have effectively displaced mitigation as the leading focus of climate research and policy in the Arctic.

In this article we critically interrogate the shift from mitigation to adaptation in Nunavut, a primarily Inuit territory in Canada. Our focus is on the translation of key concepts in climate change research and policy—including adaptation, resilience, and climate change itself—into Inuktitut. We argue that both the explicit meaning of these terms in Inuktitut and the broader intellectual, cultural, historical, and political context within which Inuktitut words make sense have played a crucial role in normalizing the shift toward adaptation and resilience in Nunavut. We do not claim that this mode of translating climate change is intentional, with a clear objective of shifting Arctic climate politics. On the contrary, those involved in developing Nunavut's glossary of climate-related terms (GN and Nunavut Tunngavik Incorporated [NTI] 2005) were committed to enhancing Inuit understandings of comlex phenomena and supporting informed responses to climate change. The term of elders, language, and climate specialists who developed the glossary was tasked with producing terms that would make sense across radically different languages and epistemologies, and they themselves understood this to be a momentous challenge, with inevitable impacts on meaning (G. Ljubicic, personal communication, 19 September 2014). Our focus, here, is not on the production of the glossary itself but rather on the later effects of particular translations and the broader discursive and epistemological context within which climate-related terms make sense in Nunavut today. As Foucault (1980) and others have made clear, it is not the intentionality of an act that creates a relationship of power; the production of knowledge has power-laden effects regardless of intent, and neither do individual knowledge producesrs bear responsibility for the broader discursive context within which they operate and within which their work is taken up. Nevertheless, how an issue is framed by a dominant discourse can be intensely political: it positions certain elements as natural and incontestable and others as the proper targets of political debate or policy development. A number of geographers have developed this point in relation to climate change in particular (e.g., Baldwin 2009; Liverman 2009; Hulme 2009; Chatterton, Featherstone, and Routledge 2013; Arnall, Kothari, and Kelman 2014).

We argue that the official Inuktitut translations of climate change are political in this sense: They frame climate change in ways that naturalize it as a biophysical phenomenon (rather than human-influenced) and thus encourage Inuit to evaluate and respond to climate change according to one epistemological frame (one that valorizes adaptation) as opposed to others (that valorize restitution, relational justice, and healing). This choice is not neutral. As Thornton and Manasfi (2010), Cameron (2012), and others have argued, adaptation and resilience shift the burden of responding to climate change from those who cause it to those who are most affected. Emphasis is placed on localized capacities to adapt to "changes" whose causes are excluded from the frame of analysis. The rise of adaptation and resilience can be understood as "postpolitical," moreover, insofar as these frameworks bracket out political economic systems and structures, contribute to the disappearance of a political subject of climate change, and support the reduction of the political to policymaking (Swyngedouw 2010). "Doing politics" in post-political times, Swyngedouw (2010) argued, is "reduced to a form of institutionalized social management" (225); in the face of the perceived inevitability of capitalism, climate postpolitics is "structured around dialogical forms of consensus formation, technocratic management, and problem-focused governance" (215). Although we find this line of argument to be resonant in the case of Nunavut, we are most interested here in distinguishing between two distinct responses to climate change in Inuktitut—one focused on adaptation and the other on relational justice—and in calling attention to how different climate-related terminology might support different climate politics in the Arctic.

Our analysis also draws on critiques of the interface between Indigenous knowledges and scientific and policy frameworks. There is a rich literature specifically engaging this issue in the circumpolar north (e.g., Cruikshank 2004; Ellis 2005; Nadasdy 2009; Haalboom and Natcher 2012), where efforts to document Indigenous knowledges are now routine. In spite of these critiques, and perhaps because so much research and policy attention is focused on translating Indigenous knowledges into non-Indigenous scientific and policy frameworks, there has been little critical attention directed to the translation of English words and concepts into Indigenous languages and the associated impacts on debate, discussion, and mobilization within Indigenous communities (but see Rudiak-Gould 2012). As Bravo (2009) notes, although there is a great deal of enthusiasm among scholars for integrating Indigenous knowledges into science, there has been almost no examination of how scientific research is received and debated in northern Indigenous communities. Meanwhile, translation has become a key site of political contestation in Nunavut, particularly in the context of environmental assessment of major resource development projects. Not only are important documents and educational resources simply unavailable in Inuktitut (Isuma 2012) but the translation of key concepts (e.g., "radiation" in the context of a proposed uranium mine in Nunavut) has been called into question by Nunavummiut (Bernauer 2012). At stake is the availability and quality of information and resources that are central to evaluating and responding to pressing contemporary issues and the resulting capacity for Inuit to govern their homelands.

We begin with a brief overview of Inuktitut and some of the particularities of Inuktitut translation. We then examine three key climate-related terms (climate change, adaptation, and resilience) in relation to Inuktitut frameworks of being–knowing–doing–accounting (McGrath 2011).[2] We go on to briefly consider what kind of knowledges and practices might emerge if Inuktitut customary law, conflict resolution, healing, and reconciliation frameworks were brought to bear on climate change. We conclude with reflection on the kind of climate politics that might result from such a shift.

Translating Inuktitut

Inuktitut is one of five primary distinct Inuit dialects spoken in Canada, and it is widely spoken in Nunavut.[3] In 2006, 91 percent of Inuit in Nunavut could hold a conversation in Inuktitut (Statistics Canada 2009) and the vast majority of elders are unilingual Inuktitut speakers. As such, interpretation and translation between English and Inuktitut is a part of everyday life in the territory and is essential for ensuring informed and meaningful public education, discussion, and decision making.

Translating unfamiliar words and concepts into Inuktitut involves distinctive challenges. Inuktitut is highly verb-based, contextual, and relational, rendering the translation of Western scientific concepts much more than a question of ontological correspondence (Laidler 2006; Pulsifer et al. 2011). According to Mallon (1993, 27), although direct word borrowing is common in English, foreign words tend to be translated into Inuktitut through "functional descriptions." While it is common in English for foreign terms to be taken up phonetically (the Inuktitut term *iglu* is taken up in English as *igloo*, for example, not *snow house* or an otherwise descriptive term), in Inuktitut, terms and concepts from other languages are usually translated descriptively using Inuktitut grammar and phrasing.

One word for *horse* in Inuktitut, for example, is *qimmirjuaq*, literally *big dog*. Because of the tendency to translate words descriptively and contextually, many terms can be in use in Inuktitut for the same English word. In reference to the many Inuktitut translations of the term *AIDS* that were developed in the 1980s and 1990s, Mallon observes that because of the "fertility of Inuktitut morphology . . . once an interpreter or translator knows the point of view of the client, she can produce a term expressive of that orientation, stressing the causes of the syndrome, or its effects, or . . . the fact that at the moment it is incurable or fatal" (27). Translation, then, is shaped not only by an interpreter-translator's understanding of Inuktitut but also their sense of what the client wishes to convey about the broader implications of a term.

Mallon (1993) notes, moreover, that Inuit have altered or rejected translation of particularly consequential terms. He recollects a meeting in Puvirnituq where Inuit delegates replaced the usual translation of "Land Claims" as *nunataarasuarniq* (which translates roughly as "trying to acquire land") with the term *nunaqarnirarniq* ("we have the land"), clarifying that the land is not something Inuit were trying to acquire but already had. McGrath recalls similar discussions in Taloyoak in the early 1970s in the lead-up to the Nunavut Land Claim negotiations, where the very notion of acquiring rights to land struck Inuit as absurd. Indeed, Inuit regularly revisit Inuktitut terminology, particularly politically or culturally sensitive terms, and sometimes change terminology to better reflect its broader significance and purpose. As part of the development of a territorial suicide prevention strategy, for example, Nunavummiut considered changing the Inuktitut term for suicide, *ingminiirniq* (literally, "to do it yourself"), because the term can be used to refer both to suicide and to more benign expressions of having done something oneself. The term only indirectly refers to killing oneself, noted Alexina Kublu, "because in Inuktitut we are trying to be sensitive to the people who are experiencing the loss and cushion their feelings" (Kublu, cited in CBC 2009), thus highlighting the relationship between terminology and a broader Inuktitut framework of relationships, knowledges, and values. Similar conversations have unfolded around how to translate the concepts of sovereignty and security into Inuktitut (Kablutsiak 2013).

Because translation into Inuktitut is descriptive, flexible, contextual, and relational, there is no single correct translation of terms like climate change, adaptation, or resilience. Terms become standardized through discussion, debate, and usage. Noting the need for standardization but also the challenge of ensuring comprehensibility of Western scientific concepts, the GN and NTI (the organization representing Inuit beneficiaries of the Nunavut Land Claim Agreement) partnered in the development of a comprehensive glossary of terminology on climate change in 2005 (GN and NTI 2005). The purpose of the glossary was to develop accurate and comprehensible terminology about climate change in both Inuktitut and Inuinnaqtun. Workshops were held across Nunavut with elders, language experts, and climate scientists, resulting in a glossary of 130 climate-related terms. In the following section we discuss three of these terms in detail, considering the broader context within which each word makes sense.

Inuktitut Climate Terminology

The vast majority of Inuit knowledge is not recorded in academic publications. Inuit have also repeatedly expressed concern that existing publications are inaccurate, misleading, and ultimately serve the interests of Qallunaaq[4] scholars, policymakers, and administrators rather than Inuit themselves (Flaherty 1995; Aupilaarjuk, cited in Price 2007). As such, although we draw here on a handful of relevant academic sources, we draw primarily on Mearns's and McGrath's knowledge of Inuktitut and experience as Inuktitut teachers, interpreters, and translators, as well as conversations with Inuktitut language specialists, elders, and intellectuals across Nunavut.

Climate Change: Silaup Asijjiqpallianinga

Climate change is defined in the glossary, in English, as "a difference in the usual and extreme global temperatures that is not just a short cycle, but lasts for decades" (GN and NTI 2005, 39). It is defined, in other words, in wholly biophysical terms and not as a process that has been induced by humans. Similarly, global warming is defined in the glossary as "the slow, continual increase of the temperature of the Earth's air and oceans" (70) and is translated into Inuktitut as *nunarjuap uqquusivallianinga*, literally an ongoing warming of the Earth's temperature. These are not inaccurate terms. What is notable is that, whereas standard English definitions of both climate change and global warming frequently include mention of pollution, greenhouse gas emissions, and other human contributors to these phenomena (e.g., Environmental Protection Agency 2013), the English and Inuktitut definitions provided in the glossary do not.

Efforts to translate the Western scientific concept of climate into Inuktitut have settled on the imperfect and not directly corresponding term *sila*. The Inuktitut translation of the term climate change is thus *silaup asijjiqpallianinga*, and it refers to a process of ongoing, continuous change in *sila*. Although *sila* is consistently translated back into English as weather, climate, or environment, it is in fact a much more complex concept. Qitsualik (2013, 29) explains that *sila* is "arguably the most important concept in classic Inuit thought ... occurring in senses that are intellectual, biological, psychological, environmental, locational, and geographical." It can mean air, atmosphere, sky, intellect, wisdom, spirit, earth, universe, and all. *Sila*, Qitsualik argues, is a kind of "super-concept, both immanent and transcendent in scope," deriving its meaning within Inuit cosmologies that cannot be reduced to Qallunaaq notions of outside, environmental phenomena (see also Leduc 2007).

Within Inuit cosmologies and epistemologies, uncertainty, unpredictability, and change are foundational. The world is understood to be ambiguous and to elude full comprehension and thus change is both expected and accepted (Bates 2007). To describe *sila* as being in a state of ongoing and continual change, then, is to state the obvious. The term *silaup asijjiqpallianinga* confirms what Inuit have always known: The world is in a continual state of flux. In Qallunaaq epistemologies, because fixity, certainty, and predictability are highly valued, to define climate change as a set of extreme and lasting changes to climatic patterns is to signal danger and to call up the question of who or what is causing the climate to change. In Inuktitut the term signals no such danger, and neither does it demand the question of cause. In fact, if climate change were translated as *silaup asijjiqtitauninga* (literally, "*sila* being made to change"), a term that implies causality and human intervention, the term would strike many Inuit as absurd. Within Inuktitut epistemologies and cosmologies, it is impossible and even nonsensical to imagine that humans could cause *sila* to change.

Although defining climate change as a process by which *sila* is being made to change by humans is cosmologically and epistemologically difficult in Inuktitut, the notion of unethical abuse of *sila*, McGrath notes (translated as *silaup asijjirluktauninga*), is indeed comprehensible and would likely elicit a very different response than the more benign notion of continually changing weather. While environmental change—even dramatic and extreme environmental change—is understood in Inuktitut as something to greet with patience, resilience, and creativity, abuse and harm are dealt with through Inuit frameworks of justice, relationality, and healing. Indeed, as we discuss later, it is precisely the translation of climate change as a wholly environmental phenomenon and not as a matter of social, political, or structural injustice and harm that naturalizes adaptation and resilience as appropriate responses to climatic change.

Adaptation: Sungiutivallianiq

Acceptance of change is not only fundamental to Inuit worldview but it is also central to Inuit understandings of what it means to be human, and how best to be *inummarik*, to be the highest standard of human in the Inuk way. To observe and adapt to change is one of the most valued and respected skills among Inuit. It reflects not only great knowledge, skill, and character but also a kind of alignment with the world and a refinement of spirit. Qitsualik (2013, 32–33) suggests that to be *inummarik* is to seek harmonious relations with Water–Land–Sky, and because Water, Land, and Sky are themselves continually changing and not amenable to total knowledge or control: "Harmony naturally arises from collaborative awareness, [and] disharmony results only from lack of awareness ... the True Human's potential is unlocked by adaptation to Water–Land–Sky, whether in concept or fact." In other words, careful observation of a changing world and creative, wise adaptation to those changes are foundational to what it means to be human in Inuktitut. Adaptation to changing environmental conditions is not only sensible or expedient; it is an expression of being Inuk.

The verb stem *sungiuti-* means to familiarize or to become accustomed to something unfamiliar. It is used not just to imply adaptation or familiarization but can also be used to suggest adapting to the ways of unfamiliar relatives (Ootoova and Quassa 2000). Invoked in the context of climate change this is particularly suggestive; if *sungiutivallianiq* refers to a process of becoming accustomed to a changing *sila*, then familiarization and adaptation are an expression of competency and strength, drawing on the ways of ancestors. Adaptation, in such a context, calls on Inuktitut values and traditions of observation, patience, and commitment to knowledge. "In Inuit society, one of the most important and respected characteristics of a successful person is their capacity for self-reliance and their ability to meet life's challenges with innovation, resourcefulness, and perseverance" (Pauktuutit 2006, 32). Faced with rapid and increasingly unpredictable changes not only to *sila* but also to the broader relations that sustain Inuit peoplehood, personhood, and livelihood (McGrath 2011), *sungiutivallianiq* is thus a particularly powerful concept to invoke.

This is not what adaptation means in English. In the context of climate and development policy in particular, critics argue that adaptation has come to imply a kind of "mechanistic" response to outside stimuli that can be enhanced through "social-ecological engineering" (Weisser et al. 2014, 112; cf. Thornton and Manasfi 2010). Bassett and Fogelman (2013) review the term's long-standing connections with human ecology and hazards approaches and argue that current adaptation research and policy is insufficiently attentive to structural causes and barriers. Indeed, as global climate politics and policy shift away from mitigation toward adaption there has been a corresponding change in focus from those who cause climate change toward those who are most *affected* by climate change. Understood in these terms, it is Inuit who must adapt to climate change—not those who continue to produce and benefit from the production of greenhouse gases—and questions of cause, accountability, justice, and rights are not only deferred but rendered illegible (Cameron 2012). Thus, whereas the GN's 2003 climate strategy emphasized international action to slow or stop the production of greenhouse gases, the 2011 strategy is almost entirely focused on Nunavummiut and their adaptive capacities. The strategy aims to assess Inuit resilience and adaptive capacity and to support Nunavummiut "in continuing to rise to the challenges and opportunities that climate change may present" (GN 2011, 3), opportunities that include expanded shipping and resource extraction. Inuit are explicitly called on within the strategy, moreover, to rally their "demonstrated . . . ability to adapt to rapidly and drastically changing circumstances. It is this adaptability that has allowed Nunavummiut to survive in what most people in the world consider to be a cold and hostile environment." But as Watt-Cloutier (2004) insists, Inuit cannot simply continue to adapt in the face of unprecedented environmental change: "The projected magnitude of climate change," she notes, will "stretch [Inuit] adaptive ability to the breaking point." We would suggest, moreover, that Inuit invocations of adaptability must be understood in the context of broader social suffering and as related to Inuit experiences of colonization and trauma over the last century.

Resilience: Annagunnarninga

Like adaptation and climate change itself, resilience derives its meaning in Inuktitut in relation to a broader cosmology. *Annagunnarninga* translates literally as the ability to survive or recover from difficulty or threat. Although it appears in this sense to conform to the English meaning of the term, it is essential to understand how difficulty or threat are themselves conceptualized in Inuktitut. The verb stem *annak-* implies survival and escaping death, suggesting that "resilience" involves not simply being strong in the face of difficulty but escaping very perilous circumstances. As discussed, moreover, in Inuktitut environmental difficulty or threat is, by definition, something that is neither caused nor changeable; the threat itself must be accepted, observed, and adapted to. Resilience, in such a context, refers not just to survival but to an epistemology and cosmology in which one expects to pass through suffering. The fact that things cannot be helped, that one must accept and move through suffering, is so central to Inuktitut ways of being–knowing–doing–accounting that the word *ajurnarmat* means both "things are difficult" and "it can't be helped." Difficulty, inevitability, and forbearance are thoroughly entangled in Inuktitut and thus *ajurnarmat* is offered both as a statement of fact and an expression of sympathy when someone is suffering. Indeed, the full Inuktitut definition of resilience in the glossary defines *annagunnarninga* as "kinatuinnaq, inugiaktulluunniit avativulluunniit ajurnaqtukkuurjuaraluaqtuq annagunnaqtuq": "a process by which an individual, collective, or our environment has gone through great challenge or suffering but can survive." The term thus ties immense suffering and struggle to assurance of survival. Mearns notes that similar concepts are rallied in everyday conversation among Inuit, to encourage each other through difficult times. Whether struggling with environmental challenges or other hardships, one might be told "kisulimaat anigusuut" ("everything passes by") or "anigulaarmijuq" ("in the future this experience will once again be past") to express confidence that someone will make it through suffering.

As with adaptation, our concern here is not with the accuracy of the translation of *annagunnarninga*, although the implied assurance in Inuktitut that people can survive climate change is notable and alarming. Our concern is with the ways in which Inuit epistemologies of suffering and resilience interact with resilience as a policy objective, particularly when climate change is defined as a wholly environmental phenomenon to which Inuit must adapt. Inuit continually emphasize that their ways of being–knowing–doing–accounting, developed over thousands of years, remain essential in the face of environmental change. But not only is contemporary environmental change unprecedented and taxing Inuit capacities for predicting and responding to environmental conditions (Nickel et al. 2005), Inuit capacities to adapt and be resilient in the face of environmental change are themselves undermined by decades of colonial and capitalist intervention. The call for Inuit to be resilient is issued in a context where that resilience is already strained by intergenerational trauma; inadequate housing, social, and health services; compromised food security; and poverty.

Inuit Approaches to Justice, Conflict, Healing, and Reconciliation

When translated as an environmental phenomenon of unspecified origin, climate change comes to demand particular responses in Inuktitut. If translated and widely discussed as a political–economic, ethical, and social phenomenon, however, involving unethical treatment of *sila*, climate change, we argue, would call on very different Inuktitut frameworks. We consider, here, the kinds of relationships with climate change that might emerge in Nunavut if climate change were translated as a form of harm or injustice. We suggest that in addition to Inuit capacities to adapt to changing environmental conditions, traditional practices of conflict resolution, customary law, healing, and reconciliation are relevant to climatic change, as well as Inuit experiences with colonization, oppression, and trauma.

Inuit approaches to conflict, justice, healing, and reconciliation are continually brought to bear on present-day concerns (Aupilaarjuk et al. 1999; McGrath 2005, 2011; Price 2007; Isuma 2012). Consider the recently concluded Qikiqtani Truth Commission (QTC), an Inuit-centered process led by the Qikiqtani Inuit Association to address the killing of *qimmiit* (Inuit sled dogs) by the Royal Canadian Mounted Police but also to document the impact of broader changes during the postwar era. The Commission's mandate was grounded in Inuit forms of relational justice: to "promote healing for those who suffered historic wrongs, and heal relations between Inuit and governments by providing an opportunity for acknowledgement and forgiveness. In Inuktitut terms, Qikiqtani Inuit are seeking *saimaqatigiingniq*, which means a new relationship 'when past opponents get back together, meet in the middle, are at peace'" (QTC 2010, 6). From the outset, then, the QTC understood itself to be addressing not just harm and suffering among Inuit but also the causes of that harm and the need to restore relationships between peoples in conflict.

The QTC's final report observes that the changes Inuit experienced during this time were "rapid and dramatic—this was not a gradual progression from a traditional to a modern way of life, but a complete transformation" (QTC 2010, 11) and that Inuit were exposed to forms of colonialism, racism, and oppression that they had never before experienced. As such, although Inuit drew on their traditional practices to cope with rapid change, ultimately these were insufficient, not least because of the direct and indirect ways in which Inuit beliefs and practices were targeted by Qallunaat. The changes Inuit experienced following World War II, in other words, were so unprecedented in scope and scale, and were caused by structures and forces that were so incompatible with and hostile to Inuit ways of life, that traditional practices were both undermined and insufficient, as much as they were also sources of tremendous pride and strength.

The QTC report insists that Inuit "displayed remarkable resilience in adapting to their new circumstances" (QTC 2010, 40) but also acknowledges the injustice and oppression that continually eroded Inuit capacities to be resilient and adaptable. We would argue that a similar framing is necessary in considerations of climate change. Thus far, Inuit have been encouraged to draw on their tremendous resilience and adaptability to confront the challenges of climatic change, but the broader social, political, economic, and historical context within which Inuit are called on to be resilient has not been adequately addressed, nor have steps been taken to ensure that those who cause harm restore balance and well-being. Within Inuit frameworks, healing and justice are only possible insofar as those who have caused harm acknowledge and account for their actions, and work to restore harmony. Whereas Canadian legal systems typically seek to "punish the offender and focu[s] primarily on the offence ... the priority within Inuit customary law was not necessarily to punish the offender or provide 'justice' per se but rather to ensure the community returned to a state of harmony, peace, and equilibrium" (Pauktuutit 2006, 9). Given that "there has been no harmonious relationship" between Qallunaat and Inuit over the course of centuries of "crushing colonialism, attempted genocide, wars, massacres, theft of land and resources, broken treaties, broken promises, abuse of human rights, relocations, residential schools, and so on" (Amagoalik 2008, 93), Qallunaaq calls for Inuit to be resilient and adaptable represent a profound disrespect for Inuit frameworks of law and justice. It is those who cause and benefit from ongoing harm to *sila*, and its impacts on Inuit lives and lands, who are required to restore balance and right relations.

Understanding climate change as a social, political, and economic phenomenon does not diminish the tremendous resilience and adaptability Inuit have demonstrated in the face of unprecedented environmental change. Rather, it places these responses in a larger context and positions resilience and adaptability as important but ultimately insufficient responses to the scale and scope of the problem. As Kusugak (2002, vi) notes, "Like acupuncture, [Inuit] know that the pain is much in their homelands but the needles have to be inserted in the south, since that is where the disease really is."

Conclusion

All forms of translation are imperfect. Our argument is not that climate change terms are translated into Inuktitut incorrectly but rather that current translations facilitate a broader shift from framing climate change as an ethical issue of globalized injustice, harm, and redress to a matter of localized, technocratic, participatory, and consensual adaptation. At stake here, but written out of the frame of contemporary climate policy in Nunavut, is not just the management of Inuit lives and livelihoods but also the continuation of Qallunaaq lives and livelihoods. If climate change was translated and widely discussed as unethical abuse of *sila*, it would not only mobilize a wholly different Inuktitut moral framework; it would also bring a particular set of subjects back into the frame: those who cause and profit from climatic change, those who suffer from it, and *sila* itself. It would make explicit the relational ties that bind the suffering of Inuit, over here, to the actions of particular subjects, over there, and their shared dependence on the earth.

We noted that for some Inuit, the notion that humans could change *sila* is absurd, even impossible. *Sila*, by definition, is far beyond human scale and scope. Such a response is not unique to Inuit; when the notion of human-induced climate change was first posited by Western scientists, it, too, struck many as impossible, and much political effort continues to be directed toward amplifying this sense of doubt. In a sense, moreover, Inuit who struggle to accept that humans could change *sila* are insightful: Although climate is being made to change by humans, *sila* will prevail, even if climatic change results in the end of life as we know it on this planet. But as human-induced climate change works to alter social, political, and ecological processes on the planet, it is essential that such shifts and impacts be made linguistically and epistemologically legible in Inuktitut. There are precedents for this. As discussed, the notion that Inuit should acquire legal rights to their lands by way of land claims initially struck Inuit as unfamiliar and irrational. In a short time, however, Inuit came to widely support the land claims movement and were ultimately successful in creating their own territory within Canada. As Amagoalik (2007, 91) recalls, this was made possible by a persistent effort among Inuit leaders to visit every community in Nunavut, explain the concept and purpose of land claims, and help Inuit understand the concepts of private property, constitutional law, subsurface mineral rights, and other terms that were central to the negotiation and settlement of the Nunavut Land Claim Agreement. What kind of climate politics might emerge in Nunavut if a similar effort were undertaken?

Our question is not merely technical, nor is it restricted in relevance to Nunavut. Climate change represents an acute threat to socioecological systems, one that disproportionately impacts Arctic peoples. At stake is not just the preservation of stable ice regimes but the very survival of peoples, cultures, languages, and lands. For Inuit, the relationships that have sustained their ways of being–knowing–doing–accounting have been under siege, first with colonial and capitalist forces and now with climate change. To call on Inuit to be more resilient and more adaptable to systems and structures that have consistently undermined their way of life, then, is to exacerbate unjust and exploitative relations. As long as climate change continues to be translated, communicated, and managed in ways that make its ethical, legal, and political–economic dimensions illegible, Inuit efforts to hold those who cause climate change to account will be severely compromised.

Notes

1. Inuit are an Arctic Indigenous people. The territory of Nunavut is a jurisdiction within the Canadian federation established as part of comprehensive land claim process in 1999. Its population is predominantly Inuit.
2. McGrath acknowledges the connections between her phrasing Inuit "being–knowing–doing–accounting" and Wilson's (2008) discussion of Indigenous epistemologies, methodologies, ontologies, and axiologies.
3. Two of the five Inuit dialects (Inuktitut and Inuinnaqtun) are spoken in Nunavut. Within these two primary dialects there are also regional variations in how Inuktitut and Inuinnaqtun are spoken and written. In this article we primarily focus on Inuktitut terminology, although in places we have also considered variation between dialects. Although there are differences in orthography and vocabulary across Nunavut, Inuit dialects connect to a shared structure of values, knowledges, beliefs, and practices.
4. *Qallunaaq* (plural *Qallunaat*) is an Inuktitut term for a non-Inuit, non-Indigenous person. It is typically translated as "white" or "white person" but the term does not specify skin color.

References

Amagoalik, J. 2007. *Changing the face of Canada.* Iqaluit, Canada: Nunavut Arctic College.

———. 2008. Reconciliation or conciliation? An Inuit perspective. In *From truth to reconciliation: Transforming the legacy of residential schools,* ed. M. B. Castellano, L. Archibald, and M. DeGagné, 93–97. Ottawa, Canada: Aboriginal Healing Foundation.

Arnall, A., U. Kothari, and I. Kelman, 2014. Introduction to politics of climate change: Discourses of policy and practice in developing countries. *The Geographical Journal* 180 (2): 98–101.

Aupilaarjuk, M., E. Imaruittuq, L. Joamie, L. Nutaraaluk, and M. Tulimaaq. 1999. *Perspectives on traditional law.* Ed. Frederic Laugrand, Jarich Oosten, and Wim Rasing. Vol. 2, Interviewing Inuit Elders. Iqaluit: Nunavut Arctic College.

Baldwin, A. 2009. Carbon nullius and racial rule: Race, nature and the cultural politics of forest carbon in Canada. *Antipode* 42:231–55.

Bassett, T. J., and C. Fogelman. 2013. Déjà vu or something new? The adaptation concept in the climate change literature. *Geoforum* 48:42–53.

Bates, P. 2007. Inuit and scientific philosophies about planning, prediction, and uncertainty. *Arctic Anthropology* 44 (2): 87–100.

Bernauer, W. 2012. The uranium controversy in Baker Lake. *Canadian Dimension* 46 (1). https://canadiandimension.com/articles/view/the-uranium-controversy-in-baker-lake (last accessed 15 March 2012).

Bravo, M. 2009. Voices from the sea ice: The reception of climate impact narratives. *Journal of Historical Geography* 35 (2): 256–78.

Cameron, E. 2012. Securing Indigenous politics: A critique of the vulnerability and adaptation approach to the human dimensions of climate change in the Canadian Arctic. *Global Environmental Change* 22:103–14.

CBC. 2009. Nunavut group seeks new Inuktitut word for suicide. http://www.cbc.ca/news/canada/north/nunavut-group-seeks-new-inuktitut-word-for-suicide-1.847390. (last accessed 25 November 2013).

Chatterton, P., D. Featherstone, and P. Routledge. 2013. Articulating climate justice in Copenhagen: Antagonism, the commons, and solidarity. *Antipode* 45:602–20.

Cruikshank, J. 2004. Uses and abuses of "traditional knowledge": Perspectives from the Yukon Territory. In *Cultivating Arctic landscapes: Knowing and managing animals in the circumpolar north,* ed. M. Nuttall and D. G. Anderson, 17–32. Oxford, UK: Bergahn.

Ellis, S. 2005. Meaningful consideration? A review of traditional knowledge in environmental decision making. *Arctic* 58:66–77.

Environmental Protection Agency. 2013. Climate change: Basic information. http://www.epa.gov/climatechange/basics/(last accessed 25 November 2013).

Flaherty, M. 1995. Freedom of expression or freedom of exploitation? *The Northern Review* 14:178–85.

Ford, J.D., T. Pearce, F. Duerden, C. Furgal, and B. Smit. 2010. Climate change policy responses for Canada's Inuit population: The importance of and opportunities for adaptation. *Global Environmental Change* 20:177–91.

Foucault, M. 1980. *Power/knowledge: Selected interviews and other writings, 1972–77,* ed. Colin Gorden. Brighton: Harvester House.

Government of Nunavut (GN). 2003. *Nunavut climate change strategy.* Iqaluit, Canada: Department of Sustainable Development.

———. 2011. *Upagiaqtavut—Setting the course, impacts and adaptation in Nunavut.* Iqaluit, Canada: Department of Environment.

Government of Nunavut and Nunavut Tunngavik Incorporated. 2005. *Terminology on climate change.* Iqaluit, Canada: Government of Nunavut and Nunavut Tunngavik Incorporated.

Haalboom, B., and D. Natcher. 2012. The power and peril of "vulnerability": Approaching community labels with caution in climate change research. *Arctic* 65 (3): 319–27.

Hulme, M. 2009. *Why we disagree about climate change.* Cambridge, UK: Cambridge University Press.

Inuit Circumpolar Commission (ICC). 2002. *Kuujuuaq Declaration.* Kuujuuaq, Canada: Inuit Circumpolar Commission.

———. 2005. Petition to the Inter-American Commission on Human Rights seeking relief from violations resulting from global warming caused by acts and omissions of the United States. Iqaluit, Canada: Inuit Circumpolar Council.

———. 2010a. Circumpolar Inuit to global leaders in Cancun: Strong action on Arctic climate change urgently needed. Press release, 1 December 2010.

———. 2010b. Nuuk declaraton. Nuuk, Canada: Inuit Circumpolar Council.

Isuma. 2012. Fact sheet: Angiqatigiingniq—Deciding together. http://s3.amazonaws.com/isuma.attachments/DID_Overview120504.pdf (last accessed 25 November 2013).

Kablutsiak, K. 2013. Almost lost in translation. In *Nilliajut: Inuit perspectives on security, patriotism, and sovereignty*, 4–5. Ottawa, Canada: Inuit Qaujisarvingat.

Kusugak, J. 2002. Foreword: Where a storm is a symphony and land and ice are one. In *The Earth is faster now: Indigenous observations of Arctic environmental change*, ed. I. Krupnik and D. Jolly, v–vii. Fairbanks, AK: Arctic Research Consortium of the United States.

Laidler, G. 2006. Inuit and scientific perspectives on the relationship between sea ice and climate change: The ideal complement? *Climatic Change* 78:407–44.

Leduc, T. 2007. Sila dialogues on climate change: Inuit wisdom for a cross-cultural interdisciplinarity. *Climatic Change* 85:237–50.

Liverman, D. 2009. Conventions of climate change: Constructions of danger and the dispossession of the atmosphere. *Journal of Historical Geography* 35:279–96.

Lynge, A. 2012. Plenary keynote, IPY Conference, Montreal, Canada.

Mallon, S. T. 1993. Early years with the Inuit interpreters: Recollections and comments from the sidelines. *Meta* 38 (1): 25–30.1

McGrath, J. T. 2005. Conversations with Nattilingmiut elders on conflict and change: Naalattiarahuarnira. Master's thesis, Saint Paul University, Ottawa, Canada.

———. 2011. Isumaksaqsiurutigijakka: Conversations with Aupilaarjuk towards a theory of Inuktitut knowledge renewal. PhD thesis, Carleton University, Ottawa, Canada.

Nadasdy, P. 1999. The politics of TEK: Power and the "integration" of knowledge. *Arctic Anthropology* 36:1–18.

Nickel, S., C. Furgal, M. Buell, and H. Moquin. 2005. *Unikkaaqatigiit—Putting the human face on climate change: Perspectives from Inuit in Canada*. Ottawa, Canada: Inuit Tapiriit Kanatami.

Ootoova, E., and J. Quassa. 2000. *Inuktitut dictionary: Tununiq dialect*. Iqaluit, Canada: Nunavut Baffin Divisional Education Council.

Qikiqtani Truth Commission (QTC). 2010. *QTC final report: Achieving Saimaqatigiingniq*. Iqaluit, Canada: Qikiqtani Inuit Association.

Pauktuutit. 2006. *The Inuit way*. Ottawa, Canada: Pauktuutit.

Price, J. 2007. Tukisivallialiqtakka: The things I have now begun to understand: Inuit governance, Nunavut, and the kitchen consultation model. Master's thesis, University of Victoria, Victoria, Canada.

Pulsifer, P., G. J. Laidler, D. R. Fraser Taylor, and A. Hayes. 2011. Towards an Indigenist data management program: Reflections on experiences developing an atlas of sea ice knowledge and use. *Canadian Geographer* 55:108–24.

Qitsualik, R. 2013. Inummarik: Self-sovereignty in classic Inuit thought. In *Nilliajut: Inuit perspectives on security, patriotism, and sovereignty*, 23–34. Ottawa, Canada: Inuit Qaujisarvingat.

Rudiak-Gould, P. 2012. Promiscuous corroboration and climate change translation: A case study from the Marshall Islands. *Global Environmental Change* 22:46–54.

Simon, M. 2009. Climate change, sovereignty and partnership with the Inuit. In *Northern exposure: Peoples, powers, and prospects in Canada's north*, ed. F. Abele, T. J. Courchene, F. L. Seidle, and F. St-Hilaire, 523–28. Montreal, Canada: Institute for Research on Public Policy.

Statistics Canada. 2009. 2006 Census: Inuit language: Inuktitut remains strong, but its use has declined. http://www12.statcan.ca/census-recensement/2006/as-sa/97–558/p9-eng.cfm#nt21 (last accessed 22 November 2013).

Swyngedouw, E. 2010. Apocalypse forever? Post-political populism and the spectre of climate change. *Theory, Culture & Society* 27:213–32.

Thornton, T., and N. Manasfi. 2010. Adaptation—Genuine and spurious: Demystifying adaptation processes in relation to climate change. *Environment and Society: Advances in Research* 1:132–55.

Watt-Cloutier, S. 2004. Climate change and human rights. *Human Rights Dialogue* 2 (11). http://www.carnegiecouncil.org/publications/archive/dialogue/2_11/section_1/4445.html/:pf_printable (last accessed 22 November 2013).

———. 2009. Keynote address, 2030 NORTH Conference, Ottawa, Canada.

Weisser, F., M. Bollig, M. Doevenspeck, and D. Müller-Mahn. 2014. Translating the "adaptation to climate change" paradigm: The politics of a travelling idea in Africa. *The Geographical Journal* 180:111–19.

Wilson, S. 2008. *Research is ceremony*. Halifax, Canada: Fernwood.

Environmental Politics After Nature: Conflicting Socioecological Futures

Becky Mansfield,* Christine Biermann,† Kendra McSweeney,* Justine Law,‡ Caleb Gallemore,§ Leslie Horner,* and Darla K. Munroe*

*Department of Geography, The Ohio State University
†Department of Geography, University of Washington, Seattle
‡Environmental Studies, Denison University
§Department of Geography and Environmental Studies, Northeastern Illinois University

This article is about the logic and dynamics of environmental politics when the environment at stake is profoundly socioecological. We investigate the socioecological forests of the coalfields of Appalachian Ohio, where once decimated forests are again widespread. Conceptualizing forests as power-laden relationships among various people, trees, and other nonhumans, we identify multiple distinct forest types that currently exist as both material reality and future vision. Each forest is characterized by antagonistic ideas about ideal species composition, structure, and function and about specific actions and actors deemed necessary and threatening for the forest's persistence. Each forest represents a very different vision for how socioecological relationships should be fostered. We argue, first, that broad acceptance that the environment is fundamentally socioecological does not mark the end of environmentalism. Rather, urges to environmentalism proliferate as people aim to foster the social natures they envision—and do so through interventions that are internal to what the forest is and does. Second, the proliferation of environmentalisms generates new forms of environmental conflict, which manifests over what sorts of social natures can and should exist (i.e., what they should do and for whom) and which interventions are beneficial or harmful to the survival and proliferation of the forest in the future. Ultimately, we demonstrate that socioecological futures are being shaped today through political struggle not over naturalness but over what should be done, by whom, to bring about which social natures, and to the benefit of whom (human and nonhuman).

本文内容关乎危及的环境具有深刻的社会生态性之时，环境政治的逻辑与动态。我们探讨阿帕拉契山在俄亥俄州的煤田中的社会生态森林，曾经大量灭绝的森林，又再度遍布其中。我们将森林概念化为各种人、树木以及其他非人类事物之间的权力承载关係，指认多种特殊的森林形态，这些森林形态在当下同时作为物质现实与未来的愿景而存在。各种森林，以有关理想物种组成、结构、功能，以及被视为对森林续存而言的必要或危害性之特定行动与行动者的对抗性概念进行分类。每一种森林，皆呈现出对社会生态关係应如何促进的截然不同的愿景。我们首先主张，广泛接受环境在根本上是社会生态的观点，并非标示着环境主义的终结。反之，当人们企图打造其所展望的社会自然时，环境主义的主张便开始激增——并透过内在于有关森林是什麼、以及作何用的介入进行之。再者，环境主义的增生，生产了环境冲突的新形式，并展现在什麼样的社会自然可以、且应该存在之上（例如它们应该做什麼、为谁而做），以及什麼样的介入对未来森林的生存与增长是有益或有害的。我们最终证实，社会生态的未来，在今日并非透过对自然性的政治斗争形塑之，而是透过需要做什麼、由谁来做、为了达成何种社会自然，以及为了谁的利益而做（人类或非人类）的政治斗争形塑之。

Este artículo se refiere a la lógica y la dinámica de políticas ambientales cuando el medio ambiente considerado es profundamente sociológico. Investigamos los bosques socioecológicos de los campos carboníferos de los Apalaches de Ohio, donde los que una vez fueron bosques aniquilados de nuevo crecen por doquier. Al conceptualizar los bosques como relaciones cargadas de poder entre varios pueblos, árboles y otros seres no humanos, identificamos múltiples tipos de bosque bien diferenciados que existen en la actualidad tanto como realidad material como visión futura. Cada bosque se caracteriza por ideas antagónicas acerca de la ideal composición por especies, estructura y función, y acerca de acciones específicas y actores que, para la persistencia del bosque, son considerados como necesarios o amenazantes. Cada bosque representa una muy diferentes visión sobre cómo se deberían promover relaciones socioecológicas. Argumentamos, primero, que esa vasta aceptación de que el medio ambiente es fundamentalmente socioecológico no marca el final del ambientalismo. Más que eso, los apremios por el ambientalismo proliferan en la medida en que la gente busca fortalecer las naturalezas sociales que ellos columbran—y hacen eso a través de intervenciones inmanentes a lo que el bosque es y hace. Segundo, la proliferación de ambientalismos genera nuevas formas de conflicto ambiental, que se manifiestan en cuáles tipos de naturaleza social pueden y deben existir (i.e., qué deben hacer y para quién) y cuáles intervenciones son benéficas o dañinas para la supervivencia y proliferación de los boques en el futuro. Básicamente, demostramos que ahora los futuros socioecológicos están siendo configurados por contienda política, no sobre la naturalidad sino sobre lo que debe hacerse y por quién para producir tales o cuáles naturalezas sociales, y en beneficio de quién (humano o no humano).

What do environmental politics look like when the environment at stake is understood to be fundamentally socioecological? When there is no recourse to a baseline nature such that dueling positions can no longer be "nature versus jobs" or "destruction versus protection"? To date, scholars have convincingly demonstrated the ubiquity and complexity of social natures, but relatively little attention has been paid to understanding the politics internal to these social natures (Robbins and Moore 2013; Lave et al. 2014).

To explicitly investigate the dynamics of environmental politics after nature, we distinguish two aspects of environmental politics. First is environmental politics in the sense of environmentalism—efforts to protect the environment—which long has been dominated by the idea of nature as a domain external to human society, whether as wilderness or resource (Cronon 1996; McAfee 1999; Braun 2002; Robertson 2006). Here we investigate socioecological environmentalism; that is, urges to environmental protection in the absence of pure, external nature (Shellenberger and Nordhaus 2011; Marris 2013; Wapner 2013). Second is environmental politics in the sense of conflict; that is, struggle to control how the environment is understood and used, by whom, and to whose benefit (Robbins 2012). Unfortunately, much scholarship on environmental conflict reexternalizes nature by treating it as an abstraction over which people struggle both materially and discursively, and it is the social dynamics of that struggle—the winners and losers—that take analytical center stage (Robbins 2012; Lave et al. 2014).

Addressing and linking both of these, we conceptualize people and their needs, visions, and actions as internal to what nature is and does. We reject identifying groups of people that come into conflict over an externalized nature, instead considering the inherently political process through which particular social natures are fostered and contested (Robbins and Moore 2013). Rather than framing forest politics as "people struggling over forests," our approach is to treat forests as power-laden, negotiated relationships among various people, trees, understory plants, wildlife, hydrological conditions, and so forth.

The socioecological environments we investigate lie in the coalfields of Appalachian Ohio, where once decimated forests are again widespread. Located on the edge of the central Appalachia plateau, this region has been a resource periphery since European American settlers arrived at the turn of the nineteenth century. People cleared forests by mining for coal and clay and by logging—for timber, for fuel for early industrial pig iron and brick ovens, and for urban and agricultural development (Bashaw et al. 2007). Whereas the region was 95 percent forested in 1800, by 1910 tree cover was reduced by more than 80 percent (Dyer 2001). During the same era, exploitation and extraction of capital during economic booms and abandonment during subsequent busts left the region's inhabitants in intense poverty (Bashaw et al. 2007). Yet, by the 1990s, forest cover had rebounded to close to 70 percent; although patchy and diverse, total forest cover has held steady since (Dyer 2001; Widmann et al. 2009). Although some trees have been planted, the vast majority of this regrowth is spontaneous: the return of forest ecosystems. Most of this regrowth, however, is not happening on land that has been abandoned or turned into a forest reserve or tourist playground. Rather, trees have returned in a densely populated area under very diverse ownership: public and private, wealthy and poor, long-term resident and newcomer.

As a result, the expanse of green that characterizes the landscape of Appalachian Ohio hides very different socioecologies—hides recognizably distinct relationships between people and trees. In what follows, we describe several of these socioecological forests, particularly in terms of the dynamics deemed essential to maintain them into the future, and we highlight inherent antagonisms and alliances as each forest jostles for position in the landscape. This is not simply a matter of matching forest types (oak

Table 1. Forests of Appalachian Ohio, identifying characteristic protagonists, actions, and ecologies associated with each

Forest type	Human protagonists	Management and social actions	Ecologies	Example
Silvicultural forests	Educated, wealthier owners in dialogue with state Service Foresters	Sustainable timber management, invasive control	Native, economically valuable tree species, especially oak	Medium to large parcels of private land managed for timber
Historic forests	Land trusts, conservationists working with like-minded landowners	Trusts and conservation easements, invasive control, restoration of extirpated species	Continuous forest cover, maximum diversity of native species in canopy and understory	Arc of Appalachia Preserve System
Exurban forests	Recent arrivals, absentee landowners	Some nontimber forest product harvesting; management for amenity and aesthetics	"Healthy" forest assemblages; useful species encouraged (for harvest or aesthetic enjoyment)	Small to medium parcels of private land owned by retirees from Ohio's cities
Livelihood forests	Long-term residents, nongovernmental organizations	Small-scale harvest of timber and other forest products; open access to public and corporate woodlands	Diverse; game species and revenue-generating nontimber forest products (e.g., ginseng) fostered	Private land managed with help from Rural Action
Matrix forests	Government or corporate ownership of large tracts of land	Maintain habitat mosaics through bushhogging, mowing, timbering	Patchy; diverse habitats to sustain greatest density and diversity of species whether rare or useful	Wayne National Forest
Privacy forests	Long-term smallholders	Let it grow, hands-off approaches; secondary forests preferred	Dense growth with no species preferences, invasive species okay if contribute to "green screen"	Private land passed down through generations

dominant, softwood plantation) with particular categories of forest owners, users, or managers (private landowners, the state, corporations). Rather, we look to how preferences for particular types of forest are enacted and articulated by entities who are active in the region, including in public agencies (e.g., Wayne National Forest, Ohio State University extension), nonprofit organizations (e.g., Rural Action, Athens Conservancy), and private industry (e.g., rental cabin owners, chambers of commerce). We draw from fifty interviews we conducted between 2009 and 2012, as well as written documents such as newsletters and annual reports, all of which the first author analyzed using an iterative process of coding, sorting, and grouping.

What emerged in this analysis were multiple socioecological forests that differed in terms of species composition, structure and function, and actions and actors (human and not) deemed necessary for the forest's persistence, as well as those deemed to threaten it. In the following sections, we describe six of these forests (Table 1). We chose these six both because of their prominence and because they demonstrate how, in this socioecological politics, threats in one forest are those things deemed necessary to another. These are the forests that people foster and want to propagate, and doing so pits their desires against those of others.

To be clear, the forests we describe are real, yet they also are conceptual artifacts of our analysis. As such, they are not absolute or exhaustive; for example, we do not discuss numerous recreational forests. Nor are they internally homogenous or discrete. They overlap both in the landscape and in their characteristics; in some cases, a single person espoused several of these visions. We choose to focus on distinct forests, rather than distinct people or groups, as a means to decenter the individual agent. Focusing instead on the dynamics through which social natures are produced, we argue that each forest represents a very different vision for how socioecological relationships should be fostered. Ultimately, we demonstrate that socioecological environmental politics is not over naturalness but rather over what should be done, by whom, to bring about which social natures, with what benefits.

Socioecological Forests of Appalachian Ohio

Each of the following sections is devoted to one of six forest types that emerged in our analysis: silvicultural, historic, exurban, livelihood, matrix, and privacy forests. We exemplify elements characteristic of these forests by beginning each section with interview excerpts (although edited for clarity, all words are original unless indicated by brackets). We then elaborate the key themes and highlight important comparisons with the other five forests.

Silvicultural Forests

> Under the [Ohio Forest Tax Law] program, landowners agree to manage their forest land for the commercial production of timber and other forest products. . . . "Forest land" is defined as land for which the primary purpose is the growing, managing, and harvesting of a merchantable forest product of commercial species under accepted silivicultural systems.

> My thing is just educating [woodland owners]. I would just like them to know what their timber's worth. Most of them don't, so two things happen. They don't get a fair price, and then they usually do a high-grade: they take all the valuable trees out, and they leave a mess. . . . I know the need for money is what's driving the way they're managing. When you need money for whatever reason, somebody knocks on your door and you just harvest, whatever's got value at that point in time. . . . [The forest is] gradually changing, especially with high-grading, to maple. Over time, we're going to lose a large component of oak, which is what sustains the industry in southern Ohio. . . . We want people to look at the forest and say, "Hey, if I want to keep this forest healthy, then I really need to do some things." It's not just a hands-off approach.

> [Newcomers are] coming from the more affluent areas of the state, they're going to add a little bit of money that they're willing to put into their property. . . . Say they come from a suburb of Cleveland . . . they are going to be receptive [to learning forest management]. Versus some of these [local] people, their great-granddad owned the property and it's just been handed down from them. Their feeling is, "I can do whatever I want out here on this property, and if I want to cut ten trees a year and sell them for cash, I'm going to do that and nobody's going to tell me any different. If I'm going to have fifteen junk cars out here and park them up in the woods I'm going to do that."

> A lot of the times when people own only twenty acres of forest, their big goal for management isn't timber management. So the amount of actual lumber that's coming [now, compared to what] used to be when these were managed as big forests is declining.

The *silvicultural forest* supports medium- to large-scale timber production. There are many recognized benefits to this type of socioecological forest, but the value of timber comes first: revenue to businesses, landowners, and the state and wages for local residents. Valuable hardwoods, especially oak, dominate, and trees should be spaced so that they can grow large and straight.

Although these are "naturally" regenerated forests (i.e., not tree farms), a steady supply of high-quality timber does not just grow; beneficial intervention comes in the form of active forest management, for instance, regular harvests for sustained yield, thinning, and control of invasive species such as *Ailanthus altissima*, the "tree of heaven." Although large landholdings (public or private) are easier to manage and cut, in Ohio 73 percent of forest land is "family" owned; 55 percent of family forests are in small parcels (< 50 acres), and only 25 percent are in large parcels (> 100 acres)—and the percentage in small parcels is increasing (Widmann et al. 2009); therefore, it is smallholders who hold the key to how forests will function over time. Smallholders must be educated about how to foster a timber forest, so experts such as professional foresters with Ohio State University Extension and the Ohio Division of Forestry are key actors in this forest, providing pamphlets, courses, and workshops. State Service Foresters work directly with individual landowners to prepare management plans that identify goals and a timetable of necessary actions; plans are required for tax reduction programs, and goals must include timber to receive the tax break.

Not all smallholders are considered to be equal in the silvicultural forest, however: Judged harmful are long-term residents (many of whom have inherited family land), whereas exurban newcomers (including retirees and absentee owners of weekend getaways) are regarded as largely beneficial. Although long-term residents might be beneficiaries of the silvicultural forest when it provides jobs, as landowners they are seen as harmful. Presumed to be ignorant of the value of their forest and unwilling or too poor to manage it, they are susceptible to unplanned logging and "high-grading," which is thought to be one of the largest threats to sustaining a valuable timber forest. In contrast, exurban newcomers, educated and wealthier, are perceived to be appreciative of the forest because it is what, after all, drew them to the region; they are open to being educated about forest management and have money to be active forest managers. Newcomers also create a challenge, though, contributing to the trend toward parcelization (i.e., subdivided, smaller holdings) that makes management of forest land more difficult—a point to which we return.

Historic Forests

> [Forests should be] just bigger and better than they are now. Continuous canopy, stopping the fragmentation that occurs. Trying to control nonnatives, invasive species as they become established. In ten or twenty years it won't really approach what it was in 1750, but, boy, that's really what we ought to be heading for.

> I do get a little concerned that they won't leave any older growth trees alone. In the state forest, they just don't want to leave anything alone that's of any size. They need to leave some areas that are undisturbed over a long period of time. . . . Outside people have moved into the county in the last three to four decades, people from Columbus, Cleveland, Dayton, the big cities. Usually these people—I don't want to discriminate, here—they're generally somewhat educated. They usually have money. If they bought it because of its beauty, they want to protect it. So this has been a trend, and I think it's really important in the preservation of the forest land here. . . . There has not been one local person that grew up here that put a conservation easement on their property. We've got thirty-two easements in effect and they've all been from the outside. So the locals take it for granted. They're wonderful people, but I think they take the forestland for granted.

> [A big] threat to this area is development of vacation/summer homes from people from more affluent communities. And the problem with that is dividing the land up into tiny little chunks.

The *historic forest* is characterized by reverent forest stewards attempting to mimic the biodiverse, mature, temperate hardwood forest that covered the area prior to the arrival of European American settlers. In these "wonderfully diverse forests," charismatic and rare or threatened native species are especially celebrated and encouraged, such as ephemeral spring wildflowers, migratory songbirds, and formerly extirpated mammals like bears and bobcats. Ideally, historic forests should never be cut. Today, this forest exists in pockets mainly on private lands, such as those protected with conservation easements or private reserves like Highlands Nature Sanctuary. Importantly, this forest is also about connecting people with the landscape, providing solitude, and fostering a sense of pride in the region's forest and rich history.

Although the historic and silvicultural forests seem quite different—appearing to represent the familiar opposing poles of jobs versus environment—they share three important characteristics. First, human intervention is required to achieve this forest. Historic forests are not remnants of the past—which do not exist in the region—but are instead forests that grow to be like those of the past. Influencing how forests grow requires activities such as planting desired species such as the American chestnut and removing undesired ones such as invasives and overpopulated deer. Second, beneficial intervention must be guided by educated experts from environmental organizations such as the Arc of Appalachia and Appalachian Ohio Alliance. Although these experts are different from those in the silvicultural forest, they, too, produce pamphlets and sponsor workshops and events. Analogous to professional foresters, land trusts assist landowners with conservation easements that reduce taxes in exchange for limiting, in perpetuity, future activity on the land. Third, with regard to private landowners, long-term residents are deemed harmful and exurban newcomers beneficial. Long-term residents are presumed to take the forest for granted, whereas exurban newcomers who have moved to the area because they appreciate the forest have the desire and means to intervene in the forest. Here, too, though, parcelization driven by exurban development is considered harmful, threatening the expanses of contiguous forest that are the cornerstone of historic forests.

Exurban Forests

> A lot of the [new] landowners here, a big thing is that timber isn't a priority. They like the idea of having woodlands, they like the idea of having some wildlife, and they like idea of just having recreation trails. They like the idea of just having trees that are growing. And some of them are interested in the understory nontimber forests products, the ginseng or whatever. So I think that's a little bit different. The focus isn't really strongly on timber production. Some people may have that, but for most it's just a by-product. And what they're really interested in is seeing woodland and trying to take care of it and seeing trees out there.

Although exurban newcomers play a central, beneficial role in both the silvicultural and historic forests, in so doing these newcomers are not enacting exactly the socioecologies that foresters and preservationists envision. Interested in neither a nature sanctuary nor a forest of revenue-generating silviculture, these newcomers instead nurture a new, *exurban forest*—a forest of elite appreciation and use. This forest is to provide a range of services: small-scale timber, aesthetic enjoyment (bird watching, autumn colors), clean water, wildlife, recreation, hunting, and nontimber forest products such as maple syrup and ginseng. Because individual landowners generally pursue management techniques that favor some but not all of these (and with varying success), the exurban forest is diverse, an overlapping patchwork of small-scale extraction, beauty, ecological services, and recreation. Aesthetics is a high priority, which makes this forest type anathema to the livelihood forests we discuss later, in which poorer local landowners prioritize use. Further, whereas parcelization is a threat to the historic and silvicultural forests, in the exurban forest, parcelization is a benefit because it brings in people who appreciate and care for the forest. Exurban forests are for the benefit of an elite class of recent arrivals and absentee landowners, who are themselves the primary protagonists in this forest.

Livelihood Forests

> [We're] really attacking the root cause of why this area is in the economic shape that it is. It's been hundreds of years of extraction and absentee ownership and money flowing out. . . . You have to have working forest; for the small-scale landowner, which is really what we're focused on, it's the most ecologically-based forest practices that are going to provide the most long-term benefit for that land. . . . [We work with two types of landowners]: the average proactive landowner who tends to be a little bit more economically well off, and they tend to be a little more interested in the green aspect of being a land steward, developing hobby-type activities. They might want to plant a little ginseng or a little maple syrup or something like that. And then you have the lower income demographic which is what we're really trying to target now because they have fewer resources to use. They want to know what they can do to earn income from the land sustainably, for the long term.

> Locals see these people [newcomers] are moving in here and barring [them] from doing the things that [they've] always done. So this has been a trend: people moving in, closing off their land, putting up all these stupid signs, about no trespassing and all that stuff and just putting up gates and barriers and all that.

The *livelihood forest* is one that provides not revenue and wage-paying jobs, but diversified entrepreneurial and subsistence livelihoods for rural residents. Forest goods and services—produced in complex ecosystems—

can be used or sold for income: timber, firewood, meat, carbon credits, natural gas wells, mushrooms, medicinal plants, fur (e.g., mink), maple syrup, and drinking water. Small agricultural fields can even be part of the livelihood forest. The array of activities overlaps with the exurban forest discussed earlier (although the range is wider), yet the livelihood forest serves poor, local residents. Livelihood activities occur on private land but also on state forests, national forests, and corporate lands (e.g., pulp and paper company lands), which often serve as a sort of commons. A response to the boom and bust cycles of large-scale industrial development, forest-based livelihoods make poor residents less reliant on wage labor and help them survive high unemployment without becoming long-distance commuters or full-fledged emigrants to Ohio's cities. A prominent proponent of this socioecological forest is Rural Action, a sustainable development nonprofit whose mission is "to foster social, economic, and environmental justice in Appalachian Ohio."

A key feature of this forest is that long-term residents are regarded as partners in fostering the forest, rather than antagonists. Use by and benefit to long-term residents is precisely what makes this forest valuable; this contrasts with the silvicultural, historic, and exurban forests in which rural residents are deemed to take the forest for granted. Even so, there is an important similarity to those other forests: Not only must the livelihood forest be managed (hands-off approaches constitute a threat), but educated professionals are thought to know best which interventions are necessary. For example, Rural Action's sustainable forestry program provides a wealth of resources on managing the forest for a range of timber and nontimber forest products; resources include on-site consultations, for instance, on how to encourage ginseng growth, and a Woodland Owners Toolkit that not only urges smallholders to develop forest management plans but even tells them "for advice on how to sell timber, talk to your service forester" (a central actor in the silvicultural forest).

Exurban newcomers and their "hobby-type" activities, however, are thought treacherous for livelihood forests. Not only do many newcomers look down on activities central to this forest, such as logging and hunting, but their arrival has very material effects. As newcomers buy up land, they often enclose former de facto commons by subdividing large holdings into smaller parcels, putting up no-trespassing signs, and putting land in conservation easements that limit use in perpetuity. The influx of well-off newcomers can also drive up property values and therefore taxes. Rising taxes are an economic hardship to low-income, long-term landowners, especially those who inherited their land; if they go into delinquency, counties eventually seize their land. Exurban newcomers might appreciate and want to protect their version of the forest, but in the process they are a force of dispossession that threatens the livelihood forest.

Matrix Forests

> So what we're doing is trying to create habitat on a large scale. A scale larger than anybody else in the state can do, because we're the largest landowner in the state. We have a management prescription that provides for an early successional habitat. We have a management prescription that provides for a closed canopy, less disturbance type area. We have areas set aside that have absolutely no management activities going on them. And we also have what we're calling historic forest prescription and trying to mimic those conditions that early European settlers found when they got here, according to their journal entries and everything else.

> [We had] 300 acres of really messed-up mine land. Probably the worst piece of mine land we had in the whole forest. Nobody using it. I mean absolutely no one. Hunters wouldn't go in the area. I'm trying to figure out how to make this place productive again. And my mind just races at night when I'm trying to go to sleep and I can't stop thinking about all that's going on out here in the forest. And what came to me one night was that what we really need to do is turn this really messed up piece of real estate into something that people want to come use.

> I would say that most of the longtime rural residents around here, we don't have much problem with. They've been around this stuff all their lives. They understand that if we harvest this area, it'll grow back. And during that time we'll get to pick some berries in here, and we'll get to do this and that, and we'll have more deer. And they understand all that. But the urbanites that come down, they're a little harder to deal with. They think that if you cut it, it will never come back. That we're doing something horrible and terrible that'll hurt the wildlife. And that's not the case.

The *matrix forest* is a diversified, resilient matrix of habitats and is primarily the purview of the Wayne National Forest, Ohio's only national forest. The Wayne exists not as a contiguous block of uninterrupted forest, but a patchwork of relatively small parcels intermixed with private and state land (much of which is tree-covered). Further, the Wayne's forest managers actively create and maintain various forest types and successional stages. The Wayne is especially interested in stands of mature oak and hickory but also young forests and meadows (perhaps on former strip mines). But even with habitat as the central focus, this is not just a forest of wildlife; managing for wildlife habitat is a way to maximize goods and services to people, who are also part of the forest. The matrix forest is not just for endangered species, such as the Indiana bat, but also abundant and useful species, such as deer and wild turkey. It also provides wood, locations for four-wheeling, understory habitats for nontimber forest products, beauty, and sites for oil and gas wells. The goal is to create usable forest. The Wayne's embrace of the matrix contrasts to its past as a silvicultural forest; "today it's more like, 'We gotta produce the habitat, and however much timber we happen to get out of that is just a by-product.'"

In sharp contrast to silvicultural and historic forests, even farm fields and meadows can be part of the matrix forest at the landscape level. Like the exurban and livelihood forests, the matrix forest does not have a dominant function; rather, its function is to provide many things. Also like previous forests, maintaining and improving the matrix requires active intervention by educated professionals. The forest-enhancing activities of the Wayne's land managers include planting, prescribed fire, herbicides, mowing, brush-clearing, and timber harvests to control invasives, encourage desired species, and keep young forests young. Finally, similar to the livelihood forests but not the others, long-term residents are not only beneficiaries of these diversified forests but are regarded as beneficial to them because they share the vision of a diversified, usable forest. The Wayne often partners with locals that have similar visions, including landowners, the state, and organizations such as Rural Action. Urbanites and newcomers, in contrast, generally do not share this vision. Urban tourists are still welcome; they come for the beauty and recreation, but after spending their money they return to the cities. Exurbanite landowners, on the other hand, are hard to work with because they see harm in the activities that foster the matrix.

Privacy Forests

> We did a series of meetings [about tax abatement programs] over a couple of year period and we had very, very little participation. Looking into it a little further, one of the things was a big underground economy that is going on. Some of these people, if they need to raise a little bit of money, maybe they'll cut eight or 10 trees down and sell the logs and it's all cash propositions, and they might sell a few truckloads of firewood, in cash, and they might have a marijuana patch growing. All of that is part of the underground economy that occurs and part of the fear was that if you let [people onto the land to] do forest management they're going to find all this stuff out. I think that it was just the fear that, we don't want the government out here looking and seeing what we're doing.

As the preceding discussion suggests, there is a striking contradiction at the heart of the matrix and livelihood forests. Long-term residents are seen as protagonists in these forests, but they also are judged to be inadequate, requiring the help of outside experts. Many residents, however, reject this contradiction. They treat organizations such as the Wayne and Rural Action not as benign partners but as untrustworthy outsiders that offer surveillance rather than assistance, a sentiment likely magnified by the fact that most employees of these organizations, and the experts they trumpet, are themselves exurban newcomers or government employees. This might explain the low rate of participation in tax abatement programs, which otherwise would seem to contribute to the livelihood forest by reducing the cost of living on the land. It can also explain why Rural Action, despite its vision, ends up working more with exurban newcomers than with long-term residents: "It is unfortunate that it is mostly transplants that are leading the effort."

The *privacy forest* is a rejection of these "outside" organizations and their contradictions, even as it also embraces diversification and livelihoods. The privacy forest provides both a physical shield and a range of goods and services to use while being shielded. Different from solitude, which is about emotional renewal and even spirituality, privacy is about being left alone without being surveilled and bothered. The significant feature of this forest is dense greenery, tucked into hills—yet this is not an untouched forest but a forest of getting by and asserting one's claims to the land. Livelihood activities might include any of those already discussed, but also illegal activities ("supplemental income") such as hunting, fishing, or gathering ginseng out of season or without licenses (i.e., poaching) and growing marijuana; legal or not, income generation is informal and only in-cash or in-kind, bypassing the tax system. Like the livelihood and matrix forests, this is a diversified, working forest in which poor and working-class long-term residents are the main protagonists, while exurban newcomers pose a threat. A dividing line in these forests is whether outsiders (government employees, university researchers, do-gooders from nonprofits) and their plans and regulations are welcome or not.

Conclusion

The forest primeval exists nowhere in Appalachian Ohio today, yet the six socioecological forests described here all do exist. The silvicultural, historic, exurban, livelihood, matrix, and privacy forests are part of one common landscape, overlapping in space. In this sense, although the matrix forest is the product of the Wayne National Forest, the matrix is also an apt metaphor for the entire region and perhaps for social natures more broadly: overlapping patchworks of different sorts of natures, all incorporating both human and nonhuman actors, visions, and beneficiaries. What does this mean for the forms of environmental politics identified at this article's outset?

First, broad acceptance that the environment is fundamentally socioecological does not mark the end of environmentalism. Rather, we find that urges to environmental protection proliferate as everyone is interested in fostering social natures with the characteristics and beneficiaries they envision. Crucially, in none of these forests is human intervention considered to be inherently damaging. In every case some interventions are deemed extremely harmful even as others are considered necessary. Desired forests have to be fostered through intentional actions, which are internal to what the forest is and does, alongside the growth of trees, hydrology of streams, fluctuations of wildlife populations, and so on. Our finding is that when human actions are part of the forest in this way, when humans are no longer an external disturbance, no one is against "the forest" in an abstract sense. Whose environment is more "environmental" thus becomes an open and necessarily political question.

The point here is not simply the proliferation of new sorts of social natures; the point is not that all social natures are equally "good" and cannot be adjudicated. This is the second part of our argument, which is that the proliferation of environmentalisms generates new forms of environmental conflict that are not about protecting or destroying the forest (let alone "nature" writ large), nor are they about controlling access to an externally given environment. Conflict manifests over what sorts of social natures can and should exist and over who gets to decide these key questions. This is about what forests should do and for whom, about which actors and actions are beneficial or harmful to the survival and proliferation of forests in the future. As the preceding excerpts make clear, the expansion of each of these forests requires excluding types of actors and actions (human and not) deemed a threat to that forest—and these threats might be the very actors and actions deemed necessary in another forest. Because no single forest has a lock on environmental protection, political struggle (entailing alliances as much as antagonisms) is about which types of forest will triumph and persist.

The broader implication for environmental politics after nature is that all assertions about socioecological futures must be adjudicated as no one social nature can be argued to be "naturally" better than any other. In this new environmental politics, nothing is inherently harmful or beneficial for social nature in the abstract. The challenge for scholars of social nature, then, is to push beyond merely recognizing and describing socioecological dynamics (in all their complexity) toward evaluating them. But if "nature" is no longer the baseline against which to measure and evaluate change in the world, then what is? An alternative is to parse social natures in terms of justice, attending to the multiple and uneven consequences of particular socioecological configurations. This is also a way to leverage academic insight into political practice. In a world of massive and ubiquitous socioecological change, it is time to rally not around the tired environmentalisms of "protecting nature" but around protecting and fostering the social natures that lead to the most just outcomes for humans and nonhumans alike.

Funding

This research received support from the U.S. National Science Foundation, Award No. 1010314, "CNH: Collaborative Research: Explaining Socioecological Resilience Following Collapse: Forest Recovery in Appalachian Ohio."

References

Bashaw, A., S. Landis, D. White, and J. Winnenberg. 2007. *At the glacier's edge*. Shawnee, OH: Little Cities of Black Diamonds Council.

Braun, B. 2002. *The intemperate rainforest: Nature, culture, and power on Canada's west coast*. Minneapolis: University of Minnesota Press.

Cronon, W. 1996. The trouble with wilderness; or, getting back to the wrong nature. In *Uncommon ground: Rethinking the human place in nature*, ed. W. Cronon, 69–90. New York: Norton.

Dyer, J. M. 2001. Using witness trees to assess forest change in southeastern Ohio. *Canadian Journal of Forest Research* 31:1708–18.

Lave, R., M. Wilson, E. Barron, C. Biermann, M. Carey, C. Duvall, L. Johnson, et al. 2014. Critical physical geography. *The Canadian Geographer* 58:1–10.

Marris, E. 2013. *Rambunctious garden: Saving nature in a post-wild world*. New York: Bloomsbury USA.

McAfee, K. 1999. Selling nature to save it? Biodiversity and green developmentalism. *Environment and Planning D: Society and Space* 17 (2): 133–54.

Robbins, P. 2012. *Political ecology*. 2nd ed. Chichester, UK: Wiley-Blackwell.

Robbins, P., and S. A. Moore. 2013. Ecological anxiety disorder: Diagnosing the politics of the Anthropocene. *Cultural Geographies* 20 (1): 3–19.

Robertson, M. 2006. The nature that capital can see: Science, state, and market in the commodification of ecosystem services. *Environment and Planning D: Society and Space* 24:367–87.

Shellenberger, M., and T. Nordhaus, eds. 2011. *Love your monsters: Postenvironmentalism and the Anthropocene*. Oakland, CA: Breakthrough Institute.

Wapner, P. 2013. *Living through the end of nature: The future of American environmentalism*. Cambridge, MA: MIT Press.

Widmann, R. H., D. Balser, C. Barnett, B. J. Butler, D. M. Griffith, T. W. Lister, W. K. Moser, C. H. Perry, R. Riemann, and C. W. Woodall. 2009. *Ohio forests 2006*. Newtown Square, PA: U.S. Forest Service.

The Place and Time of the Political in Urban Political Ecology: Contested Imaginations of a River's Future

Ryan Holifield and Nick Schuelke

Department of Geography, University of Wisconsin–Milwaukee

Urban political ecology (UPE) has become an important and influential paradigm for the geographic analysis of socioecological transformation. Despite considerable progress in its empirical and theoretical sophistication, however, what it means to analyze the specifically political dimensions of change in UPE accounts remains largely unspecified and underdeveloped. One option receiving attention is to confine analysis of the "properly political" to the disruption of prevailing orders by egalitarian challenges. As an alternative, we propose and elaborate a pragmatist approach to political analysis that has emerged in science and technology studies. Through accounts of two efforts to imagine the socioecological future of an urban river, we aim to demonstrate the potential of such an approach. We argue that in addition to local variation and the deployment of knowledge, analyses of the political trajectories of issues should address historical variation and the mobilization of desire. We contend that such an approach provides a methodology for tracing connections between conventional political processes and extraordinary moments of disruption and that it is also compatible with multiple perspectives on the "political" within UPE.

城市政治生态学（UPE），成为对社会生态变迁的地理分析而言，重要且具有影响力的范例。儘管其中的经验及理论的复杂性有长足的进展，但在城市政治生态学的解释中，何谓分析改变的特别政治面向，却大幅欠缺详细说明，并且未有充分的发展。其中一个受到关注的选项，是将对"适当的政治"之分析，限缩于平等主义所带来的打破盛行秩序之挑战。我们则提出并阐述一个政治分析的务实取径，作为另类方案，该取径已在科学与科技研究中浮现。我们透过想像一条城市河流的社会生态未来的两种努力之记述，旨在证实此一取径的潜能。我们主张，除了在地变异与知识部署之外，对于议题的政治轨迹之分析，必须处理历史的差异和慾望的动员。我们宣称，此一取径提供了追溯传统政治过程与扰乱的特殊时刻之间的连结，且可同时与城市政治生态学中多重的"政治"视角相容。

La ecología política urbana (EPU) se ha convertido en un paradigma importante e influyente en el análisis geográfico de la transformación socioecológica. Sin embargo, a pesar del considerable progreso logrado en su sofisticación empírica y teórica, lo que significa analizar las dimensiones específicamente políticas del cambio, en los recuentos de la EPU, en gran medida sigue subdesarrollado y sin especificar. Una opción que está recibiendo atención es confinar el análisis de lo "propiamente político" a la alteración de los órdenes dominantes por retos igualitarios. Como alternativa, proponemos y elaboramos un enfoque pragmático para el análisis político que ha surgido en los estudios de ciencia y tecnología. A través de los registros de dos esfuerzos para imaginar el futuro socioecológico de un río urbano, buscamos demostrar el potencial de tal enfoque. Argüimos que además de la variación local y el despliegue del conocimiento, los análisis de las trayectorias políticas de los asuntos deben abocar la variación histórica y la movilización del deseo. Sostenemos que ese enfoque proporciona una metodología para trazar las conexiones entre los procesos políticos convencionales y los momentos extraordinarios de alteración, y que eso es también compatible con perspectivas múltiples sobre lo "político" dentro de la EPU.

The geographic subfield of urban political ecology (UPE) has produced a rich and growing body of scholarship challenging conventional conceptions of urban environmental change. Drawing conceptual inspiration from Marxist political economy, feminist theory, actor–network theory, science studies, and Foucauldian governmentality approaches, UPE explains socioecological transformations as products of contested, multiscalar processes, shaped by flows of capital and uneven relations of power (Keil 2003; Swyngedouw and Heynen 2003; Heynen 2014). Its analyses also often advance a practical and normative program: to identify conditions of possibility for imagining and enacting more democratic and egalitarian socioecological futures (e.g., Loftus 2012) and, as Swyngedouw and Heynen (2003, 914) put it: "to enhance the democratic content of socioenvironmental construction." Although it still lacks the visibility of systems-based urban ecology (e.g., Alberti 2007), the UPE paradigm has generated an impressive array of empirical studies and theoretical innovations.

As its name suggests, UPE emphasizes political dimensions of socioecological change. Many accounts, for example, skillfully trace the dynamics of hegemonic projects for socioenvironmental transformation, often highlighting deployments of electoral, legislative, deliberative, and administrative processes within institutions of government and the state (e.g., Njeru 2010; Loftus 2012; Perkins 2013). These and other studies also attend to political dimensions of change beyond such institutions: actions of resistance, inequalities of power, or the normalization of socioecological processes and subjects (e.g., Véron 2006). But UPE has begun wrestling with the argument that the "properly political" lies not in the familiar institutional domains of politics or in power relations of all varieties but specifically in egalitarian demands that interrupt and disrupt established orders (e.g., Swyngedouw 2009). Currently, there appears to be no consensus on what the "political" or the "democratic" means for a UPE analysis. Should the political in UPE pertain exclusively to occasional moments of egalitarian disruption? Or should UPE address both the "properly political" and conventional, institutionalized "politics as usual"?

We argue that one promising analytical approach for connecting these disparate conceptions of the political in UPE has emerged recently in science and technology studies. This pragmatist approach recasts both conventional politics and disruptions of the *demos* as distinct moments in a nonlinear trajectory, encompassing phenomena as disparate as deliberation and governmentality (Latour 2007; Marres 2007). But it also requires empirical tracing, accounting for how issues move from one mode to another. Through accounts of two contested efforts to imagine the socioecological future of an urban river, we aim to demonstrate the potential of such an approach for analyzing political dimensions of urban socioecological transformation. We also seek to add to an evolving conceptual framework by incorporating two new elements: historical variation and the aesthetic articulation and mobilization of desire. We contend that this framework is not only compatible with multiple forms of UPE but also capable of bringing greater empirical detail and methodological coherence to specifically political analysis within UPE.

Locating the Political in Urban Political Ecology

The argument that what passes for the political today is in fact postpolitical has become highly influential, both within and beyond geography (Swyngedouw 2009, 2011). Rather than celebrating the proliferation of mechanisms for public participation in decision making as a sign of increasing democratization, critics suggest the opposite: Participation, carefully managed and typically oriented toward building consensus, is symptomatic of the absence of the political (Rancière 2001; Dikeç 2005). Institutionalized participation is part of what Rancière (2001) describes as the *police*, or the activities that order, place, and partition what and whom can be seen and heard. "Properly political" moments occur only when this police order is disrupted by the performance of egalitarian demands. Consequently, analysis of the properly political processes shaping urban environmental futures would require looking beyond the established institutions through which socioecological demands and desires are funneled into what Rancière calls the "distribution of the sensible," to the extraordinary moments in which egalitarian demands disrupt the prevailing order and call for a different one (see Dikeç 2005).

But this "ontological turn" in political theory, of which Rancière is but one representative, also has its critics. Barnett (2012, 677), for instance, argues that an exclusive focus on foundational disruptions and antagonisms "squeezes out any serious consideration of the plural rationalities of ordinary political action," such as formal deliberation. He argues:

> Rather than continuing to resort to a priori models of what is properly political or authentically democratic, geographers would do well to acknowledge the ordinary dynamics and disappointments which shape political action. (Barnett 2012, 677; see also Barnett and Bridge 2013)

A different critique comes from Bennett (2005), who argues that Rancière's conception of the political is anthropocentric. She contends that political analysis must also recognize the diverse, surprising ways in which nonhumans participate in political agency.

We contend here that rather than focusing only on identifying and analyzing "properly political" moments of disruption, UPE would benefit from adopting the pragmatist approach to the political that has evolved in science and technology studies (Latour 2007; Marres 2007).[1] Instead of restricting the scope of the political from the outset—whether to the familiar institutions of conventional political science or to the constitutive ground of difference that emerges only in moments of disruption—Latour (2007, 813) proposes conceptualizing the political as a trajectory, oriented toward the "progressive composition of the common world." Political trajectories are never abstract or empty of content; they start with and unfold around issues or matters of concern (Marres 2007). The aim of political analysis is to follow the trajectories of such issues empirically, tracing the steps through which they "generate a public around them" (Latour 2007, 814).

Although the disruption of the prevailing order—an event no longer reserved exclusively to human actors—can constitute part of an issue's trajectory, the movement of the political also includes the conventional activities others relegate to the police or postpolitics. Latour (2007) identifies five key moments through which the political trajectory of an issue might travel. The first is the traditional territory of science studies: the emergence of new associations of humans and nonhumans, often as disputed issues or "matters of concern." A second moment occurs when such issues generate a provisional new public and a third when established institutions of government attempt to translate this problematic public and its imputed will into a "common good." A fourth moment consists of deliberation, and a fifth—which Latour associates with governmentality—takes place as matters of concern are provisionally "solved," normalized, naturalized, and transformed into routines, institutions, and procedures.

Within geography the influence of this approach remains limited, but there are important exceptions (Barnett and Bridge 2013; Fuller 2013). Donaldson et al. (2013), for example, apply Latour's concept to the political trajectories of flood risks, showing not only how flood events and responses to them can move issues to new moments or modalities but also how this movement can vary among localities. Their account emphasizes that whether local flood risk management moves from "hotter" moments of more intense, active controversy to "cooler" moments of consensus and routinization depends in part on the production and deployment of particular forms of knowledge.

We build on this methodological framework by highlighting two different dimensions. First, rather than examining variation in political trajectories between two localities, we investigate two moments in the history of a single (although ever-changing) place. This historical variation, we argue, is as important as local particularity for understanding how issues move between controversy and consensus. Second, rather than emphasizing the production of knowledge in shifting the political modality of an issue, we highlight the role of desires and their translation into aesthetic representations of imagined socioecological futures. Just as the mobilization and deployment of knowledge claims can drive the trajectory of an issue, so, too, can circulating visions of a common world yet to come.

The Kinnickinnic River: A Tale of Two Urban Socioecological Futures

Our analysis compares two moments in the history of one ten-block stretch of Milwaukee's Kinnickinnic River, which runs through the city's south side. At just under 16 km (10 miles) long, the Kinnickinnic, frequently called "Milwaukee's forgotten river," is the shortest of the Milwaukee River Basin's three major rivers. It is also the basin's most heavily urbanized, with a watershed contained almost entirely within industrial and residential neighborhoods of Milwaukee and neighboring suburbs. The segment we examine flows within an artificially straightened channel between 6th Street and 16th Street, in the neighborhood of Lincoln Village. This short segment of the river has become a matter of concern more than once. The first moment we trace is in the early 1930s, when controversy erupted over a Common Council resolution to cover the stretch of river in a box sewer tunnel; the second is the recent introduction of a plan to remove a concrete channel from the stretch, deconstruct dozens of nearby houses, and restore a wider flood plain.

Because we do not have the space here to present the cases in detail, we analyze one key document for each time period, tracing the involvement of each in the trajectory of the political. Both documents respond to matters of concern—in one case contamination, in the other flood risk—in ways that assemble and mobilize the "voices of the public" but in strikingly different ways and for different purposes. The first articulates a vision of a beautified future Kinnickinnic River, forming part of a county-wide system of parks and parkways. The second deploys a remarkably similar imagined future for this river, which has now been channelized in concrete for five decades. We aim to show how the imagined socioecological futures that circulate through these documents have played central roles in shifting their respective matters of concern among moments within their political trajectories.

How the Kinnickinnic Should Look: The Box Tunnel Controversy

The first document we consider is a fourteen-page pamphlet issued in 1931 by Charles Whitnall, Socialist head of Milwaukee's Public Land Commission. This pamphlet, entitled *How the Kinnickinnic Should Look* (Whitnall 1931), was produced for a specific political purpose: to overturn a legislative proposal by Alderman Max Galasinski to enclose the stretch of river underneath a box sewer tunnel. In 1929, Galasinski, the nonpartisan representing the Fourteenth Ward, introduced a resolution to the Milwaukee Common Council to construct a box tunnel over the river, on the grounds that the stretch presented a nuisance and a public health hazard to residents of the surrounding neighborhood. In short, the increasing contamination and stagnation of the segment of river had become a matter of concern. At the end of 1930, the nonpartisan majority of the Common Council voted to support Galasinski's resolution, and the same majority overrode a subsequent veto by Socialist Mayor Daniel Hoan. Legislative procedures had ostensibly translated the box tunnel proposal into the "will of the people," and it appeared well on its way to establishment as a technical solution that, over time, would become a normal part of everyday routines.

The box tunnel resolution directly threatened an important corridor within Whitnall's grand plan for a metropolitan system of parks and parkways, however, which he had introduced in 1923 and was in the process of implementing (Platt 2010). To maintain the plan's integrity, *How the Kinnickinnic Should Look* argues that the goal for the stretch should be to "restore as nearly as practicable the country-like atmosphere that natural streams maintain" (n.p.).[2] The importance of the "country-like atmosphere" stems primarily from the benefits it conveys for ordinary people, relieving the stress of congested urban life and improving their health and disposition. As the pamphlet puts it, the Fourteenth Ward—at the time the predominantly working-class heart of Polish Milwaukee—"is in greater need of park influences than almost any other part of the city." The pamphlet's frontispiece depicts "a photograph of a parkway quite similar to what the fourteenth ward parkway would be when complete with a pedestrian walk close by" (Figure 1).

Figure 1. *How the Kinnickinnic Should Look. Source:* Image courtesy of the City of Milwaukee.

It presents a romantic vision of the future of the Kinnickinnic River: meandering and tree-lined, fit for leisurely boating, and abutted by graceful walks and parkways.

The pamphlet also contends that such a river, supplemented by human assistance, could address the city's problems conveying sewage and pollution better than a box tunnel. For instance, it cites a recent health department report that suggests cutting the floor of the river into a V shape to avoid stagnation. Moreover, it suggests that the river's low flow rates, cited as a concern in the box tunnel resolution, could be addressed by "utilizing the water from a natural spring while there is sufficient, and amplifying it from the hydrant occasionally when the spring supply is short."

How the Kinnickinnic Should Look articulates and assembles the voices and desires of an interested public in several distinctive ways. First, on its last page it features reprints of two unsigned local newspaper editorials. The second of these, entitled "Aldermen—Give Our Children a Chance," is from the *Milwaukee Leader*, a Socialist daily near the height of its circulation. The editorial does not claim to speak as the people, but it speaks on behalf of them: The aldermen voting for the box tunnel "not only showed utter lack of vision, but also gave the people of Milwaukee and especially the children of the section affected a very unfair deal." It invokes a working-class identity, arguing that the "sections in the city that need especial attention toward beautification" include those "in which live those of us who cannot go to Florida in winter or to the mountain lakes in the summer, but must stay on the job—or looking for one—all the year around." It also appeals explicitly to a sense of fairness and justice: "Why must all the beauty go only to those who have the money to pay for it? Why can't at the very least the creeks be saved for the children of hard working fathers and mothers?" In contrast, the first editorial, from the more conservative *Milwaukee Daily Journal*, is entitled "Milwaukee—Business Men Speak." This editorial expresses support for the position of the nearby Layton Park Business Men's Association that the river should be "cleaned, parked, made attractive," instead of buried beneath a box tunnel. The pamphlet thus assembles a public will that purportedly reflects not only the needs of working-class south-siders for "a bit of nature's beauty" along the Kinnickinnic but also the desires of a wider public—including the businessmen who typically supported the nonpartisans over the Socialists.

The *Journal* editorial concedes that "Mr. Galasinski seems to have the support of many south siders; at least many signed a petition favoring his box tunnel." The main text of the pamphlet also acknowledges this petition, but it suggests that the petition does not reflect the true will of the people, because it is based on a misunderstanding and a deception: "The people there [in the Fourteenth Ward] have been misled and do not understand the benefit to be derived from the parkway." In addition, the river's potential to serve as a mechanism for conveying sewage, by "amplifying" flow rates artificially, "appears to baffle these fourteenth warders' understanding." Whitnall's pamphlet condemns the Common Council vote for the box sewer not just as bad but as "erroneous," based as it was on "one of their number [complaining] of its being stagnant and creating a stench at certain times." The pamphlet recognizes the stretch of river as a matter of concern, but it aims to interrupt and disrupt the issue's trajectory, by calling into question both the public that its contamination ostensibly mobilized and the legislative effort to translate the will of this public into the construction of a box tunnel.

For the pamphlet to function not only as a means of delegitimizing one "will of the people" and articulating another but also as a tool for mobilizing popular opposition to the box tunnel, it had to circulate. A January 1931 article in the *Leader* on the pamphlet's release (it later reprinted the text in full) reported that "Whitnall proposes to carry the fight against the box tunnel to the people" ("Will continue fight against boxing of river" 1931, 7). And carry the fight to the people he did, through a whirlwind lecture tour to promote the pamphlet throughout the city. His audiences ranged from Socialist Party branches in wards near the river, where he spoke about how the parkway would "beautify many neighborhoods in the interests of the common people," to civic associations more likely to be constituents of the nonpartisans ("Whitnall to talk on box tunnel fight" 1931, 5). Many of the associations Whitnall addressed immediately filed statements with the Common Council to protest the box tunnel resolution, and when the Council's Sewers and Alleys Committee met in late February, the *Leader* reported "an estimated crowd of more than 250 persons present at the meeting, the greater part of whom appeared to protest against the creation of a box tunnel" ("Many protest box tunnel" 1931, 3). Civic association representatives continued to file petitions and speak at meetings throughout the subsequent year, and after the Socialists secured a majority coalition in the 1932

Common Council elections, they succeeded in killing the box tunnel plan entirely. Over the next decade, Whitnall's parkway was constructed instead, and his vision became the basis—temporarily, at least—for recomposing the river as part of the common world.

The Kinnickinnic River Corridor Neighborhood Plan: River Restoration and Neighborhood Redevelopment

The second document we consider concerns the same stretch of river, this time in very different circumstances. In the early 1960s, city engineers modified the Kinnickinnic in the interest of flood control, by further straightening and lining the channel with concrete—ending its brief era as a beautified part of the parkway system. Although this was another controversial historical moment, the controversy subsided rapidly as the quickly reengineered channel became the new "normal." Over the following decades, however, the project failed to alleviate flooding, often exacerbating its damage. New concerns emerged about safety, health, and overall decline in the condition of the surrounding neighborhood, the population of which had become increasingly low-income, with a growing number of Latin American immigrants. Due to the unexpected effects of both the concrete and increasing upstream development, the river segment again became a matter of concern. Serious attention to addressing these problems began in the mid-2000s, culminating in the 2009 *Kinnickinnic River Corridor Neighborhood Plan*.

Initially, deliberation on the new problems generated by channelization was confined primarily to technical experts. In 2007, the poor condition of the concrete lining and continued flooding prompted the Milwaukee Metropolitan Sewerage District (MMSD) to contract an engineering firm to study options. Subsequently, in collaboration with the nonprofit Sixteenth Street Community Health Center (SSCHC), the agency established a Technical Review Committee to review proposed solutions. The Committee, initially composed of "representatives from government agencies, educational organizations, and nonprofit organizations with expertise in water-related environmental issues," ultimately proposed modifying the channel extensively and clearing a wider flood plain, which would require the removal of more than eighty homes along the river (MMSD and SSCHC 2009, 3).

By late 2007, however, the political trajectory of the issue threatened to shift, as the question of who should compose the public of deliberation began emerging as a problem. Initially, the Technical Review Committee did not include residents of the adjacent neighborhoods. According to a leading Committee member, the technical stakeholders felt it would be more effective to discuss the range of feasible alternatives for channel design with the engineering firm before involving residents (personal interview). Given the necessity of acquiring and deconstructing homes in the flood plain, the Committee sought to resolve as many technical questions and reduce as much uncertainty as possible before opening the deliberations to a wider public.

Nonetheless, the Committee recognized that the proposed modifications would have a substantial impact on the surrounding community, and the political trajectory changed after attention shifted from channel redesign to the future of the neighborhood as a whole. In Spring 2008, the Committee was expanded to include a broader set of stakeholders, and MMSD contracted an environmental planning firm to spearhead the creation of a plan incorporating stakeholder and resident involvement in articulating a vision for the neighborhood's future. To assemble and mobilize this expanded public will, SSCHC's contractor conducted interviews early in 2009 with community leaders and local officials to provide preliminary contributions. Subsequently, the Committee convened three public meetings to present progress and to solicit additional resident input.

After the first two public meetings, the Committee and the planning firm held workshops to determine how to incorporate neighborhood concerns into the plan. Along with the interviews and survey results from a neighborhood organization, the resident feedback (classified as "Strengths, Weaknesses, Opportunities, and Threats") encompasses six pages of the document's appendix, and it reveals mixed reactions to the plan recommendations. Although most comments were supportive or neutral, others articulated opposition to aspects of the project: "Architectural images are not like what one would find in the neighborhood, look too much like for the wealthy," "Amphitheater is not appropriate for our neighborhood," "Like the old character of Pulaski Park—trees," and "No change in river width at this time" (MMSD and SSCHC 2009, 95–96).

One representation of the vision for the river's socioecological future constructed through the process appears on the document's cover (Figure 2). In this image, the river corridor is engineered with terraced

Figure 2. *Kinnickinnic River Neighborhood Plan*. *Source*: Image courtesy of SmithGroupJJR. (Color figure available online.)

banks on which people sit to relax and converse, and pedestrian and automobile bridges connect the two sides, which feature an amphitheater and an overlook. Several individuals fish as others walk, bike, or enjoy the scenic view from the banks. A woman sits at the river's edge, cooling her feet in its waters; others cross the river on stepping stones. In this vision, the river no longer serves as a barrier dividing and endangering the neighborhood but as a gathering place to socialize and connect with urban nature.

But not everyone felt that this vision reflected the desires of the public, and the trajectory of the issue shifted from the less controversial moment of deliberation toward a contested reopening of the question of the public will. During the planning process, a community organization headquartered within the neighborhood argued that resident participation was insufficient and asked for formal representation on the Technical Review Committee (personal interviews; Lackey 2009). The organization, although involved in projects throughout Milwaukee, serves as the de facto neighborhood association in the absence of an officially designated one, but it was not initially identified as a stakeholder in the project and thus was not included in the Committee until later in the planning process. The organization characterized the process as the "development of the Lincoln Village neighborhood by those that do not reside here," and it condemned "the refusal of the planning committee to invite residents to serve on this [Technical Review] committee" (Urban Anthropology Incorporated 2009a). It mobilized a protest outside one of the public meetings, featuring signs with slogans like "Being told is not the same as helping decide" (Urban Anthropology Incorporated 2009a).

The Technical Review Committee rejected this characterization, maintaining that its efforts to involve residents in the planning process were adequate and appropriate. It also held that the neighborhood organization's specific request for participation—for "residents to comprise a majority of the planning committee"—was not feasible for a project of this nature (Lackey 2009). Meanwhile, other neighborhood residents questioned the organization's articulation of the public will, composed from the vote of a group of "18 residents, some living along the banks of the KK" (Lackey 2009). One self-identified "10-year resident" of the neighborhood, for instance, posted an online comment that the claims of the organization to represent the neighborhood "[do] not qualify [its leader] to protest this process on my behalf" (Lackey 2009).

In the controversy's wake, the Committee scheduled an additional meeting that would include the sewerage district, the engineering firm, residents, and the neighborhood organization. Here, the issue's trajectory shifted again, as another conventional process of democratic election provisionally settled the question of the will of the public and nudged the project back to the moment of deliberation. Specifically, the meeting provided residents an opportunity to vote on the proposals to widen the river and acquire residential property, and in fact a majority of residents present voted in favor of both (Urban Anthropology Incorporated 2009b). At the time of this writing, the controversy has largely died down, and MMSD has begun designing the new channel and deconstructing homes in the flood plain. Although the political trajectory of this matter of concern could change directions again at any time—and several years of work remain—the process of normalizing a new version of the everyday life of the Kinnickinnic is now underway.

Convergences and Divergences

In both of these situations, documents designed to circulate an imagined socioecological future for the Kinnickinnic influenced the political trajectory of a matter of concern. Despite striking similarities between the documents' idyllic visions, however, their political roles diverge in important ways. In the case of the fight against the box tunnel, Whitnall's imagined future Kinnickinnic emerged within a historical assemblage in which decisions were entrusted to elites, with few institutions or expectations for public participation in design and planning. This vision of the future originated not from popular desires or demands but from intellectual perspectives influenced by town planner Ebenezer Howard and the Garden City movement (Platt 2010).

Nonetheless, the document functioned to disrupt a trajectory that appeared headed toward a moment of normalization. Its language and images aimed to delegitimize dissent to the parkway and the truth of the public will that Alderman Galasinski had assembled, arguing that the apparent desire for a box tunnel was based on misunderstanding and deception. It recast the box tunnel solution as problem, and around this new problem it assembled a new, larger oppositional public—both within the neighborhood and throughout the city, ranging from businessmen to working-class children—united by the desire for a beautiful parkway. As it circulated, it helped Whitnall mobilize this new public to interrupt and delay the legislative and administrative processes that would have established the tunnel as a translation of the general will.

In the more recent case, the plan document—although not designed with an explicit political purpose—helped contain and reduce controversy and conflict resulting from perceptions that outside elites were imposing a particular vision on the river without resident participation. On the one hand, the plan is not a populist document; it embodies discourses and practices of landscape designers, civil engineers, public health experts, and other river restoration professionals. Unlike Whitnall's pamphlet, however, the 2009 document legitimizes dissent by including a range of public input in its appendix, including criticism of the planners' visual representation of the Kinnickinnic's imagined future. It emerged within a historical assemblage in which the expectation of meaningful public participation has been fully institutionalized.

In contrast to the disruption that the circulation of Whitnall's pamphlet helped mobilize, the process that generated the 2009 plan effectively ended disruption of the restoration project, by opening the project to a democratic vote. Indeed, we could easily interpret this consensus-seeking process and the planning document's role as symptomatic of the present postpolitical condition. Instead, we argue that the debate over resident participation represents a key moment in the political trajectory of the flood control project as matter of concern. Specifically, the plan document and the process that produced it helped keep the restoration moving toward the moment of normalization, rather than reopening the public and its desires as matters of controversy.

Conclusion

In this study, we have introduced new elements for a pragmatist conceptual framework to guide analysis of the political in UPE. Two moments in the history of a short river segment obviously cannot provide a representative sample of the many ways in which the political trajectories of matters of concern might unfold. We contend, however, that they suffice to demonstrate ways to develop the framework that Donaldson et al. (2013) introduced in their study of flood risk management. Specifically, we have argued that in addition to local variation and the role of knowledge production, accounts of the political trajectories of issues should also attend to the mobilization of aesthetic desires—here, for particular imagined futures—within specific historical configurations.

We argue against conceptualizing the *demos* and its desires as a given, preexisting public will, which is then simply reflected or represented, adequately or inadequately, in formal articulations of socioecological futures. Instead, we contend that UPE needs to incorporate within its accounts the mundane steps involved in assembling a public and articulating its will and desire, such as gathering petitions and conducting interviews, translating voices into editorials or "public input" appendices, arranging straw votes, or mobilizing people to disrupt or legitimate formal decision-making processes. Instead of exiling these "ordinary dynamics and disappointments" outside the purview of the properly political, our analyses should trace how such activities change the particular ways that issues are political. Indeed, this need to trace also applies to the moments of disruption in prevailing distributions of the sensible—what practical steps are taken, what forms and documents circulate for such exceptional moments to occur (cf. Barnett and Bridge 2013; Fuller 2013)?

As Donaldson et al. (2013, 606) suggest, a pragmatist framework of this kind is not "opposed to other approaches to political analysis," and we contend that it offers a methodology more widely applicable within UPE. For example, analyses emphasizing the role of hegemonic projects in shaping socioenvironmental change could deploy such a framework to trace the specific means—documents, practices, events, and so on—through which such projects articulate and assemble a public interest and translate it into normalized routines of everyday life. One challenge for future research will be to identify additional means through which issues travel between the most and least controversial points within their trajectories; another will be to conceptualize the relationships among these means, along with the historically and geographically specific configurations that enable them to have effects. Under what circumstances do circulating translations of knowledge or desire—or both in combination—generate disruption of a legislative process or consolidate consensus in a process of deliberation? Are there predictable patterns in issue trajectories, or is each one irreducibly unique? As for the normative aspiration of much UPE to "enhance the democratic content of socioenvironmental construction," a pragmatist approach can lend specificity to what this "enhancement" might require under different sets of conditions (Fuller 2013).

To be sure, analyzing the political dimensions of UPE also requires continued engagement with existing research on the democratization of knowledge production, including the relationship between expert and lay knowledges (e.g., Eden and Tunstall 2006). But if the aim is to include not only a more diverse range of knowledges but also a wider range of desires in the shaping of future urban environments, then we must ask this: Where do these desires come from? How are they formalized and circulated, and how do they alter trajectories of socioecological change? And, to ask a timeless question, when desires diverge or conflict, how do we select among them? By pursuing such questions, we can not only develop richer, more complex analyses of how socioecological construction works in practice but also help build the basis for identifying better ways for it to happen.

Acknowledgments

The authors thank three anonymous reviewers, Bruce Braun, and audiences at University of Wisconsin, University of California–Davis, and the Association of American Geographers annual meeting in Los Angeles for insightful critiques and questions. They also thank the Sixteenth Street Community Health Center's Department of Environmental Health, Urban Anthropology, Milwaukee Metropolitan Sewerage District, and the Milwaukee Central Library, and SmithGroupJJR and the City of Milwaukee for permission to reproduce its artwork.

Notes

1. We cannot do justice here to the rich, growing literature on pragmatism and geography. See, for example, the 2008 special issue of *Geoforum* 39 (4). For the distinctive contributions of "French pragmatism," see Fuller (2013).
2. The pamphlet has no page numbers. All subsequent quotes are from unnumbered pages.

References

Alberti, M. 2007. *Advances in urban ecology: Integrating humans and ecological processes in urban ecosystems*. New York: Springer.

Barnett, C. 2012. Situating the geographies of injustice in democratic theory. *Geoforum* 43 (4): 677–86.

Barnett, C., and G. Bridge. 2013. Geographies of radical democracy: Agonistic pragmatism and the formation of affected interests. *Annals of the Association of American Geographers* 103 (4): 1022–40.

Bennett, J. 2005. In parliament with things. In *Radical democracy: Politics between abundance and lack*, ed. L. Tønder, L. and L. Thomassen, 133–48. Manchester, UK: Manchester University Press.

Dikeç, M. 2005. Space, politics and the political. *Environment and Planning D: Society and Space* 23:171–88.

Donaldson, A., S. Lane, N. Ward, and S. Whatmore. 2013. Overflowing with issues: Following the political trajectories of flooding. *Environment and Planning C: Government and Policy*, 31 (4): 603–18.

Eden, S., and S. Tunstall. 2006. Ecological versus social restoration? How urban river restoration challenges but also fails to challenge the science–policy nexus in the United Kingdom. *Environment and Planning* C 24 (5): 661–80.

Fuller, C. 2013. Urban politics and the social practices of critique and justification: Conceptual insights from French pragmatism. *Progress in Human Geography* 37 (5): 639–57.

Heynen, N. 2014. Urban political ecology I: The urban century. *Progress in Human Geography* 38 (4): 598–604.

Keil, R. 2003. Urban political ecology 1. *Urban Geography* 24 (8): 723–38.

Lackey, J. 2009. Lincoln Village's UrbAn displeased with KK River plans. Bay View Compass.com. http://bayviewcompass.com/lincoln-village-residents-displeased-with-kk-river-plans/ (last accessed 1 May 2014).

Latour, B. 2007. Turning around politics: A note on Gerard de Vries' paper. *Social Studies of Science* 37 (5): 811–20.

Loftus, A. 2012. *Everyday environmentalism: Creating an urban political ecology*. Minneapolis: University of Minnesota.

Many protest box tunnel; defy hecklers. 1931. *Milwaukee Leader* 20 February:3.

Marres, N. 2007. The issues deserve more credit: Pragmatist contributions to the study of public involvement in controversy. *Social Studies of Science* 37 (5): 759–80.

Milwaukee Metropolitan Sewerage District and Sixteenth Street Community Health Center (MMSD and SSCHC). 2009. *Kinnickinnic River Corridor neighborhood plan*. Madison, WI, and Milwaukee, WI: JJR, LLC and PDI/Graef.

Njeru, J. 2010. "Defying" democratization and environmental protection in Kenya: The case of Karura Forest reserve in Nairobi. *Political Geography* 29 (6): 333–42.

Perkins, H. 2013. Consent to neoliberal hegemony through coercive urban environmental governance. *International Journal of Urban and Regional Research* 37 (1): 311–27.

Platt, L. 2010. Planning ideology and geographic thought in the early twentieth century: Charles Whitnall's progressive era park designs for socialist Milwaukee. *Journal of Urban History* 36 (6): 771–91.

Rancière, J. 2001. Ten theses on politics. *Theory & Event* 5 (3). http://muse.jhu.edu/journals/theory_and_event/toc/tae5.3.html (last accessed 11 December 2014).

Swyngedouw, E. 2009. The antinomies of the postpolitical city: In search of a democratic politics of environmental production. *International Journal of Urban and Regional Research* 33 (3): 601–20.

———. 2011. Interrogating post-democratization: Reclaiming egalitarian political spaces. *Political Geography* 30 (7): 370–80.

Swyngedouw, E., and N. Heynen. 2003. Urban political ecology, justice and the politics of scale. *Antipode* 35 (5): 898–918.

Urban Anthropology Incorporated. 2009a. 84 homes to be lost in KK flood management project. *Lincoln Village Voice* September: 1.

———. 2009b. Residents vote on the KK project. *Lincoln Village Voice* October: 1.

Véron, R. 2006. Remaking urban environments: The political ecology of air pollution in Delhi. *Environment and Planning* A 38 (11): 2093–2109.

Whitnall, C. 1931. *How the Kinnickinnic should look*. Milwaukee, WI: Public Land Commission.

Whitnall to talk on box tunnel fight. 1931. *Milwaukee Leader* 27 January:5.

Will continue fight against boxing of river. 1931. *Milwaukee Leader* 20 January:7.

Toward an Interim Politics of Resourcefulness for the Anthropocene

Kate Driscoll Derickson* and Danny MacKinnon†

*Department of Geography, Environment, and Society, University of Minnesota
†Centre for Urban and Regional Development Studies, Newcastle University

Based on the need for meaningful political responses to socionatural change, in this article we develop an interim politics of resourcefulness as a strategy for addressing the limitations of postpolitical environmental governance. Drawing on political and epistemological insights of third-world feminism as well as an ongoing collaborative with environmental justice organizations in West Atlanta, we argue that visions for just socionatural futures must necessarily be generated in conversation with historically marginalized communities. We offer an interim politics of resourcefulness as one way of forging those kinds of engagements between academic researchers and communities, and describe the forms that such engagements have taken in our own research.

我们根据对社会自然变迁做出有意义的政治回应之需要，于本文中发展出资源丰沛性的临时政治，作为处理后政治环境治理限制的策略。我们运用第三世界女性主义的政治与认识论洞见，以及与亚特兰大西部环境正义组织进行中的合作，主张公义的社会自然之未来愿景，必须从和历史上受到边缘化的社群对话中生成。我们提供资源丰沛性的临时政治，作为打造上述学术研究者和社群之间相互参与的一种方式，并描绘此般参与在我们自身的研究中所採取的形式。

A partir de la necesidad de respuestas políticas significativas al cambio socionatural, en este artículo desarrollamos una política interina de ingeniosidad como estrategia para abocar las limitaciones de la gobernanza ambiental pospolítica. Apoyándonos en las estrategias políticas y epistemológicas del feminismo tercermundista, lo mismo que en la colaboración en curso con las organizaciones de justicia ambiental del sector occidental de Atlanta, sostenemos que las visiones de futuros socionaturales justos deben generarse necesariamente en la conversación con comunidades históricamente marginadas. Proponemos una política interina de ingenio como una de las maneras de forjar tales tipos de compromiso entre investigadores académicos y comunidades, y describimos las formas que han tomado esos compromisos en nuestra propia investigación.

Although climate scientists reached consensus on the relationship between human activity and a warming planet decades ago, the necessity for both elite and popular imaginaries in Western liberal democracies to contend with the inevitability of climate change is a more recent phenomenon. Indeed, climate change was on the agenda in Davos in 2014 (Confino et al. 2014) at a "high level private session" of the World Economic Forum, and the World Bank has put climate change at the center of the Bank's mission (World Bank 2014). Far from Davos, in living rooms and church basements, everyday people organize themselves to address and mitigate carbon emissions and the impact of global warming. Thus, it seems that we can no longer imagine futures, capitalist or otherwise, without thinking about climate change specifically or socionatural transformation more broadly. This is the condition of the Anthropocene. Geologists can quibble as to whether we are really in a new geological era, the irreversible consequence of human activity of a certain kind. What is clear enough is that we are in a new political era, in which futurity is conditioned by the consequences of a changing planet.

The fact of climate change and the consequences it reaps might be a problem at the planetary scale, but neither the causes nor the consequences can be understood as evenly distributed. Like crises of capitalism, those who stand to suffer most from it did not precipitate this crisis. *Climate justice* is the term meant to signal these uneven causes and consequences of climate change, both "geographically and socially" (Chatterton, Featherstone, and Routledge 2013, 2). In their recent piece "Climate Leviathan," Wainwright and Mann (2013) ask, "Do we have a theory of climate justice?" and answer a resounding "No." We want to suggest not only that we do not have a theory of climate justice but that we cannot have a theory of climate justice; not yet.

Vulnerability to climate change is not the only thing that is unevenly distributed—so, too, is the ability to meaningfully influence climate futures and contribute to the process of imaging and enacting alternative futures. This uneven capacity is shaped and conditioned along persistent axes of sedimented social difference. The margins are where climate change will be most acutely experienced (Intergovernmental Panel on Climate Change 2014), where it has been least produced, and where the barriers to imagining and engendering alternative futures are highest. The claim that the capacity to envision and engender alternative socionatural futures is unevenly distributed is not a claim about the essential nature of the marginalized but rather an observation about the present nature of social formation—an observation about the margins themselves. The challenges that historically marginalized communities face in producing and enacting visions of socionatural futures are material, cultural, and political. What might seem to some like the banalities of poverty in the United States present meaningful, material barriers to their capacities to simply be together in space to reflect on their concerns and develop strategies for the future.

These challenges must be remediated if we are to develop just theories of socionatural futures and climate justice. Here we are drawing on the epistemological tradition that emerges largely from feminist and postcolonial scholars to argue that knowledge is always partial and situated, both geographically and in relation to social and political power structures (Anzaldua 1987; Haraway 1988; Harding 1991; Rose 1997). Work in this tradition holds that knowledge is situated not only by the social and geographical location of the knower but also by the methods by which it travels (i.e., through academic journals, community-engaged projects, or policy circles) and the strands of thought with which it is engaged (i.e., continental philosophy, subaltern studies, the Frankfurt School, third world feminism, etc.). This argument has been made through philosophical and theoretical critique of critical and mainstream epistemologies and the sorts of representational regimes they reify and politics they engender, as well as through the observed and lived experiences of political movements.

In particular, third-world feminists and feminists of color mobilized forceful critiques of second wave feminist theory and practice, demonstrating the ways in which its failure to engage substantively with the lives and political desires of women of color, poor women, and women beyond the Western world rendered much second-wave feminist theory and practice not only inadequate but harmful to the degree that it reproduced marginalization (hooks 1984; Lorde 1984; Mohanty 1988, 2003; Ong 1988). This powerful observation and critique shifted the horizons of much feminist theory and practice away from universalizing narratives regarding the substance and subjects of justice and toward diverse politics of epistemology that focused on how and by whom knowledge and associated visions of the future can and should be produced. This epistemological posture has focused on the production of knowledges that can learn from other knowledges (rather than contest or silence them); the processes of "achieving" various standpoints that do not reinforce universalizing subject positions; and the creation of space for the "view from the margins" (see, e.g., Haraway 1988; Nagar 2006; Peake and Rieker 2013).

The transformative possibilities that inhere in experiences, worldviews, and knowledges that are marginalized or rendered invisible are echoed in a more recent set of observations about politics in general and environmental governance in particular. Ranciere and others (Mouffe 2005; Swyngedouw 2007; Paddison 2009) have used the concept postpolitical to highlight how that which is understood as political in the present often does little to substantively challenge the larger social and political order. Ranciere (2010) uses the term "the part of those that have no part" to refer to the modes of life that are obscured and marginalized by the dominant social order. For Ranciere, moments that can be understood as "properly political" are those in which the "part of those that have no part" are rendered visible and in this sense bring about a rupture in the social order. "Politics" he argues, "before all else, is an intervention in the visible and sayable" (37).

Swyngedouw (2007) has extended this analysis to environmental governance, arguing that much of the public discussion about environmental futures favors technological fixes in the same register as the damage wrought (e.g., cutting CO_2 emissions) rather than the substantively political question of the kinds of natures we want to inhabit. The notion of the postpolitical turns on a specific and somewhat counterintuitive use of the term *political*, yet we find it useful for identifying and understanding the ways in which environmental governance is intensely circumscribed with implications for the capacity of historically marginalized communities to meaningfully engage or transform environmental governance processes in accordance with their own visions. Crucially, this reading of postpolitics does not assert that politics are no longer relevant or possible (see McCarthy 2013) but rather identifies an approach to governance that actively marginalizes or constrains antagonisms that would meaningfully transform or challenge the social and political order and proceeds as though the questions that these thinkers consider "properly political" are not valid or even possible questions to consider.

Following on from these debates, we are proposing an "interim politics of resourcefulness" (MacKinnon and Derickson 2013) as an approach and epistemological posture for social science inquiry that aims to produce knowledge about the form that just socionatural futures might take. Given that the causes and consequences of climate change and socionatural transformation are unevenly distributed, social science inquiry must, we argue, necessarily substantially engage and actively resource those who are most vulnerable. We understand this vulnerability to be largely socially produced along persistent and sedimented access of social difference. We are proposing an approach that does not seek to produce a theory of climate justice but rather a politics that seeks to produce the conditions in which just theories of climate justice can emerge. As such, resourcefulness is a political and epistemological posture aimed at remediating the conditions that produced and reproduce the uneven capacity to engender alternative futures. It is an interim politics in that it prioritizes the act of cultivating the conditions in the immediate term that are conducive to full participation in knowledge production and visioning practices, over and above working toward the realization of predetermined, philosophically deduced conceptions of climate, environmental, and social justice.

In the following sections, we describe an ongoing collaboration that takes resourcefulness as its guiding principle between Derickson and community-based environmental justice organizations in West Atlanta. We engage Ranciere's political ontology to interpret the efforts of a nascent struggle around environmental politics in West Atlanta as a struggle for historically marginalized publics to "take part" in bringing about alternative socionatural futures. We note, however, what we consider to be critical shortcomings in Ranciere's interpretation of politics, insofar as it emphasizes spontaneous and ephemeral rupture and provides little comment on the possibilities of forming political solidarities, particularly between those who are not recognizable within the dominant social order and those who are. We turn then to Mouffe's conception of chains of equivalence to consider how such forms of solidarity might be conceived. We conclude that an interim politics of resourcefulness is an epistemological and political strategy for forging solidarities that seek to redress the everyday challenges historically marginalized communities face as they seek to articulate and realize alternative socionatural futures in the context of postpolitical environmental governance.

Resourcefulness in Two Registers

Resourcefulness, as we have practiced it and described it elsewhere (MacKinnon and Derickson 2013; Derickson and Routledge 2014), is a political posture and an epistemological approach to collaborative research with historically marginalized communities in two registers. First, it raises a set of empirical questions regarding the ways in which communities are coconstituted with the social formation, with important implications for their varying capacities to shape environmental futures. We have argued that resourcefulness should be understood as relational in the sense communities themselves cannot and should not be understood to be resourceful as a characteristic in their own right but rather their capacities for mobilizing resources are in relation to the social formation—the political, economic, and cultural practices that interact to create our social world.

Second, the notion of resourcefulness can serve as a normative ideal and an ethical practice of scholarly research (Derickson and Routledge 2014). As a normative ideal, resourcefulness can serve as one condition (among many) that political action and public policy can aim to bring about. Elsewhere we have argued that as a normative ideal, resourcefulness is far more compelling than the currently fashionable resilience, because it is expressly concerned with the capacity of communities to articulate and realize their own visions of the future (MacKinnon and Derickson 2013). As an ethical practice of scholarly research, resourcefulness is aimed at designing research questions, processes, and practices in ways that are always informed by the concerns, desires, objectives, and needs of historically marginalized communities. In other words, research should resource the capacity of historically marginalized communities. This is not to supplant intellectual and scholarly questions, nor to suggest that scholars should serve communities uncritically (see The Autonomous Geographies Collective

2010), but rather to suggest that whether and how the act of scholarly research resources historically marginalized communities should always be a substantive consideration in research design.

Taken together, as set of empirical questions, a normative ideal, and an ethical practice of scholarly research, resourcefulness can be understood as an "interim politics." Rather than a politics that has in mind a particular future that it seeks to call into being (i.e., carbon neutral, socialist/anticapitalist, antiracist, etc.), an interim politics seeks to proliferate the capacity of all groups to cultivate and work toward a range of competing visions.

Invisibility and Environmental Politics in West Atlanta

In this section, we describe some ongoing work that Derickson is doing in West Atlanta to offer an example of the challenges facing historically marginalized communities engaging in environmental politics in the context of postpolitical environmental governance, and to illustrate both the potential and necessity of an interim politics of resourcefulness. The project underway in West Atlanta is a collaboration among Derickson, two community-based nonprofits, and their networks and local residents. The goal of the project is to resource the capacity of residents of the Proctor Creek watershed to develop and engender visions for the watershed in and against the context of postpolitical environmental governance.

In collaboration with the West Atlanta Watershed Alliance and Eco-Action, in her capacity as a faculty member at Georgia State University and the University of Minnesota, Derickson has attempted to practice resourcefulness as an approach to scholarly research in West Atlanta. This has largely taken the form of a three-year project to support the development of a Watershed Stewardship Council of residents in the Proctor Creek watershed.

The watershed is located in the northwest quadrant of the City of Atlanta and historically served as the channel for raw sewage from downtown to the Chattahoochee River. The watershed's environmental condition and demographic features are well predicted by decades of research on the alarming correspondence between low-income communities of color and environmental degradation. Primarily home to low-income African Americans, the watershed contains more than twenty-nine "hot spots" identified by the Environmental Protection Agency (EPA). The creek itself floods often due to poor stormwater runoff management in the region. The floods often breach low-lying homes in the watershed, carrying sewage, chlorine disinfection by-products, and untreated stormwater runoff, and leave behind disease-causing pathogens, mold, and higher incidences of mosquitoes infected with West Nile virus (Vazquez-Prokopec et al. 2010). The neighborhoods in the watershed are also notable for their high levels of vacant housing.

The challenges that residents of the Proctor Creek watershed have faced as they attempt to engage with state agencies to change the environmental conditions in the watershed are emblematic of postpolitical environmental governance and provide insight into the applicability of resourcefulness as an interpretive frame as well as a normative ideal and ethical practice of scholarly research. As residents have sought to establish a community-based resident board of "stewards" to represent community concerns, communicate local knowledge, and influence the processes and agencies that produce and govern the watershed, they have found a startling refusal on the part of the EPA, the city Watershed department, and other environmental nonprofits to engage with their analysis and objectives. Invisibility is a common theme that arises in interviews with stewards, both in terms of the spaces they are concerned with and the concerns they have raised.

One steward jokingly suggested that the failure of the city officials to see thousands of discarded tires during what they claimed was an exhaustive survey of the creek indicated that Klingons (a species from the science fiction series *Star Trek*) must have used their powers of invisibility to hide the tires from view. Recounting his conversation with city officials who claimed that there were no discarded tires along the creek, he said, "We were out there also, but we saw signs of the Klingons. And you know, they're notorious for their cloaking devices, and we're sure that's what happened with these pictures, and why you didn't see this."

The steward further elaborated on a sense of invisibility in later comments, when he described the work of the stewardship council as similar to an adolescent girl biding her time and waiting to be noticed:

> [What we're doing is] helping people in their vision so they have a chance to see. It's there, but you have to help them take those veils." [He described advising his granddaughter that a boy she was interested in would someday notice her.] And what I said was, you just take

> your time, you develop your skills, and do all the things you need to do, and all of a sudden, one day, out of nowhere, it will hit him, he'll smell the perfume, it might be an a-ha moment, and he'll say, "Where have you been all my life?" And that's what will happen to you. And this is what occurs here [referring back to the work of the Stewardship Council].

Residents have also been met with a direct refusal by the EPA to engage in inquiry about the systemic production of environmental degradation and its uneven manifestations, even when using the language of environmental justice or working in historically marginalized communities. When approached by the EPA for a discussion of environmental protection in the watershed, residents indicated a strong interest in linking discussions of poor environmental quality and vacant housing and disinvestment in the neighborhood. They were particularly interested in drawing connections between stormwater management and associated flooding that had led to public health concerns in the neighborhood and contributed to the high levels of vacant housing. Despite framing their engagement in terms of "environmental justice," the EPA refused to facilitate such a discussion, choosing instead to teach residents how to hold community-based clean-ups.

This refusal was evidenced further at a recent environmental justice conference hosted by the EPA in Atlanta. *Environmental justice* is a term and a social movement born out of a desire to politicize and problematize the uneven exposure of people of color to environmental pollutants and locally undesirable land uses (United Church of Christ Commission for Racial Justice 1987; Bullard and Johnson 2000). In addition to a desire to politicize this racist distribution of exposure, a fundamental platform plank of the movement has been that no one should be exposed to environmental toxins. By extension, then, environmental justice activism can be understood as challenging the hegemony of racist, state-supported, industrial capitalism that has produced, sanctioned, and externalized the cost of environmental degradation at a nearly incomprehensible scale. Yet in practice, the EPA's engagement with the concept of environmental justice is to discuss management of the environment (although not necessarily the mitigation of environmental degradation) in poor communities and communities of color, with little acknowledgment of the broader social relations that produced these distributions.

For example, one of the sessions at the conference was titled "Achieving Environmental Justice: Best Practices and Success Stories—Collaborations Between Communities, Government Agencies, Business and Industry." The panel included four community organizers from the Southeast, all of whom were African American, and one white woman from New Jersey who represented Rhodia Chemical, a company that produces chemicals for cigarette filters and bioacumulative chemicals like solvents and surfactants. As each community organizer offered a brief set of remarks on how and why he or she became involved in environmental justice activism, a recurrent theme was premature death in their communities. One man told the audience that every single founding board member of his organization had died of a rare respiratory disease, which he attributed to living in close proximity to a chemical plant. Another woman told the audience that "death was all around us" when she started her work. But this was not the topic at hand at this particular EPA-hosted environmental justice panel. Instead, along with the businesswoman from Rhodia, which had plants in the vicinity of most of the panelists, panelists discussed finding "win–win" solutions with environmental polluting firms like Rhodia, through weekly conference calls, job placement programs, and forums designed to "foster trust" between the companies and residents living near their facilities.

The preceding vignettes are meant to illustrate the degree to which historically marginalized communities in West Atlanta and the region more broadly struggle to achieve visibility and recognition for their environmental concerns, even when the subject is environmental justice. This invisibility, along with the EPA's posture toward these communities, can be seen as an expression of postpolitical governance in terms of the effort to generate a consensus among government, industry, and local residents, actively overlooking or not acknowledging the incommensurability of environmental justice and some forms of chemical and industrial production, despite the overwhelming evidence of environmental degradation and negative health outcomes (i.e., death).

Resourcefulness and the Possibilities of Rupture

In our reading, Ranciere's (2010) conception of politics is quite useful for understanding what is at stake in West Atlanta. Although his body of work is far too extensive to do justice to its nuances here, we take his conceptualization of politics to be as follows. The social world is ordered by a "distribution of the sensible" (36) through which meaning and sense is made. This ordering, however, is never fully reflective of or sensitive to the myriad ways the lives are lived and, as such, the distribution always has an outside, or what Ranciere (2010) calls "the part of those who have no part" (33).

By way of illustrating the work that the "distribution of the sensible" does, Ranciere (2010) suggests that it has a "slogan": "Move along! There's nothing to see here . . . here, on this street, there's nothing to see and so nothing to do but move along" (37). He continues:

> It asserts that space for circulating is nothing but the space of circulation. Politics, by contrast, consists in transforming this space of "moving-along," of circulation, into a space for the appearance of a subject: the people, the workers, the citizens. It consists in re-figuring space, that is in what is to be done to be seen and named in it. (37)

Politics, for Ranciere, occurs when "the part of those who have no part" are rendered visible in ways that radically destabilize the social order. We read Ranciere as arguing that the incompleteness of the distribution of the sensible renders it inherently unstable, thus making what he considers political an ever-present possibility. He uses the language of "rupture" (98) to describe the moments of destabilization of the distribution of the sensible.

There is much in Ranciere's framework that helps us understand the environmental politics of state agencies and historically marginalized communities in West Atlanta. When city officials or EPA representatives deny the extent of pollution, claim to not see what is in plain sight, and refuse to make connections between pollution and neighborhood disinvestment, this is usefully understood as postpolitical governance, or governance that casts the watershed and the space of West Atlanta as a place where there is "nothing to see" and nothing to do but "move along." In fact, stewards have argued that state agencies are only concerned with the environmental well-being of the watershed to the degree that it impacts the cleanliness of the water that flows through the creek and empties into the Chattahoochee River, a major source of drinking water in the region. They are not nearly as concerned, they argue, with the public health concerns regarding mosquito breeding in tires, flooding, erosion, and dumping in the watershed itself.

The stewards clearly understand their efforts as an attempt to bring themselves and the space of the watershed into the realm of the seeable, the visible, and the recognizable. As the earlier quote from a steward talking about his granddaughter illustrates, he is longing for and expecting a moment in which he and the rest of the stewards and the work they have been doing will "help them with their vision" and the environmental managers will have an "a-ha" moment when they really see the stewards and the creek for the first time. The work of the stewards then can be interpreted as an attempt to transform the watershed and the creek from a space of flow and circulation where there is "nothing to see" into a space for what Ranciere calls "the appearance of the subject."

Yet there are ways in which Ranciere's framework obscures as much as it illuminates. In our reading, his emphasis on the ever present possibility of politics tells us little about why and how some ruptures happen and some do not, or as Povinelli (2011) puts it, "how and why . . . some things move from potentiality to eventfulness to availability for various social projects?" (14). Ranciere appears to consider these kinds of ruptures random. Indeed, as Corcoran notes in the introduction of a volume of Ranciere's work he edited and translated, "If Ranciere continually emphasizes the chance-like nature of politics against all the attempts to explain political events by referring to underlying causes," it is because he believes "*nothing* explains why people decide to rise up and demonstrate their equality with those who rule" (Ranciere 2010, 9, italics in original).

In our own engagement with what Paddison (2009) calls "local insurgencies" and what might be understood as "not-yet-insurgent" local initiatives, we have observed that what might appear to be a rupture is in fact a moment in a much longer process and that the "decision" to work to bring about rupture is conditioned substantively by unequal resource distributions. To bring about a rupture requires tremendous amounts of resources and labor, much of which remains unseen and unaccounted for in Ranciere's conception of politics as a result of his emphasis on the ever-present availability and possibility of politics.

In West Atlanta, residents who are working to shape environmental futures not only face symbolic hurdles but wide-ranging material hurdles as well. These include things like limited access to the means of mass communication (cell phones, Internet, photocopy machines), reliable transportation, affordable child care, space to be together, financial resources, and flexible employment schedules. These material challenges pose meaningful hurdles to the process of cultivating collective subjectivities, envisioning alternative political futures, and calling into being political rupture. As one steward put it:

> I think folks want to participate, but they themselves don't have jobs, can't pay bills, don't have a cell phone to call anybody or Internet to get on and reach out to people—it's like an everyday what do I do? I don't have the Internet, I don't have a phone, the library is about three and half miles from here, lack of transportation, MARTA doesn't come through here anymore.

Relatedly, Ranciere's emphasis on radical equality of the subject, and any politics that proceeds as though this is the case, runs the risk of minimizing the radically uneven topography of the social world. Politics is always possible, but it is also almost always very painful, difficult, risky, and costly, and the price is not evenly distributed, even among those who might constitute the part of which has no part.

Finally, in our reading, Ranciere's insistence that nothing can explain what brings about rupture precludes him from offering insight into how solidarities might be forged between political subjectivities that occupy knowable and seeable relationships to the partition of the sensible and the part of those that have no part. We get little guidance from Ranciere, for example, about how the production of academic knowledge might contribute to the kinds of ruptures he describes. As academics seek to articulate theories of environmental justice, climate justice, and just socionatural futures, this is an important consideration.

Mouffe's (2005) work provides insight into how these questions might be addressed in conversation with Ranciere. Like Ranciere, Mouffe is concerned with discerning what might be considered properly political in the context of postpolitical hegemony and conceives of antiessentialist political subjectivities as the location of politics. For Mouffe, the social world is conditioned by hegemony that functions as the grammar through which sense is made. Unlike Ranciere, however, Mouffe has a clear political strategy of cultivating solidarities that knit together various political subjectivities through what she (Laclau and Mouffe 1985) calls chains of equivalence with the goal of developing a counterhegemonic formation.

Mouffe is primarily concerned with finding methods for strategic and solidaristic political alliance that do not elide conflict but are likewise not obliterated by conflict. For Mouffe, political subjectivities are partial, socially constructed, and antiessentialist formations around which the "constitutive we" of democracy is constructed. Although this constitutive we always has an outside "they," Mouffe (2005) turns to agonism to imagine a "we/they relation where the conflicting parties, although acknowledging that there is no rational solution to their conflict, nevertheless recognize the legitimacy of their opponents" (20). She contrasts this with antagonism and consensus-based approaches, arguing that the former seek to eradicate the opponent and the latter eradicate politics by eliding unresolvable difference. By contrast, in an agonistic radical democratic political framework, affinity groups actively seek to engage one another explicitly with respect to conflicting visions and strategies for the establishment of a new hegemony.

The political posture Mouffe advocates for is a "war of position" (in Gramsci's terms) launched against "a multiplicity of sites" toward a "vanishing point" on the horizon toward a new hegemony. Mouffe uses the metaphor of a vanishing point on the horizon in contradistinction to a telos: Although a vanishing point can orient action, it can never be reached. It is in this sense that we see Mouffe's framework as compatible with an interim politics. Moreover, we see strong affinities between the feminist politics of epistemology we outlined earlier and Mouffe's objectives for working toward justice in ways that are expressly attuned to learning from and with differently situated knowledges, experiences, and points of view. Finally, we see promise in this conception of chains of equivalence that are sutured together, partially and ephemerally, as a way to conceive of the possibility and praxis of solidarity across the partition of the sensible or between those who are able to be seen and heard around questions of environmental futures and those who occupy the "part of those that have no part."

An Interim Politics of Resourcefulness

What Ranciere (2010, 33) calls the "part of those who have no part" is not only a symbolic relationship to the partition of the sensible but, we argue, also a material and social relationship to the resources necessary to make collective claims and effectively disrupt the partition of the sensible. On this basis, we propose an interim politics of resourcefulness in two registers outlined earlier as the proper political response to this uneven distribution of capacity in the face of postpolitical environmental governance.

For example, Derickson's collaborative work in West Atlanta (and elsewhere; see MacKinnon and Derickson 2013; Derickson and Routledge 2014) has sought to cultivate resourcefulness in communities as a normative ideal, and her approach to research has been informed by resourcefulness as an ethical practice of scholarly research. This has taken the form of becoming and staying engaged with the work of the Stewardship Council on its own terms and not in accordance with research schedules or driven by academic outputs. It has also included working on projects with the Stewardship Council that are not directly related to the research, including grant writing, brochure design, and meeting facilitation. More substantively, it has taken the form of channeling resources wherever possible from the academic institutions where Derickson has been affiliated, by writing grant budgets in ways that include community residents as researchers and collaborators rather than informants, using research funds to contribute to the salary of a community organizer rather than hiring a research assistant in the area, and making the university space available for computing, printing, and meeting. More subtly, resourcefulness as Derickson has practiced it entails sensitivity to, and a built-in effort to remediate, the mundane and everyday challenges residents face in becoming engaged and sustaining participation. This has entailed offering rides to interested participants, ensuring that a meal is served at every meeting (even when funders and institutions have strong aversions to food-related expenditures), and facilitating the project in accordance with resident priorities and organizational needs rather than the cycles of academic outputs.

Finally, as we have laid out elsewhere in detail (Derickson and Routledge 2014), resourcefulness as a practice of scholarly research entails a "triangulation" of the research question, to consider not only the advancement of scholarly knowledge but equally the needs and priorities of the communities with which we work, as well as the political projects that are advanced by the findings of the research (see Derickson and Routledge [2014] for a more substantive explanation of triangulation). Most important, the triangulation of research questions is a method for collaboratively producing knowledge in ways that "speak back" (Sheppard et al. 2013) and relate the work in communities to the broader intellectual and academic community, as a way to create space in intellectual projects for the concerns of historically marginalized communities to be recognized and engaged with meaningfully. We offer the preceding examples to illustrate what we believe to be one of many possible ways to enact an interim politics of resourcefulness as an academic.

Conclusion

As feminist and postcolonial scholars have argued convincingly, the act of knowing and theorizing is always situated, both in geographic places and in intersectional relations to power structures. Failure to confront the overwhelming degree to which historically marginalized communities are underrepresented in the mainstream and critical spaces of knowledge production and theory building runs the risk of reifying marginalization and universalizing from partial perspectives. In the context of postpolitical environmental governance, this focuses attention on the challenges of producing knowledge about environmental futures in ways that retain fidelity to the perspectives and capacities of historically marginalized communities. We have offered resourcefulness as a conceptual frame with multiple dimensions as a way of fostering the capacity of historically marginalized communities to conceive of and engender alternative environmental futures.

References

Anzaldua, G. 1987. *Borderlands/La Frontiera: The new mestiza*. San Francisco: Aunt Lute Press.

The Autonomous Geographies Collective. 2010. Beyond scholar-activism: Making strategic interventions inside and outside the neoliberal university. ACME 9 (2): 245–75.

Bullard, R., and G. Johnson. 2000. Environmental justice: Grassroots activism and its 8 impact on public policy issues. *Journal of Social Issues* 56 (3): 555–78.

Chatterton, P., D. Featherstone, and P. Routledge. 2013. Articulating climate justice in Copenhagen: Antagonism, the commons and solidarity. *Antipode* 45 (3): 602–20.

Confino, J., C. Holtum, L. Paddison, and J. Kho. 2014. Davos 2014: Climate change and sustainability—Day three as it happened. *The Guardian* 24 January. http://www.theguardian.com/sustainable-business/2014/jan/24/

davos-2014-climate-change-resource-security-sustain ability-day-three-live (last accessed 1 May 2014).

DeFilippis, J. 1999. Alternatives to the "new urban politics": Finding locality and autonomy in local economic development. *Political Geography* 18 (8): 973–90.

Fraser, N. 1997. *Justice interruptus: Critical reflections on the post-socialist condition*. London and New York: Routledge.

Haraway, D. 1988. Situated knowledges: The science question in feminism and the privilege of partial perspective. *Feminist Studies* 14 (3): 575–99.

Harding, S. G. 1991. *Whose science? Whose knowledge? Thinking from women's lives*. Ithaca, NY: Cornell University Press.

hooks, B. 1984. *From margin to center*. Boston: South End.

Intergovernmental Panel on Climate Change. 2014. Climate change 2014: Impacts, adaptation and vulnerability. https://www.ipcc.ch/pdf/unfccc/sbsta40/SED/1_lennartol son_sedpart2.pdf (last accessed 5 September 2014).

Laclau, E., and C. Mouffe. 1985. *Hegemony and socialist strategy*. New York: Verso.

Lake, R. W. 1994. Negotiating local autonomy. *Political Geography* 13 (5): 423–42.

Lorde, A. 1984. *Sister outsider: Essays and speeches by Audre Lorde*. Freedom, CA: Crossing.

McCarthy, J. 2013. We have never been "post-political." *Capitalism Nature Socialism* 24 (1): 19–25.

Mohanty, C. T. 1988. Under Western eyes: Feminist scholarship and colonial discourses. *Feminist Review* 30:61–88.

———. 2003. "Under Western eyes" revisited: Feminist solidarity through anticapitalist struggles. *Signs* 28 (2): 499–535.

Mouffe, C. 2005. *On the political*. London and New York: Routledge.

Nagar, R. 2006. *Playing with fire: Feminist thought and activism through seven lives in India*. Minneapolis: University of Minnesota Press.

Ong, A. 1988. Colonialism and modernity: Feminist re-presentations of women in non-Western societies. In *Feminism and the critique of colonial discourse*, ed. D. Gordon, 108–18. Santa Cruz, CA: Center for Cultural Studies.

Paddison, R. 2009. Some reflections on the limitations to public participation in the post-political city. *L'Espace Politique. Revue en ligne de géographie politique et de géopolitique* 8. http://espacepolitique.revues.org/1393 (last accessed 29 January 2015).

Peake, L., and M. Rieker, eds. 2013. *Interrogating feminist understandings of the urban*. London and New York: Routledge.

Peck, J. 2010. *Constructions of neoliberal reason*. Oxford, UK: Oxford University Press.

Pickerill, J., and P. Chatterton. 2006. Notes towards autonomous geographies: Creation, resistance and self-management as survival tactics. *Progress in Human Geography* 30 (6): 730–46.

Povinelli, E. 2011. *Economies of abandonment*. Durham, NC: Duke University Press.

Ranciere, J. 1999. *Disagreement: Politics and philosophy*. Minneapolis: University of Minnesota Press.

———. 2010. *Dissensus: On politics and aesthetics*. New York: Bloomsbury.

Rose, G. 1997. Situating knowledges: Positionality, reflexivities and other tactics. *Progress in Human Geography* 21 (3): 305–20.

Swyngedouw, E. 2007. Impossible sustainability and the post-political condition. In *The sustainable development paradox: Urban political economy in the United States and Europe*, ed. R. Krueger and D. Gibbs, 13–40. New York: Guilford.

———. 2010. Apocalypse forever? Post-political populism and the spectre of climate change. *Theory, Culture & Society* 27 (2–3): 213–32.

United Church of Christ Commission for Racial Justice. 1987. Toxic wastes and race in the United States: A national report on the racial and socio-economic characteristics of communities with hazardous waste sites. Public Data Access.

U.S. Census Bureau. 2010. Profile of selected population characteristics: Fulton County, Georgia. http://quickfacts.census.gov/qfd/states/13/13121.html (last accessed 4 June 2013).

Vazquez-Prokopec, G. M., J. Vandeng Eng, R. Kelly, D. Mead, P. Kolhe, J. Howgate, U. Kitron, and T. Burkot. 2010. West Nile Virus infection is associated with combined sewage overflow streams in urban Atlanta, Georgia. *Environmental Health Perspectives* 118 (10): 1382–88.

Wainwright, J., and G. Mann. 2013. Climate leviathan. *Antipode* 45 (1): 1–22.

World Bank. 2014. Climate change. http://www.world bank.org/en/topic/climatechange (last accessed 1 May 2014).

Climate Change and the Adaptation of the Political

Joel Wainwright* and Geoff Mann†

*Department of Geography, The Ohio State University
†Department of Geography, Centre for Global Political Economy, Simon Fraser University

In the face of climate change, along what path might we attempt transformation that could create a just and livable planet? Recently we proposed a framework for anticipating the possible political–economic forms that might emerge as the world's climate changes. Our framework outlines four possible paths; two of those paths are defined by what is called "Leviathan," the emergence of a form of planetary sovereignty. In this article we elaborate by examining the adaptive character of emergent planetary sovereignty. To grasp this, we need a theory that can see through our ostensibly "postpolitical" moment to grasp not the disintegration but the adaptation of the political. What does it mean to say the political adapts? Reduced to its essence, it is to say that if the character of political life prevents a radical response to crisis, then it is the political that must change. A materialist attempt to elaborate on this question must begin by reflecting on the manifest inequalities of power in the current mode of global political-economic regulation. After doing so, we conclude by arguing for a return to the concept of natural history.

面对气候变迁，我们能够依循什麽样的道路，达到可以创造公平且宜居的地球之转变？晚近我们提出期待可能的政治—经济形式之框架，这些形式或许会随着世界气候的变迁而出现。我们的框架概述四条可能的路径；其中两条路径以所谓的"利维坦（Leviathan）"定义之——一种地球主权形式的浮现。我们于本文中，透过检视浮现中的地球主权的调适特徵来阐述之。为了进行理解，我们需要能够透视我们显着的"后政治"时刻的理论，以领会政治的调适，而非政治的分解。政治调适意味着什麽？以简化的本质而言，政治调适意味着，如果政治生活的特徵阻碍了对危机的激进回应，那麽此种政治便必须改变。阐述此一问题的物质主义尝试，必须始于反思在当前的全球政治—经济规范模式中，显着的权力不均。此后，我们在结论中主张回归自然历史的概念。

Frente a la realidad del cambio climático, ¿de qué manera podríamos intentar la transformación que pudiese crear un planeta justo y habitable? Hace poco propusimos un marco que anticipara las formas político-económicas que podrán aparecer a medida que cambie el clima del mundo. Nuestro marco de predicción esboza cuatro posibilidades; dos de éstas se definen como "Leviatanes," la emergencia de una forma de soberanía planetaria. En el artículo nos extendemos un poco examinando el carácter adaptativo de la emergente soberanía planetaria. Para captar esto, necesitamos una teoría que pueda ver a través de nuestro momento ostensiblemente "pospolítico" que opte no por la desintegración de lo político sino por su adaptación. ¿Qué se quiere decir con que lo político se adapte? Concentrándonos en lo esencial, se quiere decir que si el carácter de la vida política previene una respuesta radical a la crisis, entonces lo político es lo que debe cambiar. Un intento materialista de elaborar alrededor de esta cuestión debe empezar reflexionando sobre las manifiestas desigualdades de poder en el modo actual de la regulación global político-económica. Después de hecho lo anterior, concluimos abogando por un regreso al concepto de historia natural.

The International Energy Agency (IEA) opened its 2012 World Energy Outlook with the following warning:

> The global energy map is changing, with potentially far-reaching consequences for energy markets and trade. It is being redrawn by the resurgence in oil and gas production in the United States. … By around 2020, the United States is projected to become the largest global oil producer. … The result is a continued fall in US oil imports, to the extent that North America becomes a net oil exporter around 2030. … [T]he climate goal of limiting warming to 2°C is becoming more difficult. … [A]lmost four-fifths of the CO_2 emissions allowable by 2035 are already locked-in by existing power plants, factories, buildings, etc. If action to reduce CO_2 emissions is not taken before 2017, all the allowable CO_2 emissions would be locked-in by energy infrastructure existing at that time. … No more than one-third of proven reserves of fossil fuels can be consumed prior to 2050 if the world

> is to achieve the 2°C goal, unless carbon capture and storage (CCS) technology is widely deployed. . . . Geographically, two-thirds [of proven reserves] are held by North America, the Middle East, China and Russia. These findings underline the importance of CCS as a key option to mitigate CO_2 emissions, but its pace of deployment remains highly uncertain. (IEA 2012, 1–3)[1]

In other words, a rapid and massive change in the geographies of energy production and consumption is presently underway. In a bid for energy security and a repatriated stream of profits, some of the world's largest consumers of energy are turning to friendlier, ideally domestic, suppliers. Big oil's gaze has turned north (to the Arctic), deeper (offshore), and dirtier (tar sands). If the Middle East still holds most of the world's oil reserves, it nonetheless accounts for only 31 percent of current global production (Bridge and LeBillon 2013, 15). These centripetal forces are reconfiguring the world's political geography.

There are at least two likely, and profoundly significant, political conclusions we can draw from these developments. The first is that the winners of this geopolitical game, already the world's most powerful states, will become even more dominant via a concentration of political and economic power, military force, and energy resources. The second and perhaps more profound consequence of this shift is that it signals the end of any hope for meaningful carbon mitigation. Unconventional hydrocarbons are much more carbon-intensive sources of energy than Saudi oil (Bridge and LeBillon 2013, 9).[2] Their development guarantees massive increases in greenhouse gas emissions. Moreover, the geographic and political–economic distribution of these resources deepens the global division of wealth and power, exacerbating geopolitical inequalities and further destabilizing what little ground international negotiations have cleared for cooperation on climate-related concerns.

The IEA does not say mitigation is no longer possible and, to be sure, some sectors, firms, and localities have reduced emissions (Intergovernmental Panel on Climate Change [IPCC] 2014). Green energy has expanded in certain instances. But the IEA's emphasis on the desperate need for CCS surely means it recognizes the insurmountable obstacles to CO_2 emissions reductions on the necessary timelines (i.e., "before 2017").[3] The possibility of rapid global carbon mitigation as a climate change abatement strategy has passed. The world's elites, at least, appear to have abandoned it—if, of course, they ever really took it seriously. Davis (2010) might prove prescient. In what he called a "not improbable scenario," mitigation "would be tacitly abandoned . . . in favour of accelerated investment in selective adaptation for Earth's first-class passengers."

> The goal would be the creation of green and gated oases of permanent affluence on an otherwise stricken planet. Of course, there would still be treaties, carbon credits, famine relief, humanitarian acrobatics, and perhaps the full-scale conversion of some European cities and small countries to alternative energy. But worldwide adaptation to climate change, which presupposes trillions of dollars of investment in the urban and rural infrastructures of poor and medium income countries, as well as the assisted migration of tens of millions of people from Africa and Asia, would necessarily command a revolution of almost mythic magnitude in the redistribution of income and power. (Davis 2010, 38)

What does this mean for how we conceive the political today? This question is the focus of what follows. The momentous socioecological transformations to which Davis refers, and against which the global climate justice movement might enact a "revolution of almost mythic magnitude," is best grasped as a moment of transition in the planet's natural history. This is in no way to suggest it is beyond politics. On the contrary, in the midst of these changes the urgent questions concern not merely a transformation in politics—more representative proceduralism, for example, or more precautionary environmental policy-making—but a transformation of the political. To ask via what paths might we undertake political transformations adequate to something like a just and livable planet is necessarily to ask not only what political tools, strategies, and tactics might achieve this but also what conception of the realm of the political might render adequate tools, strategies, and tactics imaginable. What conceptions of the political legitimate the warming norm, and what alternatives can provide some grounds for genuine alternatives?

Such questions preoccupy many geographers today (e.g., Yusoff 2009; Johnson 2010; Swyngedouw 2010; Bond 2011; Chatterton, Featherstone, and Routledge 2012; Labban 2012; Dempsey 2013; Braun 2014). We recently proposed a framework for anticipating the possible political–economic forms that might emerge as the world's climate changes (Wainwright and Mann 2013), to broadly outline the paths that humanity faces. Two of those paths are defined by an emergent form of planetary sovereignty capable of deciding how to save the world—a "climate Leviathan"—that "exists to the precise extent that some sovereign exists who can decide on the exception, declare an emergency, and decide who may emit carbon and who cannot . . . for the sake of *life on Earth*" (Wainwright and Mann 2013, 5).

As Lohmann (2012) notes, however, an ambiguity troubles our argument. In our focus on the question of who may emit carbon and who cannot, we seemed to premise the possibility of Leviathan on the necessity of carbon mitigation, whereas the failure of global efforts makes it clear that any emergent Leviathan will be principally a beast of adaptation. Our only mention of adaptation hardly addresses the problem: "The elite transnational social groups that dominate the world's capitalist nation-states certainly desire to moderate and adapt to climate change—not least to stabilize the conditions that produce their privileges" (Lohmann 2012, 4). Although we emphasize throughout the *emergent* character of planetary sovereignty—that is, we do not conceive of it as operated by an on–off switch—we did fail to emphasize its *adaptive* character.[4] With the tacit acceptance of runaway climate change, Leviathan might be expected to (1) enable efforts to profit from it (Funk 2014), for example via newly accessible resources in the Arctic, while (2) organizing cross-territorial forms of adaptation that augment elite social groups' power and security (e.g., military-coordinated geoengineering). Neither of these tendencies are new: they only intensify existing dynamics. To come to grips with them, we must see through our ostensibly postpolitical moment (Swyngedouw 2010).[5] The problem is neither the disintegration nor the terminal crisis of the political but its distinctive adaptation.

Leviathan and the Adaptation of the Political

> It is all about politics. Climate change is the hardest political problem the world has ever had to deal with. It is a prisoner's dilemma, a free-rider problem and the tragedy of the commons all rolled into one. . . . [Hum]ankind has no framework for it. ("Getting warmer" 2009)

What does it mean to say that the political adapts? It clearly presumes the political has a history—perhaps even a natural history—but it also presumes it has a specificity. To speak of the political, and of its adaptation, is to say the political constitutes an analytically distinct region of the social. There is no shortage of debate to which one could turn to enrich this discussion. Virtually every prominent radical philosopher these days has written about the political, and most tell the story of its demise or the onset of the postpolitical (Mouffe, Rancière, Badiou, Žižek, and others). As Žižek (2011, ix) puts it, we witness everywhere the emergence of "a new bipolarity between politics and post-politics." Like Mouffe, Badiou, and Rancière (in his "hatred of democracy"), Žižek (2011, xv) demands we endorse an "agonism" that revels in opposition and struggle for their own sakes: "to engage in struggle means to endorse Badiou's formula *mieux vaut un désastre qu'un désêtre*."[6]

Here, the political refers neither to a particular political condition or set of institutions (e.g., liberal democracy or the parliamentary system) nor to the existential fact of struggle (although this is always implicated) but to the very grounds on which such conditions, institutions, or struggles arise and are formulated. The political is not, therefore, a relational concept in the way "relational" is typically used by geographers.[7] Nor is it merely the realm of the clash of interests, nor of agonistic confrontation and collective or individual self-actualization. Rather, it defines a relation *tout court*; that is, the relationship between the dominant and the dominated. The political is definitely not the arena in which hegemony imposes its interests and the subaltern resists but the grounds on which the relation between the dominant and dominated takes form. (*Grounds* is thus an apposite term because, as geographers know, implicit in any mode of the political is a spatiotemporal context in which it unfolds and helps shape.)

On this point, we find the writings of Poulantzas especially important. In a discussion of hegemony, Poulantzas ([1965] 2008) makes the crucial decision to found his analysis in the historical separation (or "regionalization") of the political and the effect this process has had on modern state formation. Although an early work, already in this essay Poulantzas emphasizes a point most often associated with his influential later debate with Ralph Miliband, namely, "the state crystallizes the *relations* of production and *class relations*. The modern political state does not translate the 'interests' of the dominant classes at the political level [as is often suggested in economistic or instrumentalist accounts], *but [rather] the relationship between those interests and the interests of the dominated classes*—which means that it precisely constitutes the 'political' expression of the interests of the dominant classes" ([1965] 2008, 80). For Poulantzas, the "*specifically political character* of the capitalist state" does not lie in the state's domination by capital but is in fact constituted in the very "separation between state and civil society" (83). This legitimacy of this separation is thus both founded on and represents a seemingly natural result of "the characteristic of universality assumed by a particular set of values" (83). What are these values? They are "the 'universal' values of formal abstract liberty and equality":

> In societies based on expanded reproduction and generalized commodity exchange [i.e., capitalist societies], we observe a process of privatization and autonomization of men as producers. Natural human relations, founded on a hierarchy involving the socio-economic subordination of producers (witness slave and feudal states), are replaced by *"social" relations between "autonomized" individuals*, located in the exchange process. *Marx and Lenin underscore this evolution of natural relations into social relations* ... that underlies the constitution of commodity-value and labor-value and exploitation in capitalist, exchange-based society. ... This appearance of social relations in the capitalist system of production in fact presupposes, as a necessary precondition, the characteristic atomization of civil society and goes hand in hand with the advent of specifically political relations. (Poulantzas [1965] 2008, 83)[8]

Any politics assumes and asserts a historical and geographical terrain to which it lays claim. As the "specifically political character" of the capitalist nation-state is constituted in the separation of the state and civil society, these are the grounds on which the legitimacy of the nation-state rests. Its hegemony in the contemporary political imagination underwrites our assertion that if climate Leviathan is to emerge, it will do so as a transformation of the existing form of sovereignty, enabling the world's most powerful states to engage in planetary management. Yet we now recognize this claim sidesteps a crucial and difficult question that Leviathan must answer: How could we get from the present Westphalian world to planetary management? And might we get there in a way that preserves the territorial nation-state?

Any materialist approach to these questions must reflect on the manifest inequalities of power in a mode of global political–economic regulation currently constituted to a significant extent by liberal capitalism (United Nations conventions, Bretton Woods institutions, free trade agreements, the European Union, etc.). This matrix thus far has failed to produce a coordinated response to climate change; to generalize, climate change is framed as a scientific–technical problem, best addressed by fuel-efficient cars, tradable permits, and flexible adaptation "governance" (IPCC 2014).[9] Although planetary warming accelerates ecological transformation and human suffering apace, for capitalist states it nevertheless does not yet signify a fundamental transformation of the grounds of the political. In the wealthy world, climate change still does not matter—or, alternatively, its "mattering" is refused. Instead, the buildup of anthropogenic greenhouse gases is confronted as mere market failure, for which various market-mending policies are proposed: cap-and-trade, carbon offsets, catastrophe bonds, mandatory risk disclosure, flood and hurricane insurance, and so on (Johnson 2013).

Consider the debacle that is the international United National Framework Convention on Climate Change (UNFCCC) negotiations on climate change, a process slowed by both affluent sabotage and developing-world resistance. As justifiable as the latter might be, it shares with the former a futility rooted in fidelity to the conventional economic thinking by which the nation-state-centered liberal capitalist matrix operates, as it relies just as heavily on the essentially technical distribution of costs and benefits. Negotiators seek to solve an optimization problem whose terms must include coefficients for colonialism, underdevelopment, massive historical displacement, and impoverishment. This is to say nothing of inequalities internal to developing nation-states. Climate change cannot be addressed by liberal economic reason, which, denying itself a conscious politics—indeed, denouncing all "politics" as a distortion of economic rationality—cannot think about history and hysteresis (i.e., the irrepressible ways that history continues to matter). On orthodox economic terms, a global solution is not merely politically unlikely but logically impossible. There is no market-based solution to a massive problem whose causes took place before it was possible to price their repercussions. In short, there is no Coasian solution to climate change (Coase 1960; cf. Gilbertson and Reyes 2009; Lohmann 2009), no way for self-interested actors to address the problem of social cost when the very ground on which the problem must be addressed—the political—is disavowed.

This is emphatically not to deny the global environmental debt. That the luxurious life of the capitalist Global North is desertifying West Africa and inundating rural Pakistan is impossible to deny—but it is just as impossible to price. If, as we are often told, the market is by definition apolitical, then it is ridiculous to suggest it as a solution to what is in many ways today's defining political issue: Whose lives will pay the cost of adaptation to a warming planet?

This failure endlessly frustrates all market-based efforts to allocate a global pool of emittable greenhouse gases (and even the powers that be know it cannot be anything less than global [Stiglitz 2013]). The constant intrusion of the pesky politics of the unpriceable history of the present—inequality, colonialism, and underdevelopment—simultaneously legitimates southern resistance and explains affluent nations' shirking of historical and moral accountability. For the south, it justifies the rejection of petty payments to forget the crimes of history. For elites of the north, for whom the ways and means of liberal capitalism are presumed, the way forward is via the erasure of the record of past wealth-producing emissions, and the declaration of an atmospheric blank slate. "Save our global village," "we're all in this together": This is the political adaptation proclaimed by the Global North. No mention is made of assisted mitigation. It is adaptation *qua* consolidation.

This program suppresses—as it must—the fact that adaptation to climate change will not be cheap and many will suffer. To broach the question in a manner that recognizes this truth is categorically impossible in the liberal capitalist nation-state framework. This is where political struggles over the form and character of planetary management come clearly into view. For if Mitchell (2009, 401) is right that "the political machinery that emerged to govern the age of fossil fuels may be incapable of addressing the events that will end it," what will follow?[10]

Any climate Leviathan will be predicated on the assumption that the future requires the consolidation of present forms of juridical subjectivity so as to reproduce a classical liberal world, at least in the logic of rule. In the contemporary capitalist state, this means that climate change is addressed by merely tweaking citizen-subjects' juridico-scientific status to include a role as emission source. In other words, this form of adaptation necessarily invokes the nation-state, or subnational units (states, provinces, municipalities) under its direction, taking the nation as the obvious mode through which rule is exercised. Isn't every citizen, however global or local, ultimately subject to some nation-state? The effectiveness of this program is thus premised on the simultaneous adaptation of the political qua separation of state and civil society and the refusal to adjust existing juridical and legal territoriality and power. It is a performative project to create a world in which orthodox liberal concepts—concerning the liberty of markets, property rights, and the state—actually work.

With the growing awareness that the mitigation window has closed, this project is becoming tightly linked to plans to "geoengineer" our way to safety, via massive technosocial mitigation-by-planetary-manipulation (Keith 2000; Robock 2008; Hamilton 2013). These efforts are of a qualitatively different order than projects to create resilient infrastructures or produce drought-resistant agricultural seed stock. We refer, rather, to more or less well-advanced plans for what might be called geomodification. Take, for example, the 2006 NASA-sponsored workshop on "solar radiation management"; that is, artificially increasing atmospheric albedo (Lane et al. 2007). Other commonly proposed strategies include artificially generating cold-water upwelling to lower surface temperatures, or altering ocean chemistry to absorb more carbon (Keller, Feng and Oschlies 2014).[11] As the IPCC (2013, 29) itself acknowledges, carbon capture and storage belongs here, too, as depositing gigatons of carbon in the Earth's crust for thousands of years, as imagined by the IEA (2012), will involve considerable geological engineering.

Geoengineering alone will not bring Leviathan into being, of course. But large-scale projects will involve a relatively small group of actors experimenting with global systems, in what is hard not to see as the most improbable of missions: to materially reconfigure planet Earth so as to avoid having to rework human political economies. The recognition that any means of evaluating geoengineering projects will be intensely political explains the logical appeal for a legitimate planetary authority to adjudicate the merits of experimentation and the cloaking of such authority in the white coat of technoscientific confidence: "Either we are smart enough to craft that feedback mechanism ourselves, or the Earth system will ultimately provide it" (Parson and Keith 2013, 1279). It is reason versus the state of nature. Between them stands the planetary sovereign: the one that declares the (experimental) exception in the name of life itself. Planetary sovereignty thus emerges in what might be called *Weltrecht;* that is, the arrogation of the authority and duty to remake the world to save it.

A Natural History of Our Conjuncture

It might be useful, at this juncture, to return to the concept of natural history. In a preface to *Capital,* Marx ([1867] 1976, xx) writes: "My standpoint, from which the development of the economic formation of society is viewed as a process of natural history, can less than any other make the individual responsible for relations whose creature he remains, socially speaking, however much he may subjectively raise himself above them." For the contemporary climate justice movement, this is a far more radical insight than it might first appear. It is to say that an adequate account of the current transformation—the onset of the Anthropocene, we might say—must reject the subjectivist moralism of modern environmentalism. If "the development of the economic formation of society" is "a process of natural history" in which the individual is not "responsible for relations whose creature [s]he remains," then what is to be done? We are moments in natural history, nodes inextricably enmeshed in a more-than-human world. This is perhaps most readily evident in the fact that to live in and work on the world is to be a carbon being; carbon is, after all, the second most abundant element in our bodies, one that we metabolize to power every act of living labor and exhale with every breath.

The problem is how to avoid taking this as an argument for paralysis, structural overdetermination, or helplessness. Instead, if it is true—and what useful account of climate change can rely on apportioning blame to this or that individual?—then the challenge resides not only in questioning the adaption of the political but in pursuing it. The present chapter in Earth's natural history is structured to a significant extent by capitalism as a social formation and, at least for those with the largest carbon budgets, it is on this basis that our existential relation with carbon is currently constituted. What is to be governed is the very stuff we are made of.

This helps explain why Leviathan's planetary sovereignty—and the adaptation of the political—is being shaped by struggles over the relation between carbon agency and carbon subjectivity. For if "to emit, or not to emit" is not a question, or at least not one of mere subjective choice, then the hegemony of environmental individualism in liberal climate politics is leading us into a political cul-de-sac. Indeed, it seems that we are already there. The utter failure of these efforts, and the end of any hope for meaningful mitigation, is not only evidence of their futility but stands as the most powerful argument to justify a consolidating climate Leviathan: If people will not make the right carbon choices on their own, then they must be made to do so.

The more complicated—and politically urgent—question is the emerging forms of carbon subjectivity these dynamics produce, both in the heart of the carbon-spewing north and the more varied carbon geographies of the south and the status of agency therein. Examining the manifold incipient forms of subjectivity, and their relation to the forces that produce multiscalar political–economic inertia and change, is surely one of the most important tasks facing human geography today. It demands a willingness to follow the carbon, materially and theoretically, an endeavor to which many geographers are already committed (e.g., Yusoff 2009; Johnson 2010, 2013; Rice 2010; Swyngedouw 2010; Osborne 2011; Chatterton, Featherstone, and Routledge 2012; Lansing 2012; Saldanha 2012, 2013; Huber 2013; Labban 2013); more will surely follow.

It is here, in the relation between what we might call a biopolitical subjectivity and a transformational agency, that we are bound to find our questions leading, the terrain on which the adaptation of the political must inevitably unfold. This is, of course, a very old problem. Indeed, it is the puzzle at the core of all revolutionary thought, the problem to which all revolutionary action must offer a response. The question of revolution thus always haunts the adaptation of the political, which is an attempt to forestall radical material transformation. The specter of such transformation is but one more fund on which climate Leviathan might draw.

In his wonderful book on revolution and European philosophy, Kouvelakis (2003) remarks that what Kant "finds alarming" in the prospect of revolution is "the moment of vacuum thus created, insofar as it implies the threat of regression to a state of nature" (21). Many current efforts on the left, from the liberal to the radical, are emerging from this fear of the ecological-resource scarcity precipice we perceive on the horizon. It seems clear, as we suggested earlier (Wainwright and Mann 2013), that this fear is a significant part of the reason something like climate Leviathan—like other Leviathans—can come to make so much sense.

A refusal of this position is essential. There is no state of nature to which we inevitably regress in the absence of civilization. First, because the concept of civilization emerged with the birth of liberal modernity, so any concept of its lack, at least in the Hobbesian nasty-brutish-short sense it has today, is just as modern. Second, history puts the lie to it wherever we look. The critical assertion that there is no essentialist way we all are (or must be) is not merely a political preference, but a truth. If what makes climate change terrifying is the expectation that people will "revert" to the war of all against all that is the state of nature, it is essential to recognize that this reversion would be neither a product of climate-induced scarcity, nor of nature, human or otherwise. The *bellum omnia* might very well loom on some horizons, along with a Leviathan (or a Mao) to subdue it, but in all cases they will be products of historical social relations, and are only inevitable if how we live now (and that might not be a very inclusive "we") expresses the truth of how we "really are"—some base condition to which we will forever revert in the absence of counteracting forces. There is no reason to believe this. If we are Hobbesian, it is because we make ourselves so: there is nothing inevitable or irrefutable about it.

To see this is to recognize the persistence of the political in natural history, and vice versa; it is to say that climate change is an event in our natural history. As Marx argued and many geographers have reiterated, the relations between humans and nature always reflect the prevailing relations between humans.[12] Indeed, this point remains fundamental to any political ecology worth the name. It should be paired with a second analytical point concerning the temporality of the political, one more closely associated with Gramsci. To grasp the adaptation of the political we would need to read as *conjunctural* our strange present-conditional politics, in which what might happen in the future seems to determine the present. The concept of conjuncture defines a moment, emphasizing its existence as a complex of pasts and futures. Our hypothesis is that the adaptation of the political that we might anticipate with climate Leviathan is defined by the furtive way the future bends back into the now. Just as money guarantees its social power in the present through a never-yet-realized futural promise to be worth something—to be more than a dirty scrap of paper or useless lump of metal—so will climate Leviathan secure its existence by structuring a present that realizes a certain future, one worth living. The result is a politics of emergency, one where politics is deferred. This deferral arrives from, or is premised on, a future that, increasingly, many await only with fear. This peculiarity might explain why the era in which we live, saturated in struggle, could nonetheless appear to some as postpolitical, a world in which we seek institutional and industrial technologies through which we might avoid a future necessarily presupposed by the world in which, however reluctantly or unjustly, we are all condemned to live. In sum, what we are experiencing is less after politics than other politics; that is, the adaptation of the political. This other would, of course, be the first victim of the emergency whose looming threat legitimates it—precisely because it is constructed for a disappearing world.

Notes

1. The Organization for Economic Cooperation and Development founded the IEA in 1974 (at U.S. behest) to coordinate wealthy countries' response to dependence on Middle Eastern oil.
2. Hence the new geography of energy demands increasing amounts of energy in the process of extraction relative to the energy of that extracted. During the last century, the global average fell from 1:100 to 1:30 and as low as 1:5 in some unconventional operations. In other words, whereas an average extraction project once produced one hundred times the amount of energy invested, it now produces only thirty times, and often less (Bridge and LeBillon 2013, 9).
3. The data supporting this realization—familiar to many geographers—include the following: In 2011 global CO_2 emissions reached a record high of 31.6 gigatons (Gt), a 1.0 Gt (3.2 percent) increase over 2010 (IEA 2012). The world is on track to emit ~58 Gt in 2020, the year the Durban agreement commitments are supposed to begin, ~14 Gt more than can be emitted if we are to limit warming to 2°C (United Nations Environment Program 2012). From 2004 to 2013, atmospheric greenhouse gas concentrations measured at Mauna Loa increased 2.14 percent, the fastest decadal increase yet (National Oceanic and Atmospheric Administration 2013). The rate of increase continues to accelerate. Between February 2012 and February 2013, Mauna Loa recorded a 3.26 ppm rise in CO_2, registering 400 ppm for the first time in May 2013, relative to preindustrial levels of approximately 280 ppm (Zickfeld et al. 2009; Vidal 2013). Moreover, there has been no green energy boom: "[t]he drive to clean up the world's energy system has stalled," and "the average unit of energy produced today is basically as dirty as it was 20 years ago" (IEA 2013). Finally, there is no substantive progress in international climate change negotiations, to say nothing of actual carbon mitigation. In the ruins of Kyoto, the UNFCCC lacks a coherent roadmap. The July 2013 U.S.–China agreements are narrow in scope and nonbinding.
4. Braun (2014, 50) argues we treat capitalism and sovereignty as functions of a binary "'on/off' switch" (his term).
5. We endorse McCarthy's (2013) critique of Swyngedouw (2010) and his argument that "there are . . . very substantial, significant, and ongoing struggles around the politics and politicization of climate change that are directly at odds with some of the 'post-political' dynamics that Swyngedouw sees" (McCarthy 2013, 23).
6. Žižek's translation ("better to take the risk and engage in fidelity to a Truth-Event, even if it ends in catastrophe, than to vegetate in the eventless utilitarian-hedonist survival of what Nietzsche called the 'last men'")

is improbably loose (Žižek 2010, xv). *Désêtre* is a Lacanian word game, a derivative of misreading *désire* as "not going" (*ir* is a participle of "to go" in French); hence *désêtre* is "not being" or "disbeing."

7. See, for example, Lawson and Elwood's (2014) excellent work on "relational poverty."
8. By our reading, Poulantzas places the word *social* in "'social' relations" in scare quotes to emphasize that they are also natural relations. This passage, with its emphasis on the natural history of the formation of the political in capitalist society, lends support to Bob Jessop's assertion that "were he alive today, Poulantzas would be a political ecologist" (personal communication 2013). We urgently need a study that draws on Poulantzas's thought to study climate change and the capitalist state.
9. For example, "Existing and emerging economic instruments can foster adaptation by providing incentives for anticipating and reducing impacts (medium confidence). Instruments include public–private finance partnerships, loans, payments for environmental services, improved resource pricing, charges and subsidies, norms and regulations, and risk sharing and transfer mechanisms" (IPCC 2014, 24).
10. Mitchell's analysis of the natural history of capitalist democracy shares important similarities with our project, and we have learned much from his work. Unfortunately, as Labban (2013) notes, "Mitchell eliminates capitalism altogether from the natural history of carbon democracy and replaces social relations between persons with the relations of things to persons such that, to borrow from Marx (1864), the 'definite social connections appear as social characteristics belonging naturally to things.'"
11. For a broad view of the scientific discussion of geoengineering, see the contributions to a special issue of the journal *Climate Change*, Vol. 77, No. 3–4 (2006).
12. See Karatani (2008, 571). As Labban (2012) shows, Karatani's Kantian "associationism" provides the basis for our conception of "climate X," a kind of ideal path. Labban insightfully points out that in failing to articulate a properly materialist alternative, climate X remains an essentially theological concept (in reply, see Wainwright and Mann 2012).

References

Bond, P. 2011. Carbon capital's trial, the Kyoto Protocol's demise, and openings for climate justice. *Capital Nature Socialism* 22 (4): 3–17.

Braun, B. 2014. A new urban dispositif? Governing life in an age of climate change. *Environment and Planning D: Society and Space* 32:49–64.

Bridge, G., and P. LeBillon. 2013. *Oil*. London: Polity.

Chatterton, P., D. Featherstone, and P. Routledge. 2012. Articulating climate justice in Copenhagen: Antagonism, the commons, and solidarity. *Antipode* 45 (3): 602–20.

Coase, R. 1960. The problem of social cost. *Journal of Law and Economics* 3:1–44.

Davis, M. 2010. Who will build the ark? *New Left Review* II/61:29–46.

Dempsey, J. 2013. Biodiversity loss as material risk: Tracking the changing meanings and materialities of biodiversity conservation. *Geoforum* 45 (1): 41–51.

Funk, M. 2014. *Windfall: The booming business of global warming*. New York: Penguin.

Getting warmer: A special report on climate change and the carbon economy. 2009. *The Economist* 5 December: 1–15.

Gilbertson, T., and O. Reyes. 2009. Carbon trading: How it works and why it fails. Occasional Paper No. 7, Dag Hammerskjöld Foundation, Uppsala, Sweden.

Hamilton, C. 2013. *Earthmasters: The dawn of the age of climate engineering*. New Haven, CT: Yale University.

Huber, M. 2013. *Lifeblood: Oil, freedom, and the forces of capital*. Minneapolis: University of Minnesota.

Intergovernmental Panel on Climate Change (IPCC). 2013. Summary for policymakers. In *Climate change 2013: The physical science basis. Contribution of Working Group I to the fifth assessment report of the Intergovernmental Panel on Climate Change*, ed. T. F. Stocker, D. Qin, G.-K. Plattner, M. Tignor, S. K. Allen, J. Boschung, A. Nauels, Y. Xia, V. Bex, and P. M. Midgley. Cambridge, UK: Cambridge University Press.

———. 2014. Working Group II: Impacts, adaptation, and vulnerability: Fifth assessment report technical summary. http://ipcc-wg2.gov/AR5/images/uploads/WGIIAR5-TS_FGDall.pdf (last accessed 31 March 2014).

International Energy Agency (IEA). 2012. World energy outlook 2012. http://www.iea.org/publications/freepublications/publication/English.pdf (last accessed 22 July 2013).

———. 2013. Progress toward clean energy has stalled. http://www.iea.org/newsroomandevents/pressreleases/2013/april/name,36789,en.html (last accessed 10 November 2014).

Johnson, L. 2010. The fearful symmetry of Arctic climate change: Accumulation by degradation. *Environment and Planning D: Society and Space* 28 (5): 828–47.

———. 2013. Catastrophe bonds and financial risk: Securing capital and rule through contingency. *Geoforum* 45 (1): 30–40.

Karatani, K. 2008. Beyond capital-nation-state. *Rethinking Marxism* 20 (4): 569–95.

Keith, D. W. 2000. Geoengineering the climate: History and prospect. *Annual Review of Energy and the Environment* 25:245–84.

Keller, D. P., E. Y. Feng, and A. Oschlies. 2014. Potential climate engineering effectiveness and side effects during a high carbon dioxide-emission scenario. *Nature Communications* 5:3304. doi:10.1038/ncomms4304

Kouvelakis, S. 2003. *Philosophy and revolution: From Kant to Marx*. New York: Verso.

Labban, M. 2012. Beyond behemoth. http://antipodefoundation.org/2012/07/19/symposium-on-geoff-mann-and-joel-wainwrights-climate-leviathan/ (last accessed 10 November 2014).

———. 2013. Book review—Mazen Labban on Timothy Mitchell's "Carbon democracy: Political power in the age of oil." http://antipodefoundation.org/2013/03/19/

book-review-mazen-labban-on-timothy-mitchells-carbon-democracy/ (last accessed 10 November 2014).
Lane, L., K. Caldeira, R. Chatfield, and S. Langhoff. 2007. Workshop report on managing solar radiation. NASA/CP-2007-214558, National Aeronautics and Space Administration, Washington, DC.
Lansing, D. 2012. Performing carbon's materiality: The production of carbon offsets and the framing of exchange. *Environment and Planning A* 44 (1): 204–20.
Lawson, V., and S. Elwood. 2014. Encountering poverty: Space, class and poverty politics. *Antipode* 46 (1): 209–28.
Lohmann, L. 2009. Toward a different debate in environmental accounting: The cases of carbon and cost-benefit. *Accounting, Organizations and Society* 34:499–534.
———. 2012. Commentary on "Climate Leviathan." http://antipodefoundation.org/2012/07/19/symposium-on-geoff-mann-and-joel-wainwrights-climate-leviathan/ (last accessed 10 November 2014).
Marx, K. 1864. Results of the direct production process: The process of production of capital. In *Marx–Engels collected works: Karl Marx economic works 1861–1864*. Vol. 34, trans. B. Fowkes. New York: International Publishers. https://www.marxists.org/archive/marx/works/1864/economic/ch02a.htm (last accessed 10 November 2014).
———. [1867] 1976. *Capital*. Vol. I. New York: Penguin.
McCarthy, J. 2013. We have never been post-political. *Capitalism, Nature, Socialism* 24 (1): 19–25.
Mitchell, T. 2009. Carbon democracy. *Economy and Society* 38 (3): 399–432.
National Oceanic and Atmospheric Administration. 2013. Concentrations of CO_2 in the Earth's atmosphere (parts per million) derived from in situ air measurements at the Mauna Loa observatory, Hawaii. http://co2now.org/images/stories/data/co2-atmospheric-mlo-monthly-scripps.pdf (last accessed 10 November 2014).
Osborne, T. 2011. Carbon forestry and agrarian change: Access and land control in a Mexican rainforest. *The Journal of Peasant Studies* 38 (4): 859–83.
Parson, E., and D. Keith. 2013. End the deadlock on governance of geoengineering research. *Science* 339:1278–79.
Poulantzas, N. [1965] 2008. Preliminaries to the study of hegemony in the state. In *The Poulantzas reader: Marxism, law, and the state*, ed. J. Martin, 74–119. London: Verso.
Rice, J. 2010. Climate, carbon, and territory: Greenhouse gas mitigation in Seattle, Washington. *Annals of the Association of American Geographers* 100 (4): 929–37.
Robock, A. 2008. 20 reasons why geoengineering may be a bad idea. *Bulletin of the Atomic Scientists* 64 (2): 14–18, 59.
Saldanha, A. 2012. Aestheticism and post-humanism. *Dialogues in Human Geography* 2 (3): 276–79.
———. 2013. Some principles of geocommunism. Unpublished manuscript, Lancaster Environment Centre. http://www.tc.umn.edu/~saldanha/geocommunism.html (last accessed 10 November 2014).
Stiglitz, J. 2013. Sharing the burden of saving the planet: Global social justice for sustainable development. Lessons from the theory of public finance. In *The quest for security: Protectionism without protectionism and the challenge of global governance*, ed. J. Stiglitz and M. Kaldor, 161–204. New York: Columbia University Press.
Swyngedouw, E. 2010. Apocalypse forever: Post-political populism and the specter of climate change. *Theory, Culture & Society* 27 (2–3): 213–32.
United Nations Environment Program (UNEP). 2012. *The emissions gap report 2012*. Nairobi, Kenya: UNEP. http://www.unep.org/pdf/2012gapreport.pdf (last accessed 10 November 2014).
Vidal, J. 2013. Large rise in CO_2 emissions sounds climate change alarm. *The Guardian* (London) 8 March 2013. http://www.guardian.co.uk/environment/2013/mar/08/hawaii-climate-change-second-greatest-annual-rise-emissions (last accessed 10 November 2014).
Wainwright, J., and G. Mann. 2012. Solving for X. http://radicalantipode.files.wordpress.com/2012/08/authors-reply1.pdf (last accessed 10 November 2014).
Wainwright, J., and G. Mann. 2013. Climate Leviathan. *Antipode* 45 (1): 1–22.
Yusoff, K. 2009. Excess, catastrophe, and climate change. *Environment and Planning D: Society and Space* 27 (6): 1010–29.
Zickfeld, K., M. Eby, H. D. Matthews, and A. J. Weaver. 2009. Setting cumulative emissions targets to reduce the risk of dangerous climate change. *Proceedings of the National Academy of Sciences* 106 (38): 16129–34.
Žižek, S. 2010. *Living in the end times*. London: Verso.
———. 2011. *Living in the end times*. London: Verso.

A Manifesto for Abundant Futures

Rosemary-Claire Collard,* Jessica Dempsey,[†] and Juanita Sundberg[‡]

*Department of Geography, Planning and Environment, Concordia University, Montreal, Canada
[†]School of Environmental Studies, University of Victoria, Victoria, Canada
[‡]Department of Geography, University of British Columbia, Vancouver, Canada

The concept of the Anthropocene is creating new openings around the question of how humans ought to intervene in the environment. In this article, we address one arena in which the Anthropocene is prompting a sea change: conservation. The path emerging in mainstream conservation is, we argue, neoliberal and postnatural. We propose an alternative path for multispecies abundance. By *abundance* we mean more diverse and autonomous forms of life and ways of living together. In considering how to enact multispecies worlds, we take inspiration from Indigenous and peasant movements across the globe as well as decolonial and postcolonial scholars. With decolonization as our principal political sensibility, we offer a manifesto for abundance and outline political strategies to reckon with colonial-capitalist ruins, enact pluriversality rather than universality, and recognize animal autonomy. We advance these strategies to support abundant socioecological futures.

人类世的概念，对于人类如何介入自然的问题，创造了崭新的契机。我们于本文中，处理人类世正在推进剧烈变革的一个领域：环境保育。我们主张，主流的环境保育中逐渐浮现的路径，便是新自由主义及后自然。我们则对多物种的丰富性，提出一条另类路径。我们所谓的丰富性，意味着更多差异及自主的生命形式，以及共生的方式。在考量如何展现多物种的世界时，我们受到全球各地的原住民运动和农民运动，以及去殖民和后殖民学者的启发。去殖民作为我们的主要政治敏感度，我们以此提出丰富性的宣言，并概述政治策略，以清算殖民—资本主义的毁坏，并展现多重世界性，而非单一世界性，以及承认动物的自主性。我们推动这些策略以支持丰富的社会生态之未来。

El concepto del Antropoceno está creando aperturas nuevas alrededor del interrogante sobre el modo como los humanos deben intervenir en el medio ambiente. En este artículo abocamos un campo en el que el Antropoceno está incitando a un cambio marino: la conservación. La ruta que emerge en la corriente principal de la conservación es, sostenemos, neoliberal y posnatural. Proponemos una ruta alternativa para la abundancia en diversidad de especies. Por abundancia significamos formas de vida y maneras de vivir juntos más diversas y autónomas. Al considerar cómo representar mundos diversos en especies, nos inspiramos en movimientos de indígenas y campesinos a través del globo, lo mismo que en eruditos versados en descolonización y lo poscolonial. Con la descolonización como nuestra principal sensibilidad política, ofrecemos un manifiesto en pro de la abundancia y del esquema de estrategias políticas para lidiar con las ruinas colono-capitalistas, representar la pluriversalidad más que la universalidad, y reconocer la autonomía animal. Promovemos estas estrategias en apoyo de futuros socioecológicos abundantes.

The Anthropocene, says Erle Ellis, "is a new geological era characterized by humans as a force shaping nature" (*The Economist* 2011). In an interview with *The Economist*—available on YouTube—Ellis lists indicators of the Anthropocene as he sits in a London park: cropland, domesticated species, climate change, and so on. The idea of a "pristine" baseline from which to measure disturbance and degradation is a fallacy, says the professor of geography and environmental systems. When asked whether if the Anthropocene is a source of despair or hope, Ellis responds: the Anthropocene is "great from a scientific point of view." Now that we "recognize that humans are this great causal agent ... we're putting ourselves back in the picture *intentionally*. Now we can decide—rather than just kind of assume we're not having this big impact—*how* we're going to have this impact." We agree with Ellis on this point: The Anthropocene as a concept prompts the question of how humans ought to intervene in the environment; how to live in a multispecies world.

In this article, we address one arena in which the Anthropocene is prompting a sea change: conservation. Some conservationists are beginning to speak in what would have been shocking terms a mere decade ago. As Ellis (2009) writes, if "Nature is gone" and we are "living on a used planet," then it's time for a "postnatural environmentalism" where the most important "wildlife refuges" are farms, backyards, and cities. This postnatural bent is taking root in several mainstream conservation organizations like The Nature Conservancy (TNC) and Conservation International, as well as The Breakthrough Institute (TBI). Brainchild of "Death of Environmentalism" authors Shellenberger and Nordhaus (2004), TBI is an increasingly influential think tank at the forefront of cultivating and disseminating what is fast becoming a new common sense in conservation. Past and present TBI Senior Fellows include Ellis, political ecologist Paul Robbins, Bruno Latour, and TNC Vice President Peter Kareiva; its Director of Research is mainstream conservationist Linus Blomqvist. TBI's project to "modernize environmentalism" is consistent with what Buscher et al. (2012) describe as "neoliberal conservation" guided by economic rationality and human-centered managerialism (see also Sullivan 2010). Although throwing off the shackles of Nature[1] might sound the death knell for the "Edenic sciences" like conservation biology (Robbins and Moore 2013), the economics of ecosystems and biodiversity is alive and on the neoliberal postnatural conservation path (MacDonald and Corson 2012; Dempsey forthcoming).

We respond to this path with an alternative expressed in the form of a manifesto. We choose the manifesto as a declarative format that makes a path-changing proposal "to stop going further in the same way as before toward the future" (Latour 2010, 473). In his own "Compositionist Manifesto," Latour (2010, 486; see also Latour 2013) argues that such a break requires us to "turn our back, finally, to our past, and to explore new prospects, what lies ahead, the fate of things to come." Postnatural conservation's scornful take on nostalgia for Nature and its reorientation around building futures resonate with Latour. In contrast, our manifesto urges a temporal orientation to reckon with the past. Looking back directs attention to what Stoler (2008) calls *ruination*, the discursive material processes of annihilation, displacement, and replacement driven by imperialism. Indeed, MacKinnon (2013) drew on early colonial records to suggest we inhabit a planet with only 10 percent of the biological variety and abundance it had before the mass culls and extractions that have marked imperial capitalism to present.[2] Looking back also shows us what we should strive for: a world literally filled to the brim with different creatures.

In the face of ruination, we offer a manifesto for abundant futures, by which we mean futures with more diverse and autonomous forms of life and ways of living together. In the spirit of creativity and solidarity, we leave the definition of abundance open while taking inspiration from decolonizing frameworks, politics, and ethics as articulated in contemporary settler societies such as Canada, Australia, the United States, and Latin America. Decolonizing frameworks entail recognizing how knowledge production and everyday relations (including those between humans and other sentient beings) are informed by European colonial modalities of power and propped up by imperial geopolitical and economic arrangements (Maldonado-Torres 2007). We draw from decolonial and postcolonial scholars while recognizing the diversity within and between decolonizing movements and scholarship. Our decolonizing sensibility keeps an eye on the past to reckon with how we got to this place of ruination and ecological impoverishment, acknowledging that creating conditions for abundance necessitates enacting alternatives to imperial capitalism.

We begin the article by tracing the emergence and key characteristics of neoliberal postnatural conservation. We also indicate points of convergence and divergence with decades of scholarship in political ecology, science studies, and elsewhere that has questioned conservation's traditional orientation around Nature (i.e., Haraway 1991; Cronon 1995; Latour 2004), which has led to enclosure and dispossession (i.e., Neumann 1998; Chapin 2004; West, Igoe, and Brockington 2006). Many of the same logics, we suggest, persist in neoliberal postnatural conservation. Next, we present a brief critique of postnatural conservation. We then outline our response in the form of a manifesto for abundance. The political strategies we advance are shaped by the understanding that our many privileges as members of Canada's settler society stem from the theft of Indigenous people's lands and the state's ongoing policies of assimilation and appropriation. Our profound desire to transform these conditions and to build respectful and accountable relationships with multispecies others drive this manifesto.

The Path Being Taken: Neoliberal Postnatural Conservation

In 2009, one of us sat at a peanut bar with Peter Kareiva, the Vice President and Chief Scientist of

TNC. Kareiva spoke in blasphemous terms. No one cares about biodiversity, he said, except white suburbanites; it's a dead-end concept (see Kareiva n.d.). He favors ecosystem services, which, he argued, resonates across class divides and cultures. Responding, Jessica Dempsey noted that Kareiva should probably look for a new job, given how central biodiversity is to TNC programming. He laughed. "Yeah," he said, "I might not last much longer at TNC."

Only a few years later, not only is Kareiva still in his position at TNC, he also is a key "bomb thrower" (Voosen 2012) in a heated debate about the future of biodiversity conservation. The debate has those like Karieva advocating human-centered, ecosystem-service-focused conservation, what we call *neoliberal postnatural conservation*, pitted against those holding on to traditional biodiversity-focused conservation (e.g., Soulé 2013). All is not divided, of course, as there also are calls for diverse approaches to a *Rambunctious Garden* (e.g., Marris 2011). Even so, we argue, neoliberal postnatural conservation is poised to be the path taken to guide conservation in the so-called Anthropocene. In what follows, we briefly outline three characteristics of this particular path for socioecological futures, drawing mostly from the essay "Conservation in the Anthropocene," penned by Karieva, Marvier, and Lalasz (2012). For us, that essay is the exemplar of an emerging regime of truth. Our assessment of this new regime also builds on our reading of debates in journals like *Animal Conservation*, critiques of neoliberal conservation (Sullivan 2010; Buscher et al. 2012; MacDonald and Corson 2012), and other prescient work on this turn (Robbins and Moore 2013; Robbins 2014).

Hopeful, Future-Oriented Postnaturalism

For those of us schooled in Cronon's (1995) "The Trouble with Wilderness," parts of Kareiva, Marvier, and Lalasz's (2012) essay are familiar. The authors write, "The wilderness so beloved by conservationists—places 'untrammeled by man'—never existed" (Kareiva, Marvier, and Lalasz 2012). Further echoing Cronon (and social nature geographers; e.g., Braun and Castree 2001), they call this Nature inaccurate and reflective of bourgeois desires. Thus, write Kareiva, Marvier, and Lalasz (2012), retaining "conservation's intense nostalgia for wilderness and a past of pristine nature ... [and its] focus upon preserving islands of Holocene ecosystems in the age of the Anthropocene is both anachronistic and counterproductive." It is impossible to go back to a past that never existed, they say, and this is especially the case given dramatic shifts like invasive species and climate change.

Although these shifts might be cause for despair for some environmental movements, Karieva, Marvier, and Lalasz (2012) optimistically point to the possibilities: "nature could be a garden—not a carefully manicured and rigid one, but a tangle of species and wildness amidst lands used for food production, mineral extraction, and urban life." The massive marks humans have made on the planet offer a kind of liberation for those who have been so obsessed with saving. Thus, Ellis (2013) argues, "The only limits to creating a planet that future generations will be proud of are our imaginations and our social systems. In moving toward a better Anthropocene, the environment will be what we make it."

Resilient Natures for Economic Development

Postnatural conservationists ask us to "see through the illusion" that Nature exists in a "delicate state of harmony constantly at risk of collapse from too much human interference" (Nordhaus and Shellenberger 2011). From this perspective, species and ecosystems, like humans, are usually *resilient*. Even after "major disturbances such as deforestation, mining, oil spills, and other types of pollution," Kareiva, Marvier, and Lalasz (2012) contend species abundance and "other measures of ecosystem function recover, at least partially" in most cases. At the Chernobyl nuclear facility, they say, "wildlife is thriving, despite the high levels of radiation." They also claim that Amazonian rainforests "have grown back over abandoned agricultural land," although the regrown forest only hosts "40 to 70 percent of the species of the original forests."

This perspective on resilient natures articulates with a specific view of development. Sounding positively Kuznets-curve-like (in terms of linear advancement of human societies from environmental destroyers to environmentalists), Kareiva, Marvier, and Lalasz (2012) suggest "forest cover ... is rising in the Northern Hemisphere, where 'nature' is returning to former agricultural lands. Something similar is likely to occur in the Southern Hemisphere, after poor countries achieve a similar level of economic development." Although "affluent, white, upper-middle class Americans" see conservation as inherently valuable, people in poor countries will only protect the environment if "it links to their own needs," Kareiva (n.d.) says. As such, postnatural conservationists appear to accept Rostowian visions of modernist development, which are no longer viewed as the path to destruction or the fall from purity but rather as a teleological necessity (see Ellis et al. 2013). The kind of development needed is *green* development that cultivates the natures that support "thriving economies" (Kareiva, Marvier, and Lalasz 2012). Corporate partnerships are key to this vision (MacDonald 2010). "Instead of scolding capitalism," Kareiva, Marvier, and Lalasz (2012) state, "conservationists should partner with corporations in a science-based effort to integrate the value of nature's benefits into their operations and cultures."

Optimizing Enterprising Natures

The new conservation is deeply humanist and utilitarian. "Instead of pursuing the protection of biodiversity for biodiversity's sake," write Kareiva, Marvier, and Lalasz (2012), "a new conservation should seek to enhance those natural systems that benefit the widest number of people, especially the poor." Central to the new conservation is a focus on trade-offs between competing ecosystem services, between, say, the services of water purification, carbon sequestration, pollination, and timber provided by the same patch of forested land. These trade-offs must be calculated and economically valued so that ecosystem services can be rationalized and optimized. For example, the InVEST tool, created by a collaboration between TNC (with Kareiva at the helm), WWF-US, Stanford University, and the University of Minnesota, is a computer program that calculates changes in ecosystem services based on alternative future land and marine uses (e.g., increased urban development, or more restored areas). These calculations reveal the sites most important for delivering ecosystem services, often attached to economic valuations. On this path, ethical questions about how to intervene are oriented by optimizing ecosystems for the greatest number of people, maximizing utility across these trade-offs.

Responding to Postnatural Conservation

Earlier, we traced a neoliberal postnatural conservation at the forefront of guiding interventions in the Anthropocene. As neoliberal conservation has already been critiqued (i.e., Sullivan 2010; Buscher et al. 2012), we turn our attention to its *postnatural* dimensions. Although we share excitement about political opportunities offered by the end of Nature and increasing interest in entanglement and "rambunctious gardens" (Marris 2011; Robbins and Moore 2013), we worry about what makes postnature amenable to neoliberal approaches to conservation. This is not to conflate postnatural with neoliberal natures; they are not the same. Nonetheless, we believe that the postnatural is premised on ontological claims about how the world is composed that merit additional analysis. As Braun (2009, 31) cautions, "There is no hard and fast rule that a particular ontology leads necessarily to a particular politics, but neither can any ontology be said to be neutral." Hence, the notion of postnature as composed of heterogeneous assemblages of living and inert entities is not inherently more or less life-giving in its political implications. To negotiate this slipperiness, postnatural scholars surveyed by Braun (2009, 31) direct our attention to the *processes for composing*, to the development of "institutional spaces and procedures that allow us to work through, in an agonistic matter, how this composition of common worlds should proceed." Compositionism, suggests Latour (2010, 474), "takes up the task of building a common world . . . built from utterly heterogeneous parts that will never make a whole, but at best a fragile, revisable, and diverse composite material." Compositionism directs us away from the question of whether or not things are constructed—they always are—and "toward the crucial difference between what is *well* or *badly* constructed" (Latour 2010, 474).

We agree. But we are struck by the ease with which both postnatural conservation and Latour abandon Nature and move on to composing. From our perspective, Nature was an imperial imposition, not a bad phase or an inadequate ontoepistemology that may be forgotten. Nature might be dead for Karieva and Latour, but its ruins remain. Reckoning with ruination means contending with the durability and strength of conservation assemblages, which derive from associations with imperial geopolitical institutions (Chapin 2004; Robbins 2014). Conservation organizations cannot be separated from imperial formations. This is the case even if conservation no longer relies on Nature. To face this head on, compositionism needs more political signposts or it risks becoming another future-oriented invocation of *terra nullius*, a blank slate—this time an anthropogenic or "used" slate (Ellis et al. 2013).

By signposts we do not mean a new modern constitution with universal aspirations or composing a common world (Latour 2004, 2010, 2013). Although Latour claims that we have never been modern, his invocation of a common world presupposes a "we" with sufficient ontological commonality to afford communication across communities. The modern constitution Latour describes was never universal, although imperial regimes of power certainly attempted to make it so (Blaser 2013; Sundberg 2013). As such, Latour risks treating Western thought as a universal frame of reference, which in turn negates the existence of radically different ontologies (Blaser 2009). In the face of colonial efforts to eliminate difference, we find it necessary to pursue strategies for composing that foster modes of living together with radical difference (Martin, McGuire, and Sullivan 2013). How differences are to be adjudicated and by whom remain crucial questions.

Finally, we are concerned about the easy call for entanglements and even intimacy. For Latour (2012), compositionism "describes our ever-increasing degree of intimacy with the new natures we are constantly creating." The sin, he says, "is not to wish to have dominion over Nature, but to believe that this dominion means emancipation and not attachment." As Latour is suggesting, intimacy is not free of mastery. We are with him: Domination occurs through attachment. But for us, domination should be resisted. The domination of nature and other-than-humans by particular human groups is ruinous. Acknowledging entanglement is not enough to shift us away from further animal death and exploitation.

In sum, we worry about the postnatural orientation toward a future, entangled common world, as this might be, in part, what makes the approach amenable to alliances with neoliberalism. We see such alliances as fundamental impediments to abundant futures. Our manifesto, to which we now turn, takes us in a different direction.

Another Path Is Possible! Abundant Futures Manifesto

If anything, the Anthropocene is a spark that will light a fire in our imaginaries. This is a time to think big, to dream. We dream about abundant futures. In what follows, we offer this dream in the form of a manifesto, a declaration of strategies to create the conditions for supporting diverse forms of life and ways of living.

Decolonizing frameworks, politics, and ethics guide our thinking about the conditions needed to generate abundance. Although "the desired outcomes of decolonization are diverse and located at multiple sites in multiple forms" (Sium, Desai, and Ritskes 2012, 2), our decolonizing sensibility builds from scholarship and movements in settler societies that are premised on Indigenous self-determination. In this context, we draw particular attention to the ways Nature is steeped in colonial patterns of power and knowledge. Nature, we argue, must be confronted as an artifact of empire, although not "as dead matter or remnants of a defunct regime" that can be ignored (Stoler 2008, 196). Rather, as Stoler (2008, 195) notes, imperial ruins have a political life; they "impinge on the allocation of space, resources, and on the contours of material life" in the present. Discerning how the residues of Nature are reactivated in contemporary conservation politics in ways that continue to dispossess is crucial to the practice of decolonizing.

The violence of settler colonialism is ongoing (Wolfe 2006) as "land is remade into property and human relationships to land are restricted to the relationship of the owner to his property" (Tuck and Yang 2012, 5). Anishinaabeg scholar and activist Leanne Simpson beautifully articulates this transformation of land and bodies (cited in Klein 2013):

> Extraction and assimilation go together. Colonialism and capitalism are based on extracting and assimilating. My land is seen as a resource. My relatives in the plant and animal worlds are seen as resources. My culture and knowledge is a resource. My body is a resource and my children are a resource because they are the potential to grow, maintain, and uphold the extraction–assimilation system. The act of extraction removes all of the relationships that give whatever is being extracted meaning. Extracting is taking. Actually, extracting is stealing—it is taking without consent, without thought, care or even knowledge of the impacts that extraction has on the other living things in that environment. That's always been a part of colonialism and conquest. Colonialism has always extracted the indigenous.

As Simpson suggests, colonial extraction also implies attempts to erase distinct ways of bringing worlds into being. Transforming these conditions requires political struggle grounded in decolonizing. Inspired by Simpson and others, we now turn to three concrete political strategies necessary to create conditions for generating abundance rather than postnatural conservation. These strategies are informed by transformative efforts already occurring around the globe.

Strategy 1: Reckoning with Colonial-Capitalist Ruination

Like postnatural conservationists, we do not support a conservation oriented around the colonial myth of a pristine past. Yet the tendency to relentlessly focus on the future is not the answer. When considering how to intervene responsibly and ethically, an ongoing and active reckoning with the past is crucial. We can look to the past not to provide an Edenic benchmark but to understand the discursive material infrastructure we have inherited: How did we arrive where we are today, to a world of social asymmetries and ecological impoverishment? Galeano (1973) and Davis (2002) contend that we arrived at contemporary "underdevelopment" through colonialism and imperial capitalist development. Violence was central to these processes. "Millions died," Davis (2002, 11) writes, "not outside the 'modern world system,' but in the very process of being forcibly incorporated into its economic and political structures." The Capitalocene, Haraway's (2014) counterconcept to the Anthropocene, specifically foregrounds capitalist modes of political economy (and their attachment to fossil fuels) as drivers of impoverished ecologies. To recall this violence is neither nostalgic nor anachronistic but central to understanding that any intervention today is unavoidably linked to processes of imperial ruination.

Equally, we need to pay attention to histories of nonhuman abundance and the violences that led to their diminishment. MacKinnon (2013) sees the past as a measure of possibility for what "may be again." For MacKinnon, this is not a call for "some romantic return to a pre-human Eden." Rather, he posits, "A story of loss is not always and only a lament; it can also be a measure of possibility. What once was may be again." For MacKinnon, this means taking past abundance as a marker for what might be; looking back shows us what rich socioecological worlds looked like (as in Denevan 2001; Raffles 2002; Mann 2005).

"Our systems are designed to promote more life," says Leanne Simpson about her Anishinaabeg community (cited in Klein 2013). Working with the Anishinaabeg concept of *mino bimaadiziwin*, variously translated as "the good life" and "continuous rebirth," Simpson identifies an alternative to worlds that are enacted through utilitarianism and extraction. "The purpose of life," she says, "is this continuous rebirth, it's to promote more life. In Anishinaabeg society, our economic systems, our education systems, our systems of governance, and our political systems were designed with that basic tenet at their core." The concept of promoting life differs considerably from a core aspect of sustainability and earth systems science, which focuses on figuring out the limits to development or the extent to which ecosystems may be degraded before ecological function is impaired or beyond repair. As Simpson says, her community considers "how much you can give up to promote more life" (cited in Klein 2013; also Simpson 2011).

We ally ourselves with such strategies to produce abundance. For Tewolde Egziabher (2002), the tireless Ethiopian advocate for farmers' rights and agricultural diversity, supporting conditions to create and sustain biological diversity involves refusing capitalist processes of enclosure over land, waters, and living things, including patents on life. We ally with Via Campesina (2008) and its more than 200,000 members throughout the globe in defending the "collective rights of peasant farmers to access land" from those who appropriate land "for profit." Peasant farmers affiliated with Via Campesina fight relentlessly against the status quo, against the World Trade Organization and other trade agreements that privilege corporate actors, against the governments who facilitate land grabs, and against corporate enclosures. In so doing, they are creating institutions and alliances that go far beyond national borders, including the World Social Forum, farmer–farmer exchanges, and seed-saving networks.

Strategy 2: Acting Pluriversally

Recognizing entanglement is not enough to undo colonial formations such as Nature. Hence, we ally with others fostering the capacity to act in *pluriversal* instead of universal ways (Blaser, de la Cadena, and Escobar 2014). The universe is enacted through the ontological assumption of reality or nature as singular, with different cultures offering distinct conceptions of this reality (Blaser 2013). This approach equates ontology with mental maps or culture and leaves intact the assumption that differing perspectives on the world can be understood through and reduced to Eurocentric categories. Building on Indigenous thought as well as some science studies scholarship, Blaser (2009, 2013) frames ontology in terms of practices and performances of *worlding*—of being, doing, and knowing; reality "is *done* and *enacted* rather than observed" (Mol 1999, 77). Worlding practices bring worlds into being; different stories enact different worlds that may be coemergent, partially connected, or in conflict. Blaser (2013, 552) proposes the *pluriverse* as a "heuristic proposition," a commitment to enacting ontological multiplicity, to shift us away from continuously performing the universe. If different stories perform different yet interconnected worlds, then worlding practices can be evaluated in terms of their effects; some worldings might be wrong in the sense that "they enact worlds (edifices) in which or with which we do not want to live, or that do not let us live—or lets some live and not others" (Blaser, de la Cadena, and Escobar 2014).

Creating abundant futures, we believe, means supporting already existing worlding practices that enact worlds different from those produced by European imperialism and settler colonialism. We ally ourselves with Idle No More, a Canada-wide Indigenous movement sparked by federal efforts in 2012 to enact legislative changes that weaken Indigenous sovereignty and environmental regulations. Started by four women, the movement spread like wildfire, drawing national attention to ongoing Indigenous struggles, sparking, revitalizing, and supporting decolonizing efforts in a multitude of communities. Activists and authors Simpson (cited in Klein 2013) and Glen Coulthard (2013) articulate the movement's role in supporting a "resurgence of Indigenous political thought" in relation to governance models and "Indigenous political-economic alternatives."

We respond to Idle No More's invitation "to join in a peaceful revolution, to honour Indigenous sovereignty, and to protect the land and water" (Idle No More n.d.). Enacting abundance means different ways of building relationships across vast differences, best described as solidarity or collective movement in support of conditions that enable differently situated people and other-than-humans to realize abundance, to build a world of many worlds. In thinking about how to move collectively, we take inspiration from the concept of *walking with* put forth in the Zapatista movement's Sixth Declaration of the Selva Lacandona (Zapatista Army of National Liberation 2005). In this framing of solidarity, walking with implies engaging in activism wherever one lives in support of a common struggle against neoliberalism and for democracy, liberty, and justice. As such, solidarity supports autonomous forms of worlding.

Strategy 3: Recognizing Animal Autonomy

Recognizing multispecies entanglement is not a license to intensify human control over other-than-human life. Abundant futures include nonhuman animals, not as resources or banks of natural capital that service humans but as beings with their own familial, social, and ecological networks, their own lookouts, agendas, and needs. An abundant future is one in which other-than-humans have wild lives and live as "uncolonized others" (Plumwood 1993). We follow Cronon, likely the most widely cited troubler of wilderness, who actually argues for retaining the idea of wildness. As Cronon (1995, 89) writes, "Honoring the wild" is a matter of "learning to remember and acknowledge the autonomy of the other." Whereas wilderness refers to an impossible pure Nature, *wildness* refers to the autonomy, otherness, and sentience of animals (Plumwood 1993; Collard 2014). By *autonomy* we mean the fullest expression of animal life, including capacity for movement, for social and familial association, and for work and play. These capacities have been profoundly diminished with the confinement, control, and managerialism that have come to characterize humans' relationships with the wider world in humanist colonial and capitalist regimes. In particular, animals' spatial and bodily enclosure (in public zoos and aquariums, laboratories, and factory farms) impedes their autonomy and abundance.

Of course, an autonomous life is never a discrete life. Whether enclosed or not, animals are always inescapably part of socionatural networks (as are we). So what is the difference between these networks? The wild one offers—within limits—openness, possibility, a degree of choice, and self-determination. The enclosed one is controlled, cramped, contained, and enclosed. But neither do wildness or animal autonomy mean no human intervention; in a world that has always been far too entangled to permit "stepping outside," wildness and autonomy are relational. We are not advocating a return to conservation's old misanthropy but an orientation in which wildness is understood relationally, not as the absence of humans but as interrelations within which animals have autonomy. The degree to which an animal is wild thus has little to do with its proximities to humans and everything to do with the conditions of living, such as spatial (can the animal come and go), subjective (can the animal express itself), energetic (can the animal work for itself), and social (can the animal form social networks). These are conditions of possibility, of potential, not forced states of being.

We ally ourselves with the few conservationists who make the well-being of individual animals a priority (Paquet and Darimont 2010) and with efforts such as the recent campaign by Zoocheck and other Toronto and international organizations that led to the transfer of three elephants from the Toronto Zoo to a wildlife sanctuary in California. Part of a wider movement to end elephant captivity, the release of these three elephants is a sign of growing recognition of the effects of captivity on such social creatures.

Orienting toward abundant futures requires walking with multiple forms of resistance to colonial and capitalist logics and practices of extraction and assimilation. Decolonization is our guide in this process. A profoundly unsettling process, decolonization "sets out to change the order of the world," as Fanon (1963, 36) suggested fifty years ago. As the organizations, movements, and people discussed here show, unsettlings are already taking place, pluriversally. Although never perfect, they are our best chance for abundant socio-ecological futures.

Acknowledgments

For their astute comments on earlier drafts, our thanks to Emilie Cameron and Shiri Pasternak, as well as reviewers Michael Ekers, Sian Sullivan, and two others who remained anonymous. Thank you also to editor Bruce Braun for his guidance.

Notes

1. We capitalize the word *Nature* to indicate the concept of pristine and untouched.
2. MacKinnon's figure is metaphorical, not verified by scientific methods or empirical evidence. For an overview of scientific estimates of biodiversity loss, see Butchart et al. (2010).

References

Blaser, M., M. de la Cadena, and A. Escobar. 2014. The Anthropocene and the One-world (or the Anthropos-not-seen). Unpublished manuscript.

Blaser, M. 2009. The threat of the Yrmo: The political ontology of a sustainable hunting program. *American Anthropologist* 111 (1): 10–20.

———. 2013. Ontological conflicts and the stories of peoples in spite of Europe: Toward a conversation on political ontology. *Current Anthropology* 54 (5): 547–68.

Braun, B. 2009. Nature. In *A companion to environmental geography*, ed. N. Castree, D. Demeritt, D. Liverman, and B. Rhoads, 19–36. London: Blackwell.

Braun, B., and N. Castree. 2001. *Social nature: Theory, practice and politics*. Malden, MA: Blackwell.

Buscher, B., D. Brockington, J. Igoe, and K. Neves. 2012. Towards a synthesized critique of neoliberal conservation. *Capitalism, Nature, Socialism* 23 (2): 4–30.

Butchart, S., M. Walpole, B. Collen, A. van Strien, J. Scharlemann, R. Almond, J. Baillie, et al. 2010. Global biodiversity: Indicators of recent declines. *Science* 328 (5982): 1164–68.

Chapin, M. 2004. A challenge to conservationists. *World Watch Magazine* November/December:17–31.

Collard, R.-C. 2014. Putting animals back together, taking commodities apart. *Annals of the Association of American Geographers* 104 (1): 151–65.

Coulthard, G. 2013. For our nations to live, capitalism must die. *Voices Rising* 5 November. http://nationsrising.org/for-our-nations-to-live-capitalism-must-die (last accessed 19 May 2014).

Cronon, W. 1995. The trouble with wilderness; or, getting back to the wrong nature. In *Uncommon ground: Rethinking the human place in nature*, ed. W. Cronon, 69–90. New York: Norton.

Davis, M. 2002. *Late Victorian holocausts: El niño famines and the making of the third world*. London: Verso.

Dempsey, J. Forthcoming. *Enterprising nature*. New York: Wiley-Blackwell.

Denevan, W. 2001. *Cultivated landscapes of Native Amazonia*. New York: Oxford University Press.

The Economist. 2011. Tea with Erle Ellis on the Anthropocene. http://www.youtube.com/watch?v=XLCa1njCK0E (last accessed 1 November 2013).

Egziabher, T. 2002. The human individual and community in the conservation and sustainable use of biological resources. Darwin Lecture at the Annual Darwin Initiative meeting, London. http://www.nyeleni.org/IMG/pdf/Tewolde_Darwin_Lecture2002.pdf (last accessed 1 November 2013).

Ellis, E. 2009. Stop trying to save the planet. *Wired Science* 6 May. http://www.wired.com/2009/05/ftf-ellis-1/ (last accessed 12 May 2013).

———. 2013. Overpopulation is not the problem [Op-ed]. *New York Times* 13 September. http://www.nytimes.com/2013/09/14/opinion/overpopulation-is-not-the-problem.html?smid=tw-share&_r=1& (last accessed 31 October 2013).

Ellis, E., J. Kaplan, D. Fuller, S. Vavrus, K. Goldewijk, and P. Verburg. 2013. Used planet: A global history. *Proceedings of the National Academy of Sciences* 110 (20): 7978–85.

Fanon, F. 1963. *The wretched of the earth*. New York: Grove.

Galeano, E. 1973. *Open veins of Latin America: Five centuries of the pillage of a continent*. New York: Monthly Review.

Haraway, D. 1991. *Simians, cyborgs and women*. London and New York: Routledge.

———. 2014. SF: String figures, multispecies muddles, staying with the trouble. Lecture at the University of Alberta, Edmonton, Canada. http://new.livestream.com/aict/DonnaHaraway (last accessed 14 May 2014).

Idle No More. n.d. Idle no more. http://www.idlenomore.ca (last accessed 19 May 2014).

Kareiva, P. n.d. A new type of conservation: Breaking free of two myths about nature. http://thebreakthrough.org/people/profile/peter-kareiva (last accessed 30 October 2013).

Kareiva, P., M. Marvier, and R. Lalasz. 2012. Conservation in the Anthropocene. *Breakthrough Journal*. http://thebreakthrough.org/index.php/journal/past-issues/issue-2/conservation-in-the-anthropocene/ (last accessed 1 November 2013).

Klein, N. 2013. Dancing the world into being: A conversation with Idle No More's Leanne Simpson, *YES! Magazine* 5 March. http://www.yesmagazine.org/peace-justice/dancing-the-world-into-being-a-conversation-with-idle-no-more-leanne-simpson (last accessed 29 October 2013).

Latour, B. 2004. *Politics of nature*. Cambrdige, MA: Harvard University Press.

———. 2010. An attempt at a "compositionist manifesto." *New Literary History* 41 (3): 471–90.

———. 2012. Love your monsters. *Breakthrough Journal*. http://thebreakthrough.org/index.php/journal/past-issues/issue-2/love-your-monsters (last accessed 19 May 2014).

———. 2013. Telling friends from foes in the Anthropocene. Lecture for Thinking the Anthropocene, Paris. http://www.bruno-latour.fr/node/535 (last accessed 15 May 2014).

MacDonald, K. 2010. The Devil is in the (bio)diversity: Private sector "engagement" and the restructuring of biodiversity conservation. *Antipode* 42 (3): 513–50.

MacDonald, K., and C. Corson. 2012. "TEEB begins now": A virtual moment in the production of natural capital. *Development and Change* 43 (1): 159–84.

MacKinnon, J. 2013. *The once and future world*. Toronto: Random House.

Maldonado-Torres, N. 2007. On the coloniality of being. *Cultural Studies* 21 (2): 240–70.

Mann, C. 2005. *1491: New revelations of the Americas before Columbus*. New York: Random House Digital.

Marris, E. 2011. *Rambunctious garden: Saving nature in a post-wild world*. London: Bloomsbury.

Martin, A., S. McGuire, and S. Sullivan. 2013. Global environmental justice and biodiversity conservation. *The Geographical Journal* 179 (2): 122–31.

Mol, A. 1999. Ontological politics: A word and some questions. In *Actor network theory*, ed. J. Law and J. Hassard, 74–89. Boston: Blackwell.

Neumann, R. 1998. *Imposing wilderness: Struggles over livelihood and nature preservation in Africa*. Berkeley: University of California Press.

Nordhaus, T., and M. Shellenberger. 2011. Introduction. In *Love your monsters: Postenvironmentalism and the Anthropocene*, ed. M. Shellenberger and T. Nordhaus. Oakland, CA: The Breakthrough Institute. http://thebreakthrough.org/index.php/programs/philosophy/love-your-monsters-ebook (last accessed 4 November 2014).

Paquet, P., and C. Darimont. 2010. Wildlife conservation and animal welfare: Two sides of the same coin. *Animal Welfare* 19 (2): 177–90.

Plumwood, V. 1993. *Feminism and the mastery of nature*. London and New York: Routledge.

Raffles, H. 2002. *In Amazonia: A natural history*. Princeton, NJ: Princeton University Press.

Robbins, P. 2014. No going back: The political ethics of ecological novelty. In *Traditional wisdom and modern knowledge for the earth's future*, ed. K. Okamoto and Y. Ishikawa, 103–18. New York: Springer.

Robbins, P., and S. Moore. 2013. Ecological anxiety disorder: Diagnosing the politics of the anthropocene. *Cultural Geographies* 20 (1): 3–19.

Shellenberger, M., and T. Nordhaus. 2004. The death of environmentalism: Global warming politics in a post-environmental world. Paper presented at the Environmental Granmakers Association. http://grist.org/article/doe-reprint/ (last accessed 1 November 2013).

Simpson, L. 2011. *Dancing on our turtle's back: Stories of Nishnaabeg re-creation, resurgence and a new emergence*. Winnipeg, Canada: Arbeiter Ring.

Sium, A., C. Desai, and E. Ritskes. 2012. Towards the "tangible unknown": Decolonization and the indigenous future. *Decolonization: Indigeneity, Education & Society* 1 (1): i–xiii.

Soulé, M. 2013. The "new conservation." *Conservation Biology* 27 (5): 895–97.

Stoler, A. 2008. Imperial debris: Reflections on ruins and ruination. *Cultural Anthropology* 23 (2): 191–219.

Sullivan, S. 2010. "Ecosystem service commodities": A new imperial ecology? Implications for animist immanent ecologies, with Deleuze and Guattari. *New Formations* 69 (1): 111–28.

Sundberg, J. 2013. Decolonizing posthumanist geographies. *Cultural Geographies* 21 (1): 33–47.

Tuck, E., and K. W. Yang. 2012. Decolonization is not a metaphor. *Decolonization: Indigeneity, Education & Society* 1 (1): 1–40.

Via Campesina. 2008. Via Campesina intervention to CBD, 19 February, Rome. http://viacampesina.org/en/index.php/main-issues-mainmenu-27/biodiversity-and-genetic-resources-mainmenu-37/465-small-scale-producers-key-to-the-protection-of-biodiversity (last accessed 3 November 2013).

Voosen, P. 2012. Myth-busting scientist pushes greens past reliance on "horror stories." *Greenwire* April 3. http://www.eenews.net/stories/1059962401 (last accessed 14 May 2014).

West, P., J. Igoe, and D. Brockington. 2006. Parks and peoples: The social impact of protected areas. *Annual Review of Anthropology* 35:251–77.

Wolfe, P. 2006. Settler colonialism and the elimination of the native. *Journal of Genocide Research* 8 (4): 387–409.

Zapatista Army of National Liberation. 2005. Sixth declaration of the Selva Lacandona. http://enlacezapatista.ezln.org.mx/sdsl-en/ (last accessed 1 November 2013).

The Art of Socioecological Transformation

Harriet Hawkins,* Sallie A. Marston,† Mrill Ingram,† and Elizabeth Straughan‡

*Department of Geography, Royal Holloway University of London
†School of Geography and Development, University of Arizona
‡School of Geographical and Earth Science, University of Glasgow

This article uses two artistic case studies, *Bird Yarns* (a knitting collective engaging questions of climate change) and *SLOW Cleanup* (an artist-driven environmental remediation project) to examine the "work" art can do with respect to socioecological transformations. We consider these cases in the context of geography's recent interest in "active experimentations and anticipatory interventions" in the face of the challenges posed by the environmental and social uncertainties of the Anthropocene. We propose two dimensions to the force of art with respect to these concerns. First, it provides a site and set of practices from which scientists, artists, and communities can come to recognize as well as transform relations between humans and nonhumans. Second, it encourages an accounting of the constitutive force of matter and things with implications for politics and knowledge production. Through these two dimensions, we explore how the arts can enable forms of socioecological transformation and, further, how things might be different in the future, enabling us to explore who and what might play a part in defining and moving toward such a future.

本文运用两个艺术案例研究——“鸟的编织故事”（一个涉入气候变迁问题的编织系列），以及“缓慢清理”（一个由艺术家所策划的环境矫正计画），检视艺术在社会生态变迁方面可从事的“工作”。我们在晚近地理学面对人类世的环境及社会不确定性所带来的挑战时，　对“行动实验和预防干预”产生兴趣的脉络中，考量这些案例。关于这些考量，我们提出艺术趋力的两个面向：首先，它提供科学家、艺术家与社群能够前来指认并改变人类与非人类之间的关係的场域和一系列的实践。再者，它鼓励叙述事物与东西的组构力，并对政治与知识生产带来意涵。透过上述两种面向，我们探讨艺术如何能够促成社会生态变迁的形式，并进一步探讨东西如何可能在未来有所不同，使我们能够探索谁、什麼事物可能会在界定此般未来、并朝此方向前进中起作用。

Este artículo utiliza dos estudio de caso artísticos, los de *Bird Yarns* (un tejido colectivo que aboca cuestiones de cambio climático) y *SLOW Cleanup* (un proyecto de motivación artística sobre reparación ambiental), para examinar el "trabajo" que el arte puede hacer respecto de las transformaciones socioecológicas. A estos dos casos los consideramos dentro del contexto del reciente interés de la geografía en "experimentaciones activas e intervenciones anticipatorias" frente a los retos que han puesto las incertidumbres ambientales y sociales del Antropoceno. Proponemos dos dimensiones en relación con la fuerza del arte frente a estas preocupaciones. Primero, el arte proporciona un sitio y un conjunto de prácticas desde las cuales los científicos, artistas y comunidades pueden llegar a reconocer, lo mismo que a transformar, las relaciones entre humanos y no humanos. Segundo, el arte alienta la consideración de que la fuerza constitutiva de la materia y de las cosas, con implicaciones para la política y la producción de conocimiento, sea tenida en cuenta. A través de estas dos dimensiones exploramos el modo como las artes pueden habilitar formas de transformación socioecológica y, también, las maneras como se podrían cambiar las cosas en el futuro, capacitándonos para explorar quién y qué podría tener algún juego en definir y avanzar hacia ese futuro.

What might it mean to consider socioecological transformations and futures from a perspective that deploys an expanded appreciation of the social, one that incorporates the "lively processes and impure forms coexisting in inhabited landscapes" (Lorimer 2012, 594)? Such a foundational question poses significant challenges to the politics, knowledge, and imaginations of socioecological futures. One challenge is how might we imagine and bring about a recognition of ecology and its particular temporalities—less as an interaction between preordained life forms than of their emergence and transformation in a "wider field of forces, intensities and durations that give rise to [them]" (Ansell Pearson 1999, cited in Whatmore 2013, 36). How, in short, do we develop an orientation "not towards conservation,

Figure 1. *Bird Yarns* (on the wire), Deidre Nelson (2012). *Source:* Author's own photograph. (Color figure available online.)

because the world never holds still, but to the possibilities and consequences of a 'new earth' and a 'new humanity' that is still to come" (Braun 2006, 219)? A second challenge is how to research and write about such worldly liveliness and disseminate the results among the full range of "publics" and "experts" (principally scientists) on whose knowledge the performance and enactment of socioecological futures rests.

These challenges have become increasingly urgent, and increasingly foregrounded, in discussions around the Anthropocene, that current geologic era of human-driven environmental and social uncertainty (Lorimer 2012; Johnson and Morehouse 2014). But thinking through these complex materialities and vital forces, and their temporalities, let alone making space for them within expert—often scientific—knowledge-making practices, is not easy. Indeed, it is not enough to move beyond the binary of "modern" thinking (e.g., nature–culture) to make sense and take account of such vital forces and differences—from animals, to microbes, to the liveliness of inorganic, geologic, and atmospheric forces, for example. In addition, we must develop modes of knowledge production that are able to meet the epistemological, ontological, and political challenges accompanying such accountings.

Geographers have responded to these calls with a turn toward inter- and intradisciplinarity, developing creative solutions, including "active experimentations and anticipatory interventions" in collaboration with a range of scientists and publics (Lorimer 2012, 599). In this article we present two case studies, part of a wider set of ethnographies of art–science projects, that provide complementary accounts of how artistic practices might contribute to this emerging suite of experiments and interventions.[1] In doing so we contend that such intellectual concerns with the complex materialities and forces within and through which we live possess the potential to foster political and ethical relationships that might constitute "us" and our collective futures differently.

Art has long offered geographers empirical objects through which to theorize nature and society–environment relations, through paintings, installations and land-art (Daniels 1993; Gandy 1997), and, more recently, art–science collaborations (Dixon, Hawkins, and Straughan 2013). We add to this area of inquiry by examining two socially engaged artworks, *Bird Yarns* (2012; Figure 1), a collective knitting project in Scotland, and *SLOW Cleanup* (2010; Figure 2), a Chicago-based, artist-driven environmental remediation project.

Although geographers have recognized the social force of arts practices in not only reproducing worlds but also in constituting them (Daniels 1993), the last few decades of seemingly exponential growth in socially engaged art have resulted in further interest

Figure 2. *SLOW Cleanup* (the site), Frances Whitehead (2010–). *Source:* Photo courtesy of the artist. (Color figure available online.)

and scholarship. Engaging with art as a "technology of connection," geographers have studied how it can constitute human social encounters that do particular types of "work" in the world. This work might include healing communities torn apart by conflict or connecting fractured urban communities contending with urban "superdiversity" (Hawkins 2014). Art theorists interested in this recent growth have indicated the need to investigate the form and kind of social relations these artworks constitute, to challenge their instrumentalized politics, and to reflect on their reconfiguration of aesthetics (Bishop 2004; Kester 2011). Among these discussions, however, the social has largely been mobilized as a human-centric concept, and the encounters explored chiefly those between humans. Here we seek to expand this idea of the social to include the more-than-human. We investigate how the social relations constituted in our two cases unfold between humans and nonhumans alike. We further consider how such relations might be understood to expose the liveliness of the world and in doing so assemble new types of collectives that participate in the making of our socioecological futures. Ahead of our empirical discussions, we outline the experimental and interventional strategies that geographers have recently been adopting to account for vital matter and forces in our socioecological thinking.

Toward Aesthetic Experiments and Interventions

It is, by now, understood that to specify the presence of nonhumans, or simply to acknowledge the animate matter and forces of the physical worlds in which we live, is not enough. As Paulson (2001, 112) writes, the crucial point is to learn how "new types of encounter (and conviviality) with nonhumans ... can give rise to new modes of relation ... to new political practices" and, we would add, to new knowledge-making practices. A range of responses have been offered by way of experimenting with science and arts practices and theoretical vocabularies, to begin to specify how we might formulate ethico-political conduct akin to these expanded socials. Latour (2004), for example, seeks to understand how we might "learn to be affected" to build relations with inanimate objects and nonhumans. At the same time, Haraway (2008) calls for a "response-ability," detailing a politico-ethico stance that is propagated from an acknowledgment of the force of humans and nonhumans and organic and inorganic things. Such approaches have gained momentum with the rise of Anthropocene scholarship. This work foregrounds not only the negotiation of the difficulties of alliance building among human constituencies but also requires that we account for "deliberations between multiple forms of agency, expertise and subjectivity—some of which are human, some of which require tuning into the diverse becomings of nonhuman forms and processes" (Lorimer and Driessen 2013, 3). Further, the Anthropocene framing of these discussion has emphasized questions of temporality and the practice of not just living with difference but taking account of it in the co-fabrication of immanent, indeterminable, and speculative worlds and futures to come (Lorimer 2012; Johnson and Morehouse 2014).

In response, scholars, including geographers, are evoking languages and practices of experiment, intervention, and anticipation. As such, a critical space is being opened for experimental socioecological practices and politics, wherein "anticipatory interventions" are celebrated in the exploration and bringing about of "experimental and emergent futures" (Lorimer 2012, 599, 601). These are avowedly political projects, which among other things, seek to foster us as experimental researchers through a call for "hybrid research collectives" that "un-perform dualisms" (Gibson-Graham and Rolevink 2009) and assert the potential of the sensing body as a site from which to begin addressing environmental ethics.

As practices such as citizen science (Gross 2010) and ideas such as Stengers's competency groups (Whatmore and Landström 2011) have become the normative register for such experimental endeavors and interventions, there is an emerging role for art and aesthetic practices more broadly. Born and Barry (2010), for example, note the potential of art–science practices as public experiments that, rather than translating science for an assembled public, transform art, science, and the public. In a different vein, geographers have explored forms of writing and image making as enabling encounters with nonhumans drawing on ideas of response-abilities or a learning to be affected (Braun 2005; Hinchliffe et al. 2005; Lorimer 2010). Perhaps more common, however, is geographical scholarship that turns to art to explore the place of humans within the complex worlds that constitute them. Geographical studies of land art have long been grounded by such ideas (Matless and Revill 1995; Gabrys and Yusoff 2011; Hawkins 2013), but recent studies of posthuman aesthetics (Dixon, Hawkins, and Straughan 2013) consolidate this approach as a promising direction for further study.

In making sense of bioart, Dixon quotes Whatmore to underscore how such works might cultivate relations toward all manner of "social objects and forces assembled through and involved in the co-fabrication of socio-material worlds" (Whatmore 2006, 603–04 cited in Dixon 2009, 414). Exploring arts practices as sites for the suspension of "the ordinary coordinates of sensory experience," Dixon, like Gabrys and Yusoff (2012), draws on Rancière's theories of aesthetics and politics to recognize the force of arts practices to "create new modes of sense perception that would realign and resituate natural-cultural relations" (18). In their exploration of climate change imaginaries, they assert the potential of art for making, imagining, contesting, and living in shared material and affective worlds. Although agreeing wholeheartedly with the sensibility of these studies, we find Rancière less useful in terms of exploring art's potential to cultivate imaginaries of shared worlds and collective actions. Our issue lies in the limits we find in his polity. For, as useful as his thinking might be to politics, aesthetics, and the human, it falls short in accounting for the nonhuman that is absent from his fully human-centered politics.

We resist analyzing our cases through the generalized theories of aesthetics that geographers have often deployed to understand the "work" art does. Instead, we explore the heterogeneous bits and pieces of the lively worlds of difference that compose any artwork and its polity (Marston, Woodward, and Jones 2005). Opening up the connections and redistributions of matter, objects, and expertise the artworks bring about enables us to query how they might contribute to socioecological transformation and the potential for propagating alternative futures. We begin with *Bird Yarns* reflecting on the knitting needles, patterns, wool, other yarns, and bits of plastic, as well as the production of the knitted birds and their display. Through these nonhuman objects, we probe the potential for aesthetics to engage us with lived worlds of difference. By refusing the analytic separation of spheres of life, we are forced to confront difference differently, asking not only what forms of difference matter but also who decides. We take up this point in the second case, *SLOW Cleanup*, which directs us to investigate the spaces and practices of science and the role of the scientist as representative of nonhuman constituencies.

Rematerializing Imaginaries of Climate Change

Bird Yarns is gathering supporters and followers with each and every landing, each time making connections

Figure 3. *Bird Yarns* (detail), Deidre Nelson (2012). *Source:* Author's own photograph. (Color figure available online.)

in new, exciting and unexpected ways . . . it seems that their gentle activism is engaging new audiences both for *Bird Yarns* and Cape Farewell in a cultural (and knitted) response to climate change.[2]

Discussions of the forming of social connections and the resulting "gentle activism" are common in *Bird Yarns*, an ongoing project to knit collectively a flock of Arctic terns. As artist Deidre Nelson describes, the "landing" of the knitted flock of "lost" terns in various locations around Scotland, England, and Nova Scotia produces "talk" about the disturbance in seabird migration patterns due to climate change. This participatory project, originating from the island of Mull, Scotland, and funded by the U.K.-based art–climate change organization Cape Farewell, was first displayed on wire strung along the island's harbor (Figure 3). It has evolved into an international network of knitters linked through a website as well as the global circulation of materials, patterns, and finished knitted birds.

At first glance, *Bird Yarns* appears to be a classic socially engaged arts project, mobilizing craft theories of "making-is-connecting" wherein collective material-making practices are understood to foster social bonds (Gauntlett 2013). *Bird Yarns* promoted social encounters between humans—initially Mull's Woolly Wednesdays knitting group—spreading by word of mouth across the islands, then throughout mainland Scotland, and finally going viral on the Internet, via Twitter, Facebook, and Tumblr. These virtual landings of the birds described earlier soon resulted in flock visits to various art galleries around the United Kingdom and Nova Scotia, fostering an international network of knitters.

At each of the landings, audiences encounter both knitted terns and an expert narrative, a spoken-word soundtrack written by an ornithologist and narrated by a local wildlife photographer. The observable effects of climate change are storied here through tales of changed avian lives, broken habits of nutrition and dwelling, and rerouted migration paths that take material form in the changed bodies of birds. As such, *Bird Yarns* tells a story of a charismatic species as a rallying point for environmental awareness and action (Lorimer 2006). Terns become akin to polar bears or penguins, those more classic examples of how species become icons for environmental awareness as images, stories, and myths accumulate around them.

Alongside assembling a human flock, the social encounters catalyzed by *Bird Yarns* were also of a human–nonhuman variety. Climate change is often considered a "collective experiment" that overspills the confines of the lab and requires that we acknowledge an assembly of new (human) publics, also recognizing the expanded ecologies and material conditions that are integral to its politicalities (Gabrys and Yusoff 2011). If making is indeed connecting, then the question of what is being made and what is being connected in *Bird Yarns* becomes an important one. The answer, we contend, is the production of shared material and affective environments that enables a different imaginary of earthly and atmospheric collectivities than one focused on scientific fact and scaled at the global level. In the case of *Bird Yarns*, this is a material imaginary that foregrounded situated matter, forces, and atmospheres as much as it engaged with a locally common species of bird.

The first instance in which we see these engagements with worldly matter is in how the making of the birds connected the fleshy corporealities of the knitters with the woolly bodies of the birds they produced. Working from a pattern and with wool from the island's sheep, knitters form the body of the bird, knitting the head, shaping the body, and forming its black cap from rows of stitches. They then sew them together and wash them to soften the wool, and finally stuff the knitted forms "with local fleece if possible." Local knitters were invited to beachcomb for red plastics for the wings and beaks; those not on the island were encouraged in the knitting instructions to "use red threads of yarn, or found plastics—use what you have, recycle if you can."[3]

The shaping of birds' woolly bodies, sculpted through the repetitive practices of manipulating wooden needles and spun yarn in variously experienced hands, catalyzed interactions-in-the-making, connecting the fleshy bodies of the birds, their knitted form, and the corporealities of the knitters. This connection is manifest in the care and concern the knitters expressed for the fate of their woolly terns in the harsh island weather. We want to go further, though, to suggest that these connections knitted together through matter are also ones that register a material imaginary of climate change that exceeds mundane relationships between humans and birds.

Bird Yarns offered knitters—and, in part, the local community—the chance to register a different imaginary of earthly and atmospheric collectivities than one focused on scientific fact. This imaginary propagated not only from the materialities and doings of knitting but also from the changing morphologies of the knitted bird bodies. These deformed avian corporealities were a function of the simplicity of the pattern (designed to be accessible), the varying skills of the knitters (from inexperienced novices to skilled amateurs and professionals), and the island weather conditions. But their deformation acted in two ways. First, they became part of a science-based narration concerning changes to body morphology as birds fly increased distances on less food due to climate-induced ecosystem change. As Nelson notes:

> People have . . . been like "Oh they look more like oyster catchers." . . . And I've been saying, . . . "Well you know they are changing because they are not eating the same food and they are traveling shorter" and you know "they have landed here" and . . . Dawn has been going "they have been at the chip van, they are eating rubbish that you are leaving lying around." . . . It can be a way to sort of . . . engage people in sometimes quite serious issues. (Interview 22 June 2012)

Whether the woolly birds look like Arctic terns, or whether the science was correct, was less important than how the deformed knitted morphologies stimulated discussion of "serious issues." Second, the fabricated bird forms became a barometer of sorts, attuning minds to the varying severity of local weather conditions and encouraging a conscious or unconscious registering of the interrelations of the common fleshy materiality of bird bodies, human bodies, and the elemental forces that constitute weather and climate.

Those knitting the birds worried in blog posts over "how they would fare with island weather," at the same time that local people became attached to the birds, expressing concern that they "must be cold" out there and joking about "getting a hair dryer to dry them off" (Interview 22 June 2012). Constant exposure to the elements changed their knitted forms, felting the wool, altering their materiality, as Nelson noted: "If we keep exhibiting them outdoors at different venues they will keep changing" (Interview 22 June 2012).

In the transformation of knitted stitch to felted fabric by elemental exposure, *Bird Yarns* conjures climate imaginaries based around forces and material transformations. Knitted birds become a prompt to engage with environmental change less as matters of scientific fact and more as forces affecting more-than-human collectives. The geographies of this climate change imaginary are refigured, too. In place of climate change as an "out there," distributed across a global imaginary of distant places, *Bird Yarns* localizes and materializes climate change. Epic narratives of heating and cooling, of ice and sea-level rise are replaced by a situation of climate change closer to home. These narratives are given form in the migratory birds whose annual pilgrimages have long punctuated island life, and even the perhaps more prosaic imaginaries that assert a corporeality in common that renders birds and humans alike vulnerable to dramatically changing weather conditions.

A Politics of Knowledge Production

SLOW Cleanup is a Chicago-based environmental remediation project convened by the artist Frances Whitehead in collaboration with grasses, bees, flowers, and a host of environmental science students, urban planners, and remediation scientists. Focusing on a test site on Chicago's South Side, *SLOW Cleanup* brings about environmental transformations, both the cleaning of chemical contaminants from one of the city's many abandoned gas stations and the return of the abandoned land to productive social (read here human and nonhuman) functions. The project also, we argue, realigns the political economies of the production of what counts as ecological science. We make sense of this through a grammar of science and technology studies that enables us to address sites of ontological perturbation such as Callon's (1998) "hot situations," Latour's (2005) "matters of concern," or Stengers's (2005) "things that force thought." Under these rubrics, *SLOW Cleanup* renders the environmental remediation process a generative event in which expert thinking was forced to slow down and a space for reasoning differently is opened up, enrolling those affected (human and nonhuman) in new political opportunities and associations.

If such sites of ontological perturbation are often those of techno-science or of catastrophic, controversial events (Stengers 2005; Whatmore 2013), *SLOW Cleanup*, although driven by less spectacular circumstances, is no less situated amid political economies of science. Abandoned gas station lots scattered across Chicago represent continuing challenges to that city's Brownfield Initiative, which since its inception in 1993 has encouraged partnerships with developers to clean up small, widely distributed sites and to redevelop them in ways beneficial to surrounding neighborhoods. The initiative, although successful elsewhere in Chicago—as one of Whitehead's collaborators notes, "phytoremediation is big business"—has been less so in more impoverished areas, where, without real estate investment, remediation remains elusive (Higgins 2008).

Responding to the polluted soil and social deprivation that shapes these sites, Whitehead, a professor of sculpture at the Art Institute of Chicago, submitted a proposal to the Chicago Department of Environment's Embedded Artist Project to experiment with new approaches to remediation for impoverished areas. In the course of her project, Whitehead generated a collective of nonhumans alongside artists, scientists, students, and planners. In doing so, *SLOW CleanUp*

> catalyzes a regime of thought and feeling that bestows the power on that around which there is a gathering to become a cause for thinking. . . . A presence that transforms each protagonist's relations to his or her own knowledge . . . and allows the whole to generate what each one would have been unable to produce separately. (Stengers 2005, 1002)

If *Bird Yarns* provides an example of how the "work" of art might open us to the forces and matter of the world, then Whitehead's project augments this by suggesting how taking account of nonhumans might just open us to different environmental futures. In response to the challenge of environmental remediation, Whitehead, we want to argue, creates a collective in which no single body of knowledge is sufficient. What emerges is a set of practices in which the issue—in this case, environmental remediation—gains the power to "activate thinking," a thinking that belongs to no one agent and in which no one is right. Collective thinking about environmental remediation proceeds, we argue, in the course of the project by way of acknowledging the presence of those who might otherwise be disqualified as having nothing to propose, producing as a result an emergent common account.

The workings of Whitehead's collective thus transformed the old gas-station site from a locus of persistent environmental problems into a generative space where the redistribution of knowledge and expertise occurred, and constitution of publics and the political were enlarged in practically and conceptually productive ways (Marres and Lezaun 2011; Whatmore and Landström 2011; Ingram 2012). As Paul Schwab, a soil chemist at Purdue University and one of the collaborating scientists noted:

> I've worked on a lot of remediation sites and none of them particularly nice to look at, and this one is different, so aesthetic. But that is the whole point of Frances being involved, trying to avoid the engineered straight rows and grasses that are known remediators with a different endpoint entirely. (Interview 8 September 2011)

In challenging scientists to think beyond "known remediators," disordering conditions are fostered as their expert reasoning is forced to slow down, creating opportunities to arouse "a different awareness of the

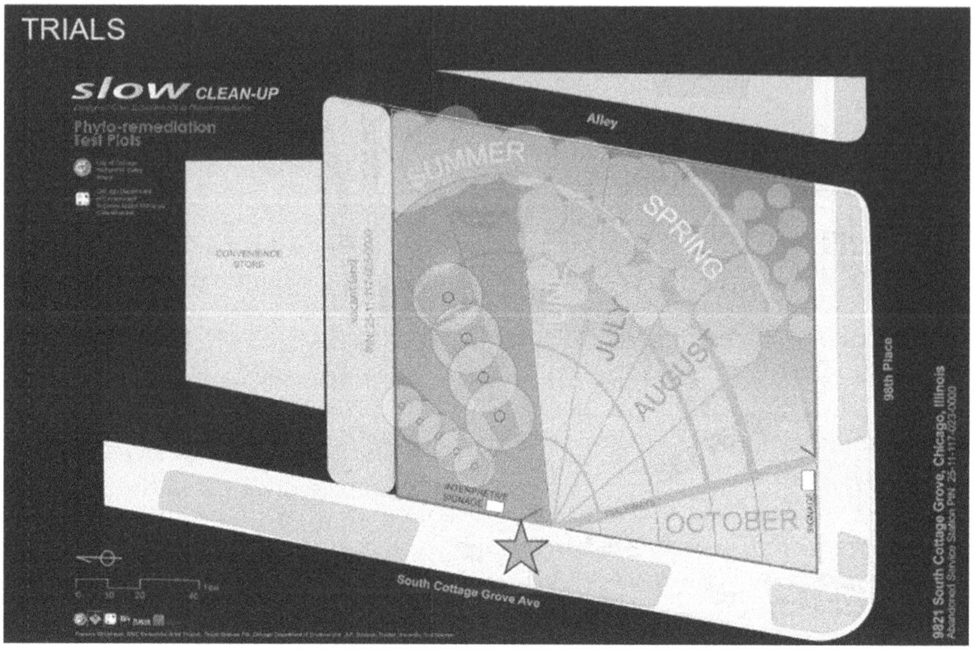

Figure 4. *SLOW Cleanup* (the plan), Frances Whitehead (2010). *Source:* Image courtesy of the artist. (Color figure available online.)

problems and situations that mobilize us" (Stengers 2005, 994). In doing so, a number of things, not least those expert knowledge claims hard-wired in the scientific practices of remediation, become the subject of political interrogation.

The urban setting of *SLOW Cleanup* challenged standard phytoremediation species, normally comprised of tall grasses and agricultural plants, hardly ideal for a small inner-city plot. Furthermore, Whitehead had a series of goals that exceeded those of remediation science. Thus began a year of lab and site-based experimentation in which the scientists went back to the drawing board after Whitehead rejected all the grasses that were their go-to choices. She turned instead to expanding the range of plants. Schwab notes:

> We worked together in terms of strategizing what should go in there in terms of plant species. On top of that [the remediation] she also has a number of scenarios she wanted to create, urban agriculture involving things that you could eat such as fruit or gardens . . . a bird sanctuary, and others that would be purely aesthetic . . . she had all kinds of things she was working on, and hundreds of species came rolling out of this. (Interview 8 September 2011)

Compiling research, assembling spreadsheets, and establishing greenhouse trials and test plots, Whitehead and Schwab, with assistance from Chicago State University environmental science students, successfully identified twelve new plant species able to manage and survive contamination and carry out other functions. Some species, for example, chemically dismantle large soil hydrocarbons in the root zone and simultaneously cocreate a garden that blooms as a seasonal clock, forming a resource and an attractive landscape for humans and nonhumans alike (Figure 4).

On the face of it, the publics and the forms of participation that *SLOW Cleanup* produces are different from those of *Bird Yarns*. In *SLOW Cleanup*, the local community—in human terms at least—is largely left out of expert discussion, as well as the larger collective that Whitehead composes. Instead, akin to "traditional" public art projects, the result is an aesthetic object (the garden) to be experienced by a (human) community that has had little involvement in its production. Significantly, however, the experiment of *SLOW Cleanup* insists that we specify more precisely than is perhaps normal with participatory arts projects, who (and what) participates in this ecological remediation and who does it serve? In forcing us to engage these questions, *SLOW Cleanup* challenges and reconstitutes ecological science and art in relation to an expanded set of participants. It develops a collective, calling on a widening circle of participants including an expanded community of plant species who bring new forms of remediative expertise to the project. What is at stake here is a redistribution of knowledge through multiple forms of expertise, whose outcomes offer a substantive challenge to conventional knowledge frameworks and technologies coded in the manuals and guides of urban brownfield remediation.

Conclusion

The environmental and social uncertainties of the Anthropocene, together with scholarship on the more-than-human, provide an impetus to expand the idea of the social in our understandings of socioecological transformations. Further, we are enjoined to take seriously the opportunities that such reconceived ideas of the socioecological present for normative understandings of spaces, practices, and participants in knowledge-making and political processes. Experimental and interventionary strategies have become an important way to approach these openings and in this article we have made a case for the value of artistic practices as part of these strategies.

In *Bird Yarns* and *SLOW Cleanup*, we considered the work art can do in bringing about forms of socioecological transformation. Further, we explored how these projects expose who and what might play a part in defining and moving toward alternative futures and how this might be accomplished. Our discussion suggests that this process of discernment happens in two ways, first, by recognizing that arts practices are able to make connections between humans and humans, humans and nonhumans, and between the matter and forces "out there" in the world and those more personal and local imaginaries. Considering intersections of the flock of birds and the distributed flock of human makers, *Bird Yarns* enabled an imaginary that localized the global collective experiment of climate change and the transcendental imaginaries of climate science. Second, by taking seriously the challenges that such worldly livelinesses pose to normative modes of politics and the practices of knowledge making, *SLOW Cleanup* provides us with an example of how artistic practices both disturb and subsequently redistribute power and knowledge. As such, normative ecological science practices are reordered to take account of multiple expertise and the enrollment of human and nonhuman actors in experiments in environmental remediation.

In addition to enabling us to consider the production of collective socioecological futures, our two cases highlight further potentially fruitful engagements between geography and art. As the introduction outlined, art theorists exploring contemporary socially engaged art such as *Bird Yarns* and *SLOW Cleanup* focus on questions concerning the type and effect of the social relations these art projects produce. Their work often assumes a human social, but as our discussion illuminates, the social of socially engaged art can be as much nonhuman as human. In our cases, the encounters that were formed brought together humans and nonhumans in manifold ways but, more than this, as *SLOW Cleanup* demonstrates, the "publics" *of* and *for* these works, brought into being by them, were as much nonhuman as they were human. This is not the place to belabor the point, but it is worth noting that geographical perspectives on nonhuman socials have much to add to, and much more to learn from, those concerned with art and aesthetics.

Reprising the claims the article opened with, and following Braun and Whatmore (2010), we contend that without an ontology that accounts for human–nonhuman collectives, we are unable to explore new capacities or to reflect seriously on how we could be different than we currently are. Our discussions have reflected on how artworks do not merely allow the nonhuman into political constituencies but also enable us to examine what is distinctive about doing so; in short, to acknowledge that "it is not only subject-bodies but other materialities that may be at work in the emergence of politicalities" (Woodward, Jones, and Marston 2010, 275). With the redistribution of the political to mundane spaces including the harbor, the knitting circle, the home, and the village hall, their disorderly operations required a rethinking of the logics of publics and participation at stake when art is "contributing to the generation of something new within scientific practice itself, challenging the boundaries of disciplinary authority" (Born and Barry 2010, 114). What is perhaps encouraging about efforts such as *Bird Yarns* and *SLOW Cleanup* is how these expanded socials allow for the possibility that unconventional collectives might give rise to the imagination and articulation of new political practices and futures.

Acknowledgments

The authors would like to thank Cape Farewell, Deirdre Nelson, the Wooly Wednesdays knitting group, Francis Whitehead, and her collaborating scientists and students for the help and assistance during the research process. They would also like to thank Danny McNally, Miriam Burke, Sasha Engelmann, and audiences at the School of Geography and Development at the University of Arizona for commenting on written and spoken versions of this text.

Funding

Research for this article was funded by an Arts and Humanities Research Council (AHRC)/National Science Foundation grant (AHRC Grant No. AH/I500022/1; NSF Grant No. 86908) and AHRC Grant No. AH/L005034/1.

Notes

1. The empirical material on which this article is based is drawn from ethnographies of art–science collaborations conducted by the authors between January 2012 and October 2013. They involved participant observation and action and interviews with those involved. See the project Web site for more details: http://artscience.arizona.edu/ (last accessed 15 December 2014).
2. Cape Farewell website: http://www.capefarewell.com/ (last accessed 21 May 2014).
3. *Bird Yarns* knitting kit instructions, authors' own.

References

Ansell Pearson, K. 1999. *Germinal life: The difference and repetition of Deleuze*. London and New York: Routledge.

Bishop, C. 2004. Antagonism and relational aesthetics. *October* 110:51–79.

Born, G., and A. Barry. 2010. Art science: From public understanding to public experiment. *Journal of Cultural Economy* 3 (1): 103–19.

Braun, B. 2005. Environmental issues: Writing a more-than-human urban geography. *Progress in Human Geography* 29 (5): 635–50.

———. 2006. Towards a new earth and a new humanity: Nature, ontology, politics, global natures in the space of assemblage. In *David Harvey: A critical reader*, ed. N. Castree and D. Gregory, 191–222. London: Blackwell.

Braun, B., and S. Whatmore, eds. 2010. *Political matter: Technoscience, democracy and public life*. Minneapolis: University of Minnesota Press.

Callon, M. 1998. An essay on framing and overflowing: Economic externalities revisited by sociology. In *The laws of markets*, ed. M. Callon, 244–69. Oxford, UK: Blackwell.

Daniels, S. 1993. *Fields of vision: Landscape and national identity in England and the United States*. Princeton, NJ: Princeton University Press.

Dixon, D. 2009. Creating the semi-living: On politics, aesthetics and the more-than-human. *Transactions of the Institute of British Geographers* 34 (4): 411–25.

Dixon, D., H. Hawkins, and L. Straughan. 2013. Of human birds and living rocks: Remaking aesthetics for post-human worlds. *Dialogues in Human Geography* 2 (3): 249–70.

Gabrys, J., and Yusoff, K. 2011. Climate change and the imagination. *Wiley Interdisicplinary Reviews: Climate Change* 2 (4): 516–634.

———. 2012. Arts, science and climate change: Practices and politics at the threshold. *Science as Culture* 21 (1): 1–24.

Gandy, M. 1997. Contradictory modernities: Conceptions of nature in the art of Joseph Beuys and Gerhard Richter. *Annals of the Association of American Geographers* 8 (4): 636–65.

Gauntlett, D. 2013. *Making is connecting*. London: Polity.

Gibson-Graham, J. K., and G. Rolevink. 2009. An economic ethics for the Anthropocene. *Antipode* 41 (1): 320–46.

Gross, M. 2010. The public proceduralization of contingency: Bruno Latour and the formation of collective experiments. *Social Epistemology* 24 (1): 63–74.

Haraway, D. 2008. *When species meet*. Minneapolis: University of Minnesota Press.

Hawkins, H. 2013. Geography and art: An expanding field: Site, the body and practice. *Progress in Human Geography* 37 (1): 52–71.

———. 2014. *For creative geographies: Geography, visual art and the making of worlds*. London and New York: Routledge.

Higgins, J. 2008. Evaluating the Chicago brownfield's initiative: The effects of city-initiated brownfield redevelopment on surrounding communities. *Northwestern Journal of Law & Social Policy* 3:240–62.

Hinchliffe, S., M. B. Kearnes, M. Degen, and S. Whatmore. 2005. Urban wild things: A cosmopolitical experiment. *Environment and Planning D: Society and Space* 23 (5): 643–58.

Ingram, M. 2012. Washing urban water: Diplomacy in environmental art in the Bronx, New York City. *Gender, Place & Culture* 21 (1): 105–22.

Johnson, E., and H. Morehouse. 2014. After the Anthropocene: Politics and geographic inquiry for a new epoch. *Progress in Human Geography* 38 (3): 439–56.

Kester, G. 2011. *The one and the many: Contemporary collaborative art in a global context*. Durham, NC: Duke University Press.

Latour, B. 2004. How to talk about the body? The normative dimension of science studies. *Body and Society* 10 (2): 205–29.

———. 2005. From realpolitik to dingpolitik or how to make things public. In *Making things public*, ed. B. Latour and P. Weibel, 14–43. Cambridge MA: MIT Press.

Lorimer, J. 2006. Nonhuman charisma: Which species trigger our emotions and why? *ECOS* 27:20–27.

———. 2010. Moving image methodologies for more-than-human geographies. *Cultural Geographies* 17 (2): 237–58.

———. 2012. Multinatural geographies for the Anthropocene. *Progress in Human Geography* 36 (5): 593–612.

Lorimer, J., and C. Driessen. 2013. Bovine biopolitics and the promise of monsters in the rewilding of Heck cattle. *Geoforum* 48:249–59.

Marres, N., and J. Lezaun. 2011. Materials and devices of the public: An introduction. *Economy and Society* 40 (4): 489–509.

Marston, S. A., K. Woodward, and J. P. Jones, III. 2005. Human geography without scale. *Transactions of the Institute of British Geographers* 30:416–32.

Matless, D., and G. Revill. 1995. A solo ecology: The erratic art of Andy Goldsworthy. *Ecumene* 2 (4): 423–48.

Paulson, W. 2001. For a cosmopolitical philology: Lessons from science studies. *Substance* 96 30 (3): 101–19.

Stengers, I. 2005. The cosmopolitical proposal. In *Making things public*, ed. B. Latour and P. Weibel, 994–1003. Cambridge, MA: MIT Press.

Whatmore, S. 2013. Earthly powers and affective environments: An ontological politics of flood risk. *Theory, Culture & Society* 30 (7–8): 33–50.

Whatmore, S., and C. Landström. 2011. Flood apprentices: An exercise in making things public. *Economy and Society* 40 (4): 582–610.

Woodward, K., J. P. Jones, III, and S. A. Marston. 2010. Of eagles and flies: Orientations toward the site. *Area* 42 (3): 271–80.

These Overheating Worlds

Kendra Strauss

The Labour Studies Program & The Morgan Centre for Labour Research, Department of Sociology & Anthropology, Simon Fraser University

In 2003 the British literary magazine *Granta* published an issue on climate change, "This Overheating World," containing reportage and essays but almost no fiction—and the claim that our "failure of the imagination" regarding socioenvironmental change is both a political and a literary one. The decade since has seen a relative burgeoning of what has been dubbed "cli-fi," dominated by apocalyptic and dystopian literary–geographical imaginations. In this article I ask this question: If these are our ways of imagining the future, what are the relationships among cultural imaginaries, theories, and politics of socioenvironmental change? Engaging the work of Frederic Jameson on utopia, and the novels of Margaret Atwood and Barbara Kingsolver, I argue that the flourishing interest in narrative, stories, and storytelling in human geography opens up opportunities for exploring political imaginaries of climate change through utopian and dystopian impulses present in its "fictionable worlds."

2003年，英国的文学杂志《格兰塔》（Granta）发表了一本气候变迁特刊——"这个过于炙热的世界"，其中包含了报导文学与论文，但几乎没有虚构小说——并主张我们对社会环境变迁"丧失想像"，同时是政治也是文学的。此后的十年，见证了被称为"气候科幻电影（cli-fi）"的相对兴盛，并充斥着天启式与去乌托邦的文学——地理的想像。我于本文中提出以下问题：如果这是我们想像未来的方式，那麽文化想像、理论与社会环境变迁政治之间的关联性为何？我透过涉入詹明信（Frederic Jameson）有关乌托邦的着作，以及玛格丽特．阿特伍德（Margaret Atwood）和芭芭拉．金索佛（Barbara Kingsolver）的小说，主张人文地理学中，对论述、故事和说故事的繁盛兴趣，开启了藉由存在于"可虚构的世界"中的乌托邦与去乌托邦之动力，探讨气候变迁的政治想像之契机。

En 2003 el magazine literario británico Granta publicó un número sobre cambio climático, "Este mundo sobrecalentado," que contenía reportaje y ensayos pero casi nada de ficción—y el reparo de que nuestra "falta de imaginación" en lo que concierne al cambio ambiental es a la vez política y literaria. La década siguiente ha visto un relativo florecimiento de lo que ha sido tildado de "*cli.fi*," una tendencia dominada por imaginaciones literario–geográficas apocalípticas y distópicas. En este artículo me pregunto lo siguiente: Si estas son las maneras como nos imaginamos el futuro, ¿cuáles son las relaciones entre imaginarios culturales, teorías y políticas de cambio socioambiental? Abocando el trabajo de Frederic Jameson sobre utopía y las novelas de Margaret Atwood y Barbara Kingsolver, sostengo que el floreciente interés en narraciones, historias y cuentos en geografía humana descorre oportunidades para explorar imaginarios políticos sobre cambio climático a través de impulsos utópicos y distópicos presentes en sus "mundos ficcionales."

> Is this the promis'd end? . . .
> Is this love
> nothing now
> or all?
> Water? Fire? Good?
> Evil? Life? Death?
>
> —W. G. Sebald (2002)

Anglo-American popular culture is besieged by monsters (Giroux 2011; McNally 2012).[1] The genre walls have come down, and the zombies, vampires, and cyborgs of fantasy and science fiction have invaded not only mass popular culture but "high" culture, too. What is striking about their permeation of the mainstream in the last decade is, on one hand, the prevalence of ecological themes and, on the other, their sheer ubiquity. Dawson (2013, 46), linking apocalypse and environmental collapse directly (like McNally's monster metaphors from Marx) to crises of capitalism, writes, "The obsession with apocalypse is shared across all levels of society, from the typical suburban multiplex cinema to war-gaming bunkers deep within the Pentagon to the obscure annals of environmental science." We are, it seems, transfixed by visions that enact, literally, Jameson's maxim that it is easier to imagine the end of the world than the end of capitalism. Yet at the same time some of the most popular genre-blurring films and books, from *The Hunger*

Games (Collins 2008) to the "speculative fiction" of Canadian novelist Margaret Atwood, turn away from straightforward apocalypse narratives to explore feminist themes through the construction of worlds that blend utopian and dystopian imaginations of sociospatial transformation.

If these are our ways of imagining of the future in the era of proclaimed climate crisis (Yusoff and Gabrys 2011), what are the relationships among cultural imaginaries, theories, and politics of socioenvironmental change? In this article, focusing on utopian and dystopian elements in the recent work by two writers (Atwood and American writer Barbara Kingsolver), I take this question as my starting point to argue that the flourishing interest in stories and storytelling in human geography opens up opportunities for examining narratives of socioecological change as a resource for exploring political imaginaries. I am interested in how theorizations of utopian and dystopian cultural themes might problematize and complicate assumptions about political–geographical imaginations and their generative potentials (Harvey 2000). Moreover, these engagements have the potential to be productive across disciplinary boundaries. I argue that a feminist geographical approach highlights the ways in which some arguments about the importance of the utopian impulse (Jameson 2005) are grounded in particular assumptions about the nature of radical difference. Furthermore, such an approach amplifies calls for ecocritical analyses grounded in a "political theory of nature" that is alive to the material inequalities and oppressions that are coconstitutive of nature–society relations (Smith 1996, 49; see also Castree and Braun 1998; Trexler and Johns-Putra 2011).

Geographies of Fictionable Worlds

In 2003 the environmentalist Bill McKibben diagnosed a "failure of imagination" at the heart of the social and political inertia about climate change, a failure that is also a literary one. "Global warming has still to produce an Orwell or a Huxley, a Verne or a Wells, a *Nineteen Eighty-Four* or a *War of the Worlds*," McKibben (2003, 11) wrote, "It may never do so."[2] These sentiments were echoed the following year by Macfarlane (2005), writing in *The Guardian:* "Where is the literature of climate change? Where is the creative response to what Sir David King, the government's chief scientific adviser, has famously described as "the most severe problem faced by the world?"

It is notable that science fiction is the genre McKibben implicitly identified with literary imaginations of climate catastrophe—and the same year also saw the publication of *Oryx and Crake* by Atwood, a Canadian novelist. For Jameson, the novel, in which "two utopias and a dystopia were ingeniously intertwined," heralded Atwood's debut as a science fiction writer (Jameson 2009, 7). Yet all of it, Atwood (2011) has claimed, is extrapolated from current conditions: It is *speculative fiction* about things that really could happen. The term speculative fiction has been taking up by those hailing the emergence of a new genre of climate fiction, dubbed "cli-fi" (Evancie 2013; Gal 2013; Glass 2013).

Atwood's claim has resonance beyond a debate about genre. The burgeoning in the last decade, in print and on film, of fantastic, apocalyptic, and dystopian visions of our collective future has happened alongside literary fiction's more widespread embrace of the issue of human-induced climate change. There has also been, during the same period, a flourishing and diverse interest in literary narratives and their reception, stories, and storytelling in human geography (see, among others, Saunders 2010; Kneale 2011; Yap 2011; Cameron 2012; Dunnett 2012; Caquard 2013; Noxolo and Preziuso 2013) even though direct engagements with eco-criticism and climate fiction are rare.

For human geographers the tradition of engagement with literary texts is grounded in the relationship between landscape and literary representation (Pocock 1988; Barnes and Duncan 1992). Around the time of the publication of "This Overheating World," however, literary geographies were being extended in new directions (see, e.g., Sharp 2000; Kitchin and Kneale 2001); they had also developed into one avenue for exploring nature–society relations. At the same time, the concept of the geographical literary imagination had evolved beyond its roots in place, landscape, and culture (Cosgrove 1979; Harvey 1990) to become a capacious framework for exploring the construction of varied sociospatial and political discourses.

Shifts in approaches to literary geographies involved new theoretical orientations and epistemologies. On one hand, geographers were exploring textual production and reception in new ways (Brace and Johns-Putra 2010; Saunders 2010; Yap 2011) and, on the other, the nature of texts and what counts as a text were being interrogated. These approaches included texts as artefacts (Sharp 2000) and moving from landscapes as texts to new readings of materialities, objects, practices, and spaces (Lorimer 2003; Whatmore 2006; Weston 2011). New (Anderson and Wylie 2009; Tolia-Kelly 2013) and vital (Braun 2008) materialisms and attention to liveliness stress immanence contra pregiven form (Bingham 2006; Braun 2006), with implications for approaches to those texts—like the literary novel—that rehearse and thus replicate form and genre (even when they seek to disrupt them). Thus the emphasis in cultural geography has shifted in the last decade to favor "small stories" (Lorimer 2003): personal accounts that engage an array of texts, narratives, and artifacts (e.g., Hill 2013).

The interest in literary geographies in general, and the novel in particular, did not disappear, of course. Johnson (2004) highlights how geographers such as Brosseau (1994) have examined novelistic and literary geographical imaginations of exploration and the city in ways that are attentive to how novels write and generate particular geographies. More recently, Saunders (2010, 2013) has distinguished between the geography of literature versus literary geography and argued for bringing these approaches together—foregrounding the importance of knowledge production. "[N]ovels are hypothetical, they can speak the unspeakable, they can say what they know is wrong, they expose the speculative nature of all knowledge and, in this, what they do know is unassailable" (Saunders 2010, 440).

The nature of knowledge production and its relationship to the political also concerns Noxolo and Preziuso (2013) in their discussion of Michael Wood's notion of fictionable worlds. The exploration of postcolonial geographies and geographies of postcolonialism are intimately related with what the authors characterize as factuality and materiality. The fictionable world is one "open to multiple interpretations from multiple located perspectives" (165). This opening up to knowledge production that is multiple in its vocality and perspective speaks also to opening up to diverse and multiple geographical imaginations, including of nature–society relations. The process motivates us not only to contemplate different social and spatial orders but also to reflect on writing as an instrument of power that helps define—and disrupt—social and spatial hierarchies. Furthermore, drawing on the work of Andrew Thacker, Saunders (2010) argues that the relationship between text and space is a recursive one: Material spaces shape the creation of literary space, but the acts of reading and writing also imagine space in ways that mediate experiences of material space and place.

Yet Cameron (2012), assessing the growth of geographical approaches to stories and storytelling, worries that recent work is less invested in connecting acts of knowing, writing, and telling with domination and control. In other words, in the fine-grained focus on recursivity described by Saunders, we lose sight of how stories relate to the exercise of power and struggles over social change. The estrangement produced by science fiction, however, can bring power and oppression sharply back into view. As Kitchin and Kneale (2001, 20) suggest, science fiction is a vital resource for exploring "the postmodern condition now prevalent in western societies, and their future visions of the new spatialities this condition will invoke."

Whereas there has been relatively little written on cli-fi by geographers, a larger body of work exists on science fiction and utopias (Harvey 2000; Kitchin and Kneale 2001; Kneale 2010, 2011; Koch 2012) that explores these future visions. Following Harvey (2000), geographies of utopia have tended to revolve around architecture, urbanism, and the built environment (see, e.g., Pinder 2005; Kraftl 2007, 2012; Firth 2012). The relationship between science fiction and utopianism is also, however, the relationship between literary narrative and political imaginations of collective social transformation, which are also deeply spatial.

Maps of Desire

Utopia's specific origins are both political and geographical, melding political tract with (fictional) travelogue. Thomas More's ([1910] 1992) text is in two parts, the first centered on what Mezciems (1992, ix–x), in her contemporary introduction, calls "sustained criticism of the state of contemporary England," grounded in an exploration of "the best state of the commonwealth," or public good. The second part is the narrative of a traveler to the island of Utopia. Utopia is More's neologism—a Latin play on the Greek *ou topos* (no place) and *eu topos* (good place; Milner 2009b); it is a society characterized by the absence of money and private property, religious intolerance, and some dimensions of patriarchal social organization. Although indebted to Plato, More's text is very much a product of the historical moment of its writing, with England on the cusp of staggering social, economic, and political change and deeply implicated in emerging forces of capitalistic, imperial, and colonial development and conquest. Jameson (2005, 15) thus characterizes the production of Utopia as the creation of a "pocket of stasis within the ferment and rushing forces of social change" that is fundamentally spatial; "Utopian space is an imaginary enclave within real social space ... the very possibility of Utopian space itself is itself the result of spatial and social differentiation."

Jameson retains, throughout his extended analysis of Utopia, the fundamental link between Utopia, what More calls commonwealth (the collective well-being of the social body) and relations of production and private property. There is thus something of a bifurcation between different approaches to the study of Utopia—and its antithesis. More generally, utopia is an ambivalent idea. It is a double referent, signalling a good but nonexistent (imagined or dreamed) place and society, both of which are therefore impossible (Levitas 2011). In contemporary Anglo-American cultures and political discourses it is often denigrated as the imagination of unattainable perfection (usually in social or economic relations but also often in the spatial organization of society and the built environment) and often so broadly defined that it is evacuated of its historical–political force. Thus, the visions of free-marketeers, neoliberals, and religious conservatives are characterized as utopian, as are institutions like marriage and the family (Jameson 2005; cf. Harvey 2000).

Writers in the field of utopian studies, however, locate the meanings and developments of utopia within an even longer genealogy of the desire for a better life. Both Levitas and Sargent have pointed to common themes present in the dreams and myths of a range of societies, at different times, including plenitude, leisure, and the deferral or banishment of illness, pain, and death. The field of utopian studies also tends to differentiate between utopia, the general form, and its attributes (including exploration, travelers, guides, island societies); eutopia, the positive variant of the form; and dystopia, a coinage of John Stewart Mill in 1868, its negative variant (Milner 2009b).

For Sargent (1994), then, utopia represents "a story of the men and women who dreamed of a better life *for all of us* and those who tried to create a better life. It is also the story of those who had differing dreams and the conflicts among them" (1, italics added). Sargent thus identifies utopianism as *social dreaming*; that is, not the individual dreaming of the collective but dreaming the collective into being. Such dreams are intimately related with geographical imaginations: Dreams of good societies are dreams of good places. In this sense utopianism is not just social dreaming but spatial dreaming.

This is not to say that space is the only constitutive dimension of the utopian dream or program. "Topos means place ... rather than time," Milner (2009b, 827) has pointed out and thus "the most important recent transformation of the genre, clearly occasioned by the European's successful mapping of the world, is the relocation of plausible good and bad places from geography to history."[3] As the utopian tradition developed, especially through the genre of science fiction (Suvin 1979), it became associated with futurity rather than historicity, as the boundaries of mappable worlds expanded into the cosmos. At the same time, the interplay of technological change and ecological despoilment have become increasingly important drivers of this expansion, whereas the latter (in ecotopias and dystopias) has refocused attention on place and impermanence.

If Sargent is attentive to the political function of utopianism and the content of utopian programs, it is Jameson who has produced the most nuanced and extended theorization of the political dimensions of utopia. Jameson distinguishes between the utopian program, which is systematic and associated with utopian texts, and the utopian impulse, which is more generalized (obscure yet omnipresent) but observable in a wide range of human activities, including in social and political theory (Jameson 2005; Milner 2009a). Following Suvin (1979), Jameson retains and reinforces the link between the utopian program and the genre of science fiction. More significant, however, has been Jameson's development of his analysis of utopia pace science fiction and his periodization of the evolution of cultural form and expression (Wenger 2007) into trenchant argument for the necessity of anti-anti-utopianism (Jameson 2005).

"The Utopians not only offer to conceive of ... alternative systems; Utopian form is itself a representational meditation on radical difference, radical otherness, and on the systemic nature of the social totality," Jameson (2005, xii) writes; the form achieves this through cognitive estrangement, which is a feature of the utopian imagination named by Marcuse. Radical difference is associated with a particular spatial concept of totality, which presides over utopian forms and differentiates the utopian impulse from the utopian program: The former relates to enclaves within our existing world, the latter replaces that world. Utopias, according to Jameson (2004, 43), emerge "at the moment of suspension of the political" when institutions and forms of social organization seem unchangeable. The resulting social forces amassed against such utopian imaginations are anti-utopian; in this sense, the market utopianism analyzed by Harvey (2000) in *Spaces of Hope* (see, in particular, Chapter 8) is not truly utopian as Jameson understands it.

Jameson's concept of utopianism has opened him up to the critique, however, that his approach elides and forecloses the analysis of critical utopian and dystopian forms, which others suggest are among the most relevant in our contemporary moment (Baccolini and Moylan 2003; Milner 2009b). Dystopia, according to Baccolini and Moylan (2003), moved from the margins of the literary and sci-fi genres to a more central or dominant imaginary in the postwar period, most significantly since the 1980s. Dystopia has been neglected relative to utopia, yet important works of science fiction are simultaneously eutopias and dystopias. Baccolini (2004) has identified this boundary disruption with the work of feminist writers, arguing that feminist criticism impels us to reconsider dystopianism in the same way that it forced us to consider anew the conventions of eutopia. This aligns with the separate but related critique that Jameson's notions of social totality and radical difference, despite his attention (at times) to questions of reproduction and gender, are grounded in the social organization of property and production in ways that privilege the latter rather than deepening a dialectical analysis of the relations between dimensions of social difference.

Archaeologies of the Present

Margaret Atwood and Barbara Kingsolver are prize-winning literary authors who write about human natures (in all senses), power, and inequality in fiction and nonfiction genres.[4] Although Atwood is perhaps better known as a feminist writer—*The Handmaid's Tale* (Atwood 1998) is now a canonical feminist text—both have written about gender and difference, with Kingsolver's (1998) Pulitzer-nominated *The Poisonwood Bible* addressing intersecting themes of patriarchy, religion, colonialism, and imperialism. They also have in common a strong interest in, and knowledge of, environmental and biological sciences, which inflect and inform their writing; in the novels I describe here, they direct these to the exploration of socioenvironmental change in the era of climate politics. Yet in neither case are ecological crises inseparable from political–economic and social crises nor gender oppression inseparable from other oppressions.

In Atwood's (2003) *Oryx and Crake*, global warming is but one of a set of interrelated processes of socioecological and socioeconomic change that represent a deepening and intensification of crisis tendencies across multiple domains. The novel proved to be the first installment of an extended exercise in imagining the North America of the near future.[5] *Oryx and Crake* opens in an unspecified future with the voice of Snowman, the last (as far as we know) fully human being left alive after the ravages of a global plague. Near Snowman, however, lives a species of "bioengineered humanoids" created by Snowman's childhood friend Crake (also, we come to learn, the originator of the pandemic or "waterless flood"). The Crakers have been designed without the human characteristics deemed most destructive by Crake. We come to understand that Snowman was once Jimmy, who grew up in the corporate-controlled Compounds where he met Crake; both become connected, through an online game, with a group of bioterrorists associated with the MaddAddam cult, which in turn is related to the green religious group God's Gardeners. The latter is the focus of *The Year of the Flood*. The narrative shifts among three female characters who come into the God's Gardeners fold, an austere intentional community and haven in the violent pleeblands outside the Compound walls. These characters, in particular the older woman Toby, continue to be central to the final book, in which all of the surviving characters—including the Crakers—encounter each other in the postplague landscape and must come together to survive the dual threats of bioengineered nature, and brutalized (and brutalizing) human enemies.

Flight Behavior (Kingsolver 2012) is also set in the near future, although a future that Kingsolver suggests is truly immanent. It is the story of Dellarobia, who marries into a small town farming family in the American South when she becomes pregnant at seventeen. On her way to a secret meeting with a man who is not her husband, she discovers a large colony of monarch butterflies clustered in the trees in the hills near her farm. It is interpreted at once as a religious miracle and a portent of climate chaos, and a team of scientists discover that the colony's deviation from its normal migratory and breeding patterns has been caused by catastrophic mudslides at the site of its Mexican breeding ground. The arrival of the scientists, and Dellarobia's interactions with them, are set against the interpretation and contestation of the miracle by the media, church, and local community. In the end, both the butterflies and Dellarobia take flight. Dellarobia, with her young children, leaves her husband and his family to resume the education that her stillborn first child cut short.

From even these brief summaries, one can discern the dystopian vision of Atwood's trilogy and utopian impulses (as opposed to program) of Kingsolver's narrative. Distinguishable from critical utopias (Moylan 1986), dystopia involves extrapolation from the present in a way that includes a warning (Sargent 2006). Atwood's fictional world is clearly just that: This, she warns us, is what could happen if science continues to be sublimated to the logic of accumulation, if we continue to worship (literally, in Atwood's world) fossil-fuel-driven growth, if the state becomes entirely elided by corporate interests, and if gender and sexuality are naturalized and commercialized in the free play of commercial desire.

In *Flight Behavior*, on the other hand, ostensibly the farther of the two texts from the conventions of science fiction, the utopian impulse is located in the two-fold emancipatory potential of science. Science illuminates the fundamental interconnectedness of human animals, their modes of economic and social organization, and the material and ecological bases of those modes. As such, it can distinguish between obfuscating and ideological anti-utopian politics and the utopian politics of critique and renewal. Science also has the potential to liberate the individual from the collective, to free her to pursue new projects of knowledge production, and thus create the basis for new forms of collectivity.

To a degree, then, these texts seem to reinforce Jameson's analysis of utopianism, and "prejudice" against dystopianism (Milner 2009b), although pace Atwood Jameson (2009, 8) circumvents the problem by positing that "the post-catastrophe situation [in *The Year of the Flood*] in reality constitutes the preparation for the emergence of Utopia itself." In fact, however, the writings of Atwood and Kingsolver suggest that feminist explorations of difference problematize certain assumptions that underwrite Jameson's critical project. Although he identifies utopia with radical difference, radical otherness, and the systemic nature of the social totality, Jameson undertheorizes (or fails to adequately elaborate) the interaction of social categories of difference, in particular gender and race, in his dialectical method and focus on systems of production. This results in blindness to the destabilizing effects on the utopian project of the problems of identity or recognition versus redistribution—or social reproduction versus production—which are alluded to but allocated a subsidiary role in Jameson's analysis.

Thus, although Jameson's overall framework remains hugely valuable, Sargent's flexible concept of social dreaming offers a way of expanding the totalizing thinking process developed in and through Jameson's work (and his method of negation) to address the question of differentiation within radical difference. Social dreaming can encompass both positive (eutopian) and negative (dystopian) variants but not anti-Utopianism (the injunction against social dreaming). Taken together with my suggested corollary of spatial dreaming, we open up the field of analysis to relationship between the present warning and Jameson's "desire called Utopia" and how this relationship is cognitively and spatially mapped.

Kingsolver and Atwood shuttle us between dystopian visions and utopian impulses and, in doing so, map the relationship between them in spatial terms. Places are redreamed, transformed: in Kingsolver through the appearance of the monarchs and in Atwood through the plague that scours the landscape of what is simultaneously a bad place (the near future United States) and no place. Spatial dreaming clears a path between Jameson's elucidation of the relationship among space, the utopian imagination, and the utopian moment and the very real critiques of his neglect, or dismissal, of dystopias. It speaks to the ways in which the contemporary interplay of utopian impulses and dystopian visions associated with socioecological transformations starts to collapse spatial–temporal distinctions. This collapsing produces estrangement through foreshortened horizons that emphasize the closeness to, rather than the distance from, the fictional worlds they portray.

The feminist dimensions of Kingsolver and Atwood's work, then, do not necessarily negate the process of totalization, but they do disrupt it. Estrangement and the encounter with radical difference does not happen in the smooth space of theory but from a topographical plane marked by preexisting privileges, inequalities, and oppressions—especially, but not exclusively, of gender and class. This means that the socioecological transformations they seek to explore, and the new worlds they dream into being, do not privilege one set of processes or one dimension of the problematic of the collectivity. They also challenge the conceptual antimonies at play in Jameson, confirming his emphasis on negation over synthesis. We can imagine the negation of sex, but what of gender? Of race? Multidimensional analyses demand conjunctural understandings of crisis and transformation, which does not negate the central problematics of Utopia (production, collectivity, the education of desire) but rather seeks to extend them.

Atwood's and Kingsolver's novels also suggest important limits that need to be confronted in their literary–geographical imaginations. In Atwood, the collectivity that is dreamed into being in a landscape washed clean by the waterless flood is held together by necessity but redeemed, for the protagonist Toby, by love. This rather contradicts the impulse, located in *Year of the Flood* by Colebrook (2014), to question the value of the maximization of life and "living on." In *Flight Behavior* (Kingsolver 2012) it is the promise of self-determination that is emphasized in the final instance over the possibility of acceptance within another kind of collectivity, a community of science. These are quite different from the social dreams of earlier feminist utopias, which sought to radically challenge relations of social reproduction.

Perhaps more fundamental, though, is the degree (as opposed to the type) of multivocality these texts produce. Both are U.S.-centric, and both largely ignore issues of race—which is not to deny, of course, that they are peopled by some characters of color (Kingsolver's novel is even, at one register, a meditation on the interaction of poverty, race, and privilege, but this is achieved by the rather blunt instrument of category inversion). Thus, although contemporary utopias and dystopias and their geographical imaginations are vital to the imperative to imagine radical alternatives to existing social orders, geographers also have a role to play in mapping their exclusions, or what they fail to dream.

Conclusion

More's *Utopia* predated the modern novel, now the preeminent form through which utopian and dystopian political–geographical imaginations are expressed. Harvey (2000) is surely right that the novel, as the vehicle for radical political imaginations, is limited by its increasingly rarefied form. But if, as Noxolo and Preziuso (2013) suggest, fictionable worlds are ones open to multivocality, the point of geographical attention to utopia and its variants is that they speak to socioecological transformations in a different form and register than most academic knowledge production. They travel in particular ways, offering a different set of opportunities for considering circulation, reader engagement, and cultural and political resonances and closures. If "utopia as a form is not the representation of radical alternatives; it is rather simply the imperative to imagine them" (Jameson 2005, 416), fictionable worlds render legible those imaginations in all of their variety. They are a resource for imagining alternative political futures.

In this sense the (re)turn to stories and storytelling in geography can provide an opportunity for engagement with the forms and varieties of spatial dreaming that shape and reflect our cultural and political hopes and anxieties. Just as important, however, are the ways in which we are all ourselves spatial dreamers. Utopian and dystopian imaginations are resources for the continuous process of renegotiating our own terms of engagement with the political, and for reimagining our own daily praxis—including as academics. Such resources are as fundamental to the politics of the everyday—like Loftus's (2012) everyday environmentalism—as to the reimagination of the commonwealth (and wealth in common) and a broader politics of hope signaled by Harvey (2000). As Jameson (2005, xii) writes: "What is crippling is . . . the universal belief . . . that no other socio-economic system is conceivable . . . one cannot imagine any fundamental change to our social existence which has not first thrown off Utopian visions like sparks off a comet."

Acknowledgments

I would like to thank Maria Kaika for her guidance and encouragement when I first started working on this theme as a graduate student almost a decade ago. Friends and colleagues, especially Kate Derickson, Katie Meehan, and Alex Loftus, provided an invaluable sounding board for the ideas as they evolved. I would also like to thank the three anonymous reviewers of the article and Bruce Braun for their perceptive, constructive comments, and suggestions—with the usual caveat that all errors and omissions are mine alone.

Notes

1. In the interests of length, this article is confined to Anglo-American print and visual culture and focuses on the literary novel. I am aware that this is a partial and selective approach, and unfortunately follows a Western bias in much ecocriticism and science fiction

studies. See inter alia Hollinger and Gordon (2002), O'Connell (2012), and Estok (2013).

2. A review I carried out at the time in fact identified dozens of examples of Anglo-American fiction engaged with themes related to climate change, spanning a range of genres (Strauss 2004).
3. As Milner also noted, this has also produced a conventional academic distinction between uchronia (no time), euchronia (good time), and dyschronia (bad time).
4. Atwood won the Booker Prize in 2000; Kingsolver won the Orange Prize in 2010.
5. *Oryx and Crake* proved to be the first of three interrelated novels, which are together known as the MaddAddam trilogy.

References

Anderson, B., and J. Wylie. 2009. On geography and materiality. *Environment and Planning* A 41:318–35.

Atwood, M. 1998. *The handmaid's tale*. New York: Anchor Books.

———. 2003. *Oryx and Crake*. London: Virago.

———. 2011. *In other worlds: SF and the human imagination*. London: Virago.

Baccolini, R. 2004. The persistence of hope in dystopian science fiction. *PMLA* 119:518–21.

Baccolini, R., and T. Moylan. 2003. Introduction: Dystopia and histories. In *Dark horizons*, ed. R. Baccolini and T. Moylan, 1–12. London and New York: Routledge.

Barnes, T., and J. S. Duncan. 1992. *Writing worlds: Discourse, text, and metaphor in the representation of landscape*. London and New York: Routledge.

Bingham, N. 2006. Bees, butterflies, and bacteria: Biotechnology and the politics of nonhuman friendship. *Environment and Planning* A 38 (3): 483–98.

Brace, C., and A. Johns-Putra. 2010. Recovering inspiration in the spaces of creative writing. *Transactions of the Institute of British Geographers* 35:399–413.

Braun, B. 2006. Environmental issues: Global natures in the space of assemblage. *Progress in Human Geography* 30 (5): 644–54.

———. 2008. Environmental issues: Inventive life. *Progress in Human Geography* 32 (5): 667–79.

Brosseau, M. 1994. Geography's literature. *Progress in Human Geography* 18 (3): 333–53.

Cameron, E. 2012. New geographies of story and storytelling. *Progress in Human Geography* 36 (5): 573–92.

Caquard, S. 2013. Cartography I: Mapping narrative cartography. *Progress in Human Geography* 37 (1): 135–44.

Castree, N., and B. Braun. 1998. Construction of nature and nature of construction. In *Remaking reality: Nature at the millenium*, ed. B. Braun and N. Castree, 3–42. London and New York: Routledge.

Colebrook, C. 2014. *Sex after life*. Ann Arbor, MI: Open Humanities Press with Michigan Publishing.

Collins, S. 2008. *The hunger games*. New York: Scholastic.

Cosgrove, D. E. 1979. John Ruskin and the geographical imagination. *Geographical Review* 69 (1): 43–62.

Dawson, A. 2013. Edward Said's imaginative geographies and the struggle for climate justice. *College Literature* 40:33–51.

Dunnett, O. 2012. Patrick Moore, Arthur C. Clarke and "British outer space" in the mid 20th century. *Cultural Geographies* 19 (4): 505–22.

Estok, S. C. 2013. Global problems, local theory: Moving beyond particularity and eco-exceptionalism to action in ecocriticism. *Foreign Literature Studies* 35:1–12.

Evancie, A. 2013. So hot right now: Has climate change created a new literary genre? *NPR Books*. http://www.npr.org/2013/04/20/176713022/so-hot-right-now-has-climate-change-created-a-new-literary-genre (last accessed 15 May 2014).

Firth, R. 2012. Transgressing urban utopanism: Autonomy and active desire. *Geografiska Annaler Series B: Human Geography* 94B (2): 89–106.

Gal, H. 2013. From Noah's ark to superstorm Sandy—The rise and rise of CliFi. *Huffington Post*. http://www.huffingtonpost.co.uk/hannah-gal/climate-change-fiction_b_4147855.html (last accessed 15 May 2014).

Giroux, H. 2011. *Zombie politics and culture in the age of casino capitalism*. New York: Peter Lang.

Glass, R. 2013. Global warning: The rise of "cli-fi." *The Guardian*. http://www.theguardian.com/books/2013/may/31/global-warning-rise-cli-fi (last accessed 15 May 2014).

Harvey, D. 1990. Between space and time—Reflections on the geographical imagination. *Annals of the Association of American Geographers* 80 (3): 418–34.

———. 2000. *Spaces of hope*. Edinburgh, UK: University of Edinburgh Press.

Hill, L. 2013. Archaeologies and geographies of the post-industrial past: Landscape, memory and the spectral. *Cultural Geographies* 20:379–96.

Hollinger, V., and J. Gordon, eds. 2002. *Edging into the future: Science fiction and contemporary cultural transformation*. Philadelphia: University of Pennsylvania Press.

Jameson, F. 2004. The politics of utopia. *New Left Review* 25:35–52.

———. 2005. *Archaeologies of the future: The desire called utopia and other science fictions*. London: Verso.

———. 2009. Then you are them. *London Review of Books* 31 (17): 7–8.

Johnson, N. C. 2004. Fictional journeys: Paper landscapes, tourist trails and Dublin's literary texts. *Social & Cultural Geography* 5 (1): 91–107.

Kingsolver, B. 1998. *The poisonwood bible*. New York: HarperFlamingo.

———. 2012. *Flight Behavior*. London: Faber and Faber.

Kitchin, R., and J. Kneale. 2001. Science fiction or future fact? Exploring imaginative geographies of the new millennium. *Progress in Human Geography* 25 (1): 19–35.

Kneale, J. 2010. Counterfactualism, utopia, and historical geography: Kim Stanley Robinson's The years of rice and salt. *Journal of Historical Geography* 36 (3): 297–304.

———. 2011. Plots: Space, conspiracy, and contingency in William Gibson's pattern recognition and spook country. *Environment and Planning D: Society and Space* 29 (1): 169–86.

Koch, N. 2012. Urban "utopias": The Disney stigma and discourses of "false modernity." *Environment and Planning A* 44 (10): 2445–62.

Kraftl, P. 2007. Utopia, performativity, and the unhomely. *Environment and Planning D: Society and Space* 25 (1): 120–43.

———. 2012. Utopian promise or burdensome responsibility? A critical analysis of the UK government's building schools for the future policy. *Antipode* 44 (3): 847–70.

Levitas, R. 2011. *The concept of utopia.* Witney, UK: Peter Lang.

Loftus, A. 2012. *Everyday environmentalism: Creating an urban political ecology.* Minneapolis: University of Minnesota Press.

Lorimer, H. 2003. Telling small stories: Spaces of knowledge and the practice of geography. *Transactions of the Institute of British Geographers* 28 (2): 197–217.

Macfarlane, R. 2005. The burning question. *The Guardian.* http://www.theguardian.com/books/2005/sep/24/featuresreviews.guardianreview29 (last accessed 15 May 2014).

McKibben, B. 2003. Worried? Us? *Granta* 83:8–12.

McNally, D. 2012. *Monsters of the market: Zombies, vampires, and global capitalism.* London: Haymarket Books.

Mezciems, J. 1992. Introduction. In *Utopia*, ix–xxiii. New York: Everyman's Library, Knopf.

Milner, A. 2009a. Archaeologies of the future: Jameson's utopia or Orwell's dystopia? *Historical Materialism: Research in Critical Marxist Theory* 17 (4): 101–19.

———. 2009b. Changing the climate: The politics of dystopia. *Continuum: Journal of Media & Cultural Studies* 23 (6): 827–38.

More, T. [1910] 1992. *Utopia.* New York: Everyman's Library, Knopf.

Moylan, T. 1986. *Demand the impossible: Science fiction and the utopian imagination.* New York: Methuen.

Noxolo, P., and M. Preziuso. 2013. Postcolonial imaginations: Approaching a "fictionable" world through the novels of Maryse Conde and Wilson Harris. *Annals of the Association of American Geographers* 103 (1): 163–79.

O'Connell, H. C. 2012. Mutating toward the future: The convergence of utopianism, postcolonial SF, and the postcontemporary longing for form in Amitav Ghosh's The Calcutta chromosome. *Mfs: Modern Fiction Studies* 58: 773–95.

Pinder, D. 2005. *Visions of the city: Utopianism, power and politics in twentieth-century urbanism.* Edinburgh, UK: University of Edinburgh Press.

Pocock, D. C. D. 1988. Geography and literature. *Progress in Human Geography* 12 (1): 87–102.

Sargent, L. T. 1994. The three faces of utopianism revisited. *The Journal of Utopian Studies* 5 (1): 1–37.

———. 2006. In defense of utopia. *Diogenes* 53 (1): 11–17.

Saunders, A. 2010. Literary geography: Reforging the connections. *Progress in Human Geography* 34 (4): 436–52.

———. 2013. The spatial event of writing: John Galsworthy and the creation of "Fraternity." *Cultural Geographies* 20 (3): 285–98.

Sebald, W. G. 2002 *After nature.* Trans. M. Hamburger. New York: Random House.

Sharp, J. P. 2000. Towards a critical analysis of fictive geographies. *Area* 32 (3): 327–34.

Smith, N. 1996. The production of nature. In *Future natural: Nature, science and culture*, ed. G. Robertson, M. Mash, J. Tichner, J. Bird, B. Curtis, and T. Putnam, 35–54. London and New York: Routledge.

Strauss, K. 2004. *Dispatches from an overheating world: Writing the geographical imaginations of global warming.* Unpublished MSc dissertation, School of Geography, University of Oxford, Oxford, UK.

Suvin, D. 1979. *Metamorphoses of science fiction: On the poetics and history of a literary genre.* New Haven, CT: Yale University Press.

Tolia-Kelly, D. P. 2013. The geographies of cultural geography III: Material geographies, vibrant matters and risking surface geographies. *Progress in Human Geography* 37:153–60.

Trexler, A., and A. Johns-Putra. 2011. Climate change in literature and literary criticism. *Wiley Interdisciplinary Reviews: Climate Change* 2 (2): 185–200.

Wenger, P. E. 2007. Jameson's modernisms: Or, the desire called utopia. *Diacritics* 37 (4): 3–20.

Weston, D. 2011. The spatial supplement: Landscape and perspective in WG Sebald's The rings of Saturn. *Cultural Geographies* 18:171–86.

Whatmore, S. 2006. Materialist returns: Practicing cultural geography in and for a more-than-human world. *Cultural Geographies* 13:600–09.

Yap, E. X. Y. 2011. Readers-in-conversations: A politics of reading in literary geographies. *Social & Cultural Geography* 12 (7): 793–807.

Yusoff, K., and J. Gabrys. 2011. Climate change and the imagination. *Wiley Interdisciplinary Reviews: Climate Change* 2:516–34.

When Horses Won't Eat: Apocalypse and the Anthropocene

Franklin Ginn

Institute of Geography, School of Geosciences, University of Edinburgh

In this article I suggest that fantasies of apocalypse are both a product and a producer of the Anthropocene. Although images and narratives of contemporary environmental apocalypse have usually been understood as politically regressive and postpolitical distractions, I demonstrate that a more hopeful reading is possible. Apocalypse tells us that the human as currently configured in the Anthropocene—an ideal universal subject who is energized through fossil fuels and who has been elevated to a position of ecological mastery—cannot continue indefinitely. This article therefore considers what apocalyptic imaginaries reveal about the limits to being human and the future of human life after the Anthropocene. It does so by analyzing a critically acclaimed film, *The Turin Horse* (2011). In this film an old farm horse refuses to eat, drink, or leave its stall, while a daughter and her father struggle on through an unspecified disaster, gnawing on raw potatoes as their world slowly unravels. *The Turin Horse* discloses the earth forces that have made Anthropocene humans along three lines: the geological, the biological, and the temporal. The film also hints at three challenges to be overcome to make humans differently: the need to surpass carbon humanity, the need for nonhuman allies, and the need to affirm agency against the inevitability of deep time. I suggest that contemporary apocalyptic visions are a core aspect of how geographers should understand socioecological transformation, as they challenge those who view them to feel the condition of the Anthropocene, and pose the question of how to respond well to unruly earth forces.

我在本文中主张，启示录的幻想，同时是人类世的产物与生产者。儘管当代环境启示录的意象和叙事，经常被理解为政治退化与后政治的干扰，但我将证明，更具希望的阅读方式是有可能的。启示录向我们揭露，人类作为当前人类世中的构成——一个由化石燃料所驱动、并上昇至生态统治者之位的理想普遍主体——是不可能永远持续下去的。本文因此考量启示录的想像，揭露了身为人类的何种极限，以及生活在人类世之后的人类未来。本文透过分析一部广受推崇的批判性电影《都灵之马》（2011）进行之。在这部电影中，一匹农场老马拒绝进食、饮水、或离开它的马厩，于此同时，一名女儿和其父亲正与一场未指明的灾难进行斗争，在他们的世界缓慢展开中，以啃食生土豆为生。《都灵之马》揭露了造就人类世的地球驱力的三条轴线：地质、生物与时间。该电影同时暗示三个必须克服才能让人类脱胎换骨的挑战：必须压制碳的人性、必须与非人类结盟、以及确认对抗深度时间的不可避免性的行动主体之必要。我主张，当代启示录的视野，是地理学者应如何理解社会生态变迁的核心面向，因为他们挑战了那些认为他们相信人类世的限制之人，并质问应如何善加回应难以驾驭的地球趋力。

En este artículo sugiero que fantasías de apocalipsis son a la vez un producto y un productor del Antropoceno. Aunque las imágenes y narrativas del apocalipsis ambiental contemporáneo usualmente han sido entendidas como distracciones políticamente regresivas y pospolíticas, demuestro que es posible una lectura más esperanzadora. El apocalipsis nos dice que lo humano como está actualmente configurado en el Antropoceno—un sujeto universal ideal que se energiza por medio de combustibles fósiles y que ha sido elevado a una posición de predominio ecológico— no puede continuar indefinidamente. Por lo tanto, este artículo se ocupa de lo que los imaginarios apocalípticos revelan sobre los límites de lo que significa ser humano, y del futuro de la vida humana después del Antropoceno. Se hace esto analizando una película aclamada por la crítica, *El Caballo de Turin* (2011). En esta película un viejo caballo granjero se rehúsa a comer, beber, o a salir de la pesebrera, mientras una hija y su padre luchan por sobreponerse a un desastre no especificado, mordiendo papas crudas a medida que su mundo lentamente se desdobla. *El Caballo de Turin* revela las fuerzas de la tierra que han moldeado los humanos del Antropoceno a lo largo de tres líneas: la geológica, la biológica y la temporal. La película también apunta hacia tres retos que deben ser vencidos para que los humanos pueda ser cambiados: la necesidad de sobrepasar a la humanidad del carbono, la necesidad de aliados no humanos y la necesidad de afirmar agencia contra la inevitabilidad del tiempo profundo. Sugiero que las visiones apocalípticas contemporáneas son un aspecto medular de cómo deben entender los geógrafos la transformación socioecológica, en la medida en que ellos retan a quienes los ven sentir la condición del Antropoceno, y plantean el interrogante de cómo responder mejor a la fuerzas rebeldes de la tierra.

Thinking of the Earth today and not, at the same time, thinking of its devastation is increasingly difficult. Popular apocalyptic stories multiply on screen: resource extraction and imperialism (*Avatar* [2009]); the vanity and depression of the rich (*Melancholia* [2011]); alien migration and hybridization (*District 9* [2009]; *Falling Skies* [2011]); zoonotic pandemics (*Contagion* [2011]); climate disaster (*The Day After Tomorrow* [2004]; *The Colony* [2013]); the end of friendship between human and animal (*After Earth* [2013]; *The Life of Pi* [2013]), as well as in text (*The Possibility of an Island* [Houellebecq 2005]; *The Windup Girl* [Bacigalupi 2009]; the *MaddAddam trilogy* [Atwood 2013]), and in landscapes (witness the popular and critical interest in ruins). The flavor and political tone varies, too, from conservative family drama, to regressive ecology, to exuberant tales of technonatures to come. Such visions are an imaginative force oriented toward the future, driven by pervasive anxiety about the prospects for life.

Of course visions of civilization's end are nothing new, and indeed every culture seems to obsess over its own ruination (Hall 2009). We must therefore historicize. "Our" current time is the Anthropocene, that new geological epoch in which humans have become a planetary force, according to analysis of the lithographic, geochemical, biological, and atmospheric records of human activities (Zalasiewicz et al. 2011). For some, the Anthropocene signals a final enclosure of politics and culture within ecology: a new geopolitics in which Earth is the sovereign authority and humans are inmates of a planet-sized camp in a permanent state of emergency. For others, it is an occasion to double down on techno-hubris and call forth more fevered bouts of rationality and management (Oxford Martin Commission 2013). Optimistic commentators hope that naming this new epoch might accelerate action on the pressing challenges of our time—that the Anthropocene is an "unprecedented opportunity" (Ellis et al. 2013, 7978), a wake-up call for "planetary stewardship" (Steffen et al. 2011), or just good to think with (Ellsworth and Kruse 2013). Critics remind us that the unitary human of the Anthropocene hides political difference, and risks elevating a particular kind of consumer to a motor of history (Malm and Hornborg 2014). For the purposes of this article, however, the very act of asking the question, "Is this the Anthropocene?" demonstrates that we have moved into an era of anxiety about the prospects for planetary life. Indeed, the Anthropocene might be defined as an emotional condition as well as a physical event (Robbins and Moore 2013). It is worth remembering that the Anthropocene arrives not with a socioecological transition (an event), but rather with our capacity to measure and to read signs of that event through scientific or artistic means (Szerszynski 2012). Only once we can measure, read, and therefore sense how the Earth has become sensitive do we enter the Anthropocene. Thus if the Anthropocene is partly formed through "affective atmospheres" (Anderson 2012) and ways of representing that constrain and enable political imaginaries, we should consider these as important components of socioecological transition.

In this article I suggest that fantasies of apocalypse are both a product and a producer of our current epoch—the Anthropocene—and that they also take us beyond this epoch by confronting what might be to come. I contend that "Anthropocene apocalypse" reveals how we have always been more-than-human in ways at once both geological and biological, ways through which earth forces have been folded within us. Against the dominant grain in social science and drawing on recent work in ecocriticism, I offer a hopeful reading of apocalypse. If contemporary apocalypse emerges as a nightmare of the Anthropocene's socioecological risks, it also produces something over and above anxiety—something escapes, and such excesses might be mined for their transformative kind of feeling, not just criticized for their politically regressive negativity. The article therefore focuses on apocalyptic cinema, examining how film offers us a way of measuring our sensitivity to the Earth (rather than measuring the Earth's sensitivity to human activities). The article analyzes one avant-garde, critically acclaimed film: *The Turin Horse* (2011), directed by Béla Tarr. Tarr's film represents a particular kind of apocalyptic vision: uncompromising, difficult, culminating in cosmic emptiness that is implied but not presented. It is far from a Hollywood blockbuster. The film nonetheless distills into an intense form many anxieties of Anthropocene apocalypse, making it a suitable vehicle through which to explore the cultural politics of how we are sensitized to the Earth.

Postpolitical Apocalypse, Ecocriticism, and Film

Can a film make someone think or act differently? Can apocalyptic cinema change the world? We can characterize two prevailing perspectives on apocalyptic cinema that respond to these questions in very different ways. The first, an eco-Marxist perspective dominant in geography, suggests that "quite simply apocalypticism is politically disabling" (Katz 1995, 277). The second, more prevalent in social theory and ecocriticism, to summarize crudely, seeks to locate cinema's potential to "reframe perception" for progressive ends (Rust, Monani, and Cubbitt 2013, 11). This article thus seeks to connect geographical debates about socioecological futures to those taking place in the interdisciplinary field of the environmental humanities.

In a series of articles, Swyngedouw (2010, 2013) has laid bare what he sees as the politically regressive logic of apocalypse. In Swyngedouw's analysis, modern ecology is a thoroughly depoliticized enterprise in which politics has been replaced by policy and governance. His paradigmatic example is climate change, in which carbon dioxide has been fetishized to stand in as "the problem," masking the underlying cause: the unequal socioeconomic and geopolitical networks behind each unit of carbon dioxide. Negotiations turn on parts per million, thresholds are set at atmospheric concentration levels, and sustainable policies, such as carbon pricing, are the only game in town. Even radical environmentalists translate their goals into scientific measurements by fixating on carbon dioxide levels. Capitalism—in particular its neoliberal variegation—remains uncontested. As Swyngedouw puts it, implicit in crisis and apocalypse is a universal threat: We are all potential victims, ultimately in it together. Swyngedouw (2010) argues that "the imaginary of crisis and potential collapse produces an ecology of fear, danger and uncertainty while reassuring the 'people' . . . that the techno-scientific and socio-economic elites have the necessary tool-kit to readjust the machine such that things can basically stay as they are" (11). The end of everything is an ongoing crisis that we are assured can be managed within the current system; apocalyptic imaginaries become a key way to sustain this postpolitical consensus. The politically regressive nature of apocalypse is therefore its tendency to entrench further forces that precipitate catastrophe in the first place.

Since the late 1990s the burgeoning field of ecocriticism has been analyzing the potential of environmental film and literature, including apocalyptic visions, for consciousness raising that might inculcate a sense of planetary care (Levene 2013; Rust, Monani, and Cubbitt 2013). The prescriptive moral tone of some progressive ecocritical readings and the assumed link between consuming a proenvironmental vision and a more enlightened planetary citizenry are, of course, naive to say the least (Berger 1999; Buell 2003; Hulme 2008). Even "progressive" environmental films, including apocalyptic films intended to shock us out of complacency, do not easily escape the postpolitical critique outlined earlier. A film like *Avatar*, for example, while showcasing the evils of an imperialist, militarized capitalism, symptomatically falls back on troubling notions of a sovereign nature, complete with a native people, the Na'vi. Other films, such as *Melancholia*, explicitly embrace passivity in the face of disaster as the only possible alternative to Hollywood bombast (wherein techno-scientific knowhow saves the day; Latour 2013a). Yet other ecocritics suggest that films, although always suffused in the prevailing postpolitical ideology of the day (eco-doom as consumerist spectacle), do not robotically follow such ideologies, but also contain ideological contradictions and excesses (Ingram 2004). Such critics are also aware that, historically, apocalypse has generated highly varied political positions, from the regressive to the revolutionary (Skrimshire 2014). For Hageman (2013), the power of ecofilms comes not from "explicit ecological programming," but "their contradictions," and "the fissures through which we may glimpse and further imagine an ecology without capital—an ecology to come" (66). Hageman thus rereads film against the grain of hegemonic ideology, and suggests that we approach ecofilm not by asking what a film does (does it change people's actions?), but rather what a film can do. In other words, instead of assessing measurable changes in people's actions, we can speculate how ecofilm might offer novel ways to think about the future and in so doing relativize the present.

To understand how film works, we can draw on ecocritic Ivakhiv's ecological approach. Ivakhiv (2013) suggests that films have internal ecologies, and that this makes cinema different from other cultural forms. The main difference, according to Ivakhiv, is film's motility and openness: Each film's internal ecology is poised to be read in a different way, to mutate, and by drawing viewers into a relationship become more than merely internal to the film, by overflowing the film and making connections to the world beyond. Film draws viewers into its world along three vectors: the spectacular, which is the immediacy of affect and response in the experience of seeing a film; the narrative, or the recognition of connection through scenes that make up the film world; and the exo-referential, which is the recognition of meaningful reference to things outside the film. Through each of these three vectors—the spectacular, the narrative and the exo-referential—films can be "affectively generative" and can change the viewer (Ivakhiv 2013, 300). The force of these vectors is neither preordained nor contained in the film alone, but emerges from the ecological relationship among viewer, world, and film. Such capacities make cinema more than programmed texts readable in only one way.

Drawing on Hageman and Ivakhiv to read the spectacular, narrative, and exo-referential ecologies of *The Turin Horse,* I contend that the apocalyptic differs from other ecofilms, which might aim to shock the viewer, or to prompt an emotional reaction to suffering, or to mourn, to bear witness, or to inspire. Although apocalyptic film can do those things, its more important function is to prompt a yearning for something different, a transformation—the beginning of a new world, not the end of an old one—as well as prompt the question of how to respond to an uncertain future. Apocalyptic cinema is not, of course, preprogrammed to do this; rather, a desire for the new emerges out of the ecological relationship among film, viewer, and world. The "new" here, I demonstrate, involves confronting the disempowering deep time horizons implicit in the Anthropocene as well as the geological and biological commitments of the anthropos of the Anthropocene.

The speculative style of this article is, of course, not intended to supplant more grounded analyses of socioecological politics and transition. We know from such work that apocalypse is already here, it is just not very evenly distributed (to paraphrase novelist William Gibson). Moreover, the capacity to be affected by film is governed by a distribution of the sensible that follows lines of privilege (Rancière 2004). Indeed, cinematic representations of Anthropocene apocalypse might be characterized as entertainment for a privileged elite. But as Castree (2014) argues, representations produced and consumed by the powerful have force and should not escape our critical attention. If, as I have suggested earlier, the Anthropocene is defined as much by anxiety as by transitions in earth systems, then it is crucial to interrogate how any cultural elite produces and consumes knowledge and imaginaries of the future as well as the physical science base.

The Turin Horse: Cosmic Unmaking and Hope in Time

Hungarian director Béla Tarr's work spans thirty years. Tarr's early films, beginning with *The Prefab People* (1982), grapple with the problems of communist Hungary, and his later works chart the hollowness of the good life under capitalism. His most famous film, the seven-hour *Satantango* (1994), follows the breakup of a collective farm after the end of communism. Tarr's films also get darker and darker in tone, culminating with the final scenes of his last film, *The Turin Horse* (2011, 146 minutes), in which the sun and all light are extinguished. Tarr describes starting his career with the desire to show how messed up the world is, but that gradually he "began to understand that the problems were not only social; they are deeper. . . . They were cosmic," which led to his style becoming ever simpler and by the end "very pure" (Tarr 2012).

The Turin Horse describes six days on a bleak, windswept, and dusty farmstead. After the opening scene, a mesmeric close-up of a horse-and-cart journey home, the horse ("played" by a horse called Ricsi) is shut away in her stable. Each day the peasant farmer Ohlsdorfer (played by János Derzsi) and his unnamed daughter (Erika Bók) repeat their routine: They get dressed, stare out the same windows, and eat the same meal (one boiled potato). The young woman has to dress her father each morning because his right hand is paralyzed; he seems barely capable even of tending to their horse. The film shows a pared down life without joy, although the audience is clearly expected to fear that worse might be to come in this antigenesis narrative arc. On the second day, despite being whipped, the horse refuses to move. We see their daily routines once again, but from slightly different angles and perspectives. On the third day, a drunken neighbor warns them of the encroaching final darkness in a long, rambling monologue. On the fourth day, the horse refuses to eat or drink. The well runs dry. On the fifth day, the farmer and daughter decide to flee with their horse. They load their possessions on their cart, and slowly the three disappear over the horizon. The camera stays fixed on a lone tree until forty seconds later we see the trio return, even more desperate and exhausted than before. The horse is shut away again, and does not reappear. On the sixth day, the storm has ended, but a great darkness has descended; their lamp gutters and dies. The film's closing shots are of the man worrying at a raw potato with one hand and the woman staring into her bowl; although we don't see the final breath, the end is inevitable.

On one level, Tarr's film is a Nietzschean vision in which the farmer and daughter fail to break out of their lives and drown in repetition as the cosmos slowly slides back to its dark origins. All Tarr's films are caught in a tension between the endless repetition of material events (the wind, the potato, a shot of pálinka) and the potential each being has for breaking out of that repetition, of exhibiting a will to power that can break with this history and bend other life to its own enhancement. The characters in *The Turin Horse* are caught between, on the one hand, as Rancière (2013) puts it in his commentary on Tarr's films, "the law of wind and misery," and on the other, "the weak but indestructible capacity to affirm 'honor and dignity' against this law" (46). For Tarr's characters, affirming "honor and dignity" would mean not surrendering to environmental or apocalyptic forces, but striving for paths out of eternal repetition and confronting cosmic dissolution. *The Turin Horse* discloses potential for honor and dignity through three themes: the geological, the biological, and the temporal.

The elements play a strong role in *The Turin Horse:* incessant wind, the well running dry, and the life-sustaining fire. The film's soundscape is dominated by howling wind, which at times makes human speech impossible, and the encroaching darkness creates creeping dread and heightens the importance of the family's firewood and paraffin supplies. Wind, water, and fire are not a backdrop against which the "action" unfolds, but are active participants. This elemental emphasis creates, as Ivakhiv would have it, a spectacle that captures physical hardship rooted in earth forces, as well as perhaps the (off-screen) politics of land tenure. As well as a spectacle, though, the narrative prominence of the pair worrying about fuel mirrors conditions outside the film world (Ivakhiv's exo-referential element). As Mitchell (2011) has shown, fossil fuels were central to the emergence of modern liberal-democratic states, first as the entrenched material networks of coal extraction bred class politics, then as oil made possible the just-in-time energy distribution networks that fuel the global economy. From Mitchell it is a small step to acknowledge that the Anthropocene is a project done not by humans alone, but done with geological forces: laid down deposits of fossilized solar energy. Others have taken this to its next logical step, which is that as well as liberal democratic capitalism, geology has made possible modern carbon humans (Clark 2011). To summarize, fossil fuels are folded into the human: materially (think of pervasive endocrine disruption), through our embedding in socio-technical systems that ultimately depend on fossilized energy (the notion that to be human is always already to be a tool user), and psychologically (our desires, hopes, and fears shaped by the geopolitics of oil). All this means that inasmuch as we might have become a geological force in the Anthropocene, "we" are also a "historically locatable capitalization on geo-power" (Grosz 2012, 975). In *The Turin Horse* it is the lamp that shows how wedded the man and woman are to the "geologic life" of the Anthropocene (Yusoff 2013). The woman gathers firewood and tends the fire each day—there is no electricity, only a supply of paraffin for their lamp. In the last days, the fire and lamp burn low, and—eventually—die. The pair is unable to relight the lamp, even though the woman assures her father that she has filled it. But faced with this mystery, rather than seek an explanation or another fuel source, the pair grope their way to bed in the dark. "What is all this?" the woman asks, as the sun vanishes; "I don't know, let's go to bed," the man replies. The sixth day's separation of life from the sun (as the great darkness descends to block the ultimate source of fossil fuels) refers to the inevitability of the end of carbon humans not just in this film, but in our own world. The characters' inability to respond to this crisis evokes the way in which imaginations of the Anthropocene have been colonized by fossil fuels.

Possibilities of a new kind of human linger at the edges of Tarr's film if only the pair can break out of the "law of rain and misery" and find other ways to harness earth energies. Hope, such as it is, lies with another character, however. In the film's opening scene, the man sits on his cart, buffeted by the wind. But he is unmoving—instead it is the horse that fights the wind, the horse that struggles to overcome the weather. The farm's residents are a knot of companion species bound together on a journey into darkness. As well as being makers of and made by Earth's geology, the horse signals how humans are made through the biological. Biophilosophers now make much of our originary relationality, the idea that the human is a strategic essentialism distilled from multispecies practices. We are, Bennett (2010, 113) puts it, "nested sets of microbiomes." Relationality goes all the way in (to the 80 percent of our DNA shared with daffodils), and all the way back (our genome replete with remainders of canine viruses transmitted in the saliva of dog companions back in deep time; Haraway 2008). Just as the "short" Anthropocene was done with geological earth forces (fossil fuels), so too the "long" Anthropocene (which takes the domestication of grain as its starting point) was never a human project (these two are the main competing versions of the Anthropocene, although they are not incompatible). The first garden plant was a blow-in weed, an opportunist landing on garbage at the edge of a camp, tended by a speculative human: a partnership of convenience, not the effect of human genius (Doolittle 2004). Since then, palm oil, wheat, cows, and the weedy foot soldiers of empire have changed earth systems along with their human kin (De Landa 1997). That is to say that as well as suffering in it, nonhumans have been fundamental to creating the Anthropocene.

The film takes its title from an apocryphal tale that has Nietzsche throwing his arms around a horse being cruelly whipped in a Turin square, before collapsing on a sofa and declaring himself "dumb." The hook for the film is curiosity: What happened to the horse in this tale? The answer to this question arrives during a remarkable one-and-a-half-minute shot in which we first zoom in to focus on the horse's face, and then zoom out as the stable door is shut forever. This shot begins as a dense moment of spectacle (in Ivakhiv's terms), creating an intense connection between viewer and animal. But the shot's length ensures that we realize that this is a real, fleshy horse, as well as a cipher for human–nonhuman entanglement. Tarr is famed for using nonactors in his films, and this horse is no exception. Tarr searched for a horse that didn't want to work, and found Ricsi in a "very ugly, shitty, miserable animal market" near the Romanian border (Tarr 2011a). The cinematographer, Fred Kelemen (2012), noted that "She had this deep sadness in her eyes." As the horse stares at us through the camera, her eyes pull us in, asking for response. The horse, then, embodies both the characters' and the viewers' debts and obligations to nonhuman kin.

Although the horse suffers a hard life, it is far from being merely a dumb beast of burden that can only accede to human demands. For the horse betrays the man and daughter: She seems to sense the coming end better than the two humans. Crucially, it is when the horse refuses to eat that any possibility of escape for the humans is extinguished. For the horse cannot then drag their cart, and the daughter must try instead (either because the man is impaired or because of a gendered division of labor). She fails. The final terror here is not any historical calamity; the final terror is that the father and daughter are deserted by their nonhuman ally. The humans are left bereft, realizing too late that Tarr's injunction to go beyond mere survival to enhance the self, expand one's capacities, affirm dignity, and break out of the "law of rain and misery" requires, well, a horse—a significant earthly other (Rancière 2013, 46). The human characters in *The Turin Horse* are caught between, on one hand, the endless repetition of the same, and on the other hand the possibility of escape, of a line of flight that leads to another life, but that without the horse's help cannot be followed. This apocalyptic vision demonstrates the need for the contemporary human subject to become more actively aware of its debts to nonhumans and to enter into progressive alliances with nonhumans as partners, not mere victims to be saved.

If the anxieties of Anthropocene apocalypse in this film take shape through the geological and biological dimensions of life, these are in turn overshadowed by the film's peculiar temporality. Time is crucial to Tarr's film. The director is famed for long shots, and indeed there are only thirty cuts in the whole film. This warps the viewer's perception of time, demanding patience and a slower form of engagement that retrains perception and flirts with boredom (MacDonald 2004). As in other apocalyptic worlds, life in *The Turin Horse* is slow. This is a double juxtaposition: first to the snappy, energetic style of Hollywood disaster movies (a contemporary Hollywood film has a cut on average every four to six seconds), and second to the accelerated, globally networked world of the Anthropocene. In *The Turin Horse* there are two main temporalities at work. The first is the repetitive time of the everyday, which wears the characters down. The second is cosmic time, as the antigenesis narrative arc moves slowly to day seven and the end of everything. The film thus juxtaposes the embodied, lived experience of apocalypse against prophetic eschatology. Much as the Anthropocene names a disaster that has already occurred (Morton 2013), the cosmic temporality in *The Turin Horse* encompasses a slow unravelling after some undisclosed, past calamity. On one hand, this could be read as a postpolitical shrug, a deep cynicism about any creatures' capacities to influence events and forces operating on vast scales (as the narrative of the film shows the wind, darkness, and end of fire overwhelming the characters, who remain unable to break of their repeated daily routines). This would be a circumscribed reading, though. The film does not encourage us to welcome the void. Rather, we want the characters, amid their geologic and biologic commitments, to act differently than they do in the film: to act, that is, despite the inevitability of the end by breaking out of their repetitive loops. In other words, if the creatures in *The Turin Horse* seem bent to their fate and the world's unravelling, their placidity makes us desire them to act otherwise. We want them to wake up, to assert their honor and dignity in the face of the wind and darkness. They do not, which challenges the audience to think about how they themselves might achieve a less fatalistic relation between the temporality of the everyday and the time of geologic inevitability. The film's ecology, in other words, overflows the film world into our own, and mirrors the predicament of the Anthropocene: Both asking how we should best respond to deep time and intractable Earth Forces (Clark 2014).

Conclusion

Tarr (2012) is convinced the world can be changed for the better, but confesses that he is "just a poor filmmaker," who wants to "show you something, some pictures, just some human eyes, something that is close to you. . . . Just listen to your heart and trust your eyes. That's enough." Tarr's film shows us characters downtrodden by forces vast and alien on the one hand (the wind and dark), intimate and fleshy on the other (unraveling bodies and animal agencies). The absence of the sun brings about the end of carbon modernity, their stockpile of oil useless and unburnable, and bereft of ideas the characters stumble to bed. An animal offers a line of escape from the encroaching gloom: an old horse that could, if it were able or wanted to, help the man and daughter escape. The horse shows the need for a desire that overflows the self and seeks connections, ways to feel more deeply our debts and obligations to nonhuman others. *The Turin Horse* is not a hopeful film. It shows the destruction of a version of the human that has been elevated into a planetary agent as the anthropos of the Anthropocene. The film enjoins us to imagine the characters doing things differently, breaking out of the law of rain and misery, seeking alternatives to their repeated daily routines.

In this article I have been suggesting that apocalyptic cinema, with its portrayals of collapse and of what might come after, is a kind of 'earth dreaming' that constructs the Anthropocene (along with scientific measurement, carbon emissions, etc.). Such earth dreaming does not work the same way as other knowledge. It is more open, its test of verification is not the "transport of indisputable necessities," but its capacity to create "beings of fiction" that are carried along and transformed by their dreaming (Latour 2013b, 112). We encounter cinema as an open ecology, a provocation to feel something different, and as a relativization of current political power; cinema does not instruct us with its knowledge claims, nor need it reinforce apathy, helplessness, and postpolitical impasse. I have stressed that, following the ecological model of Ivakhiv (2013), cinema works along spectacular, narrative, and exo-referential vectors that reconfigure the relation among audience, film, and world. We are not spectators of apocalyptic films, we are participants; their ecology is an invitation to feel the condition of the Anthropocene and what might lie beyond. If the earth-dreamers watching apocalyptic cinema are parochial, they are no less parochial than the legislators of sound science, the technocrats of earth systems governance, or the salespeople of shiny futures, and their version of the Anthropocene requires scrutiny in good faith, not just dismissal as vicarious indulgence or postpolitical passivity.

Anthropocene apocalypse does not therefore demand action or politics in the traditional sense. Instead, apocalypse undercuts the familiar modern narrative of progress. It shows that our projected future will be rudely interrupted by more-than-human forces; that, really, our collective myth of progress, of a humanity reaching ever upward toward great feats of rational management, will collapse as surely as global fish stocks. Thus the political charge of apocalypse is that it destroys the future—specifically, the future as a field in which the present human will endure unchanged. For some this is liberating: "Moderns always had a future ... but never a chance, until recently that is, to turn to what I could call their *prospect:* the shape of things to come" (Latour 2010, 486). Or as Colebrook (2012) puts it, "any truly futural future is apocalyptic, which is to say that it is destructive of the present, and certainly cannot be contained by any thought of saving, surviving, enduring, or maintaining life as cosmos or oikos" (205). The dark geographies of apocalyptic life demand possibilities for other ways of being human, for a people to come after carbon humans. Anthropocene apocalypse might not be exactly hopeful, but it demands a kind of depressing redemption: realizing that the question is not how to continue present ways of life, but the deeper challenge of crafting new ways to respond with honor and dignity to unruly earth forces.

Acknowledgments

Thanks are due to audiences at the Association of American Geographers Annual Conference in Los

Angeles, and *Apocalypse: Imagining the End* in Oxford, as well as to members of the Edinburgh Environmental Humanities network. Four reviewers and the special issue editor improved the article enormously.

Funding

This article was funded through a UK Arts and Humanities Research Council project, *Ancestral Time* (AH/K005456/1).

References

After Earth. 2013. M. Night Shyamalan, director. Culver City, CA: Columbia Pictures.
Anderson, B. 2012. Affect and biopower: Towards a politics of life. *Transactions of the Institute of British Geographers* 37:28–43.
Atwood, M. 2013. *MaddAddam*. London: Bloomsbury.
Avatar. 2009. J. Cameron, director. Los Angeles, CA: Twentieth Century Fox.
Bacigalupi, P. 2009. *The windup girl*. San Francisco: Night Shade Books.
Bennett, J. 2010. *Vibrant matter: A political ecology of things*. Durham, NC: Duke University Press.
Berger, J. 1999. *After the end: Representations of post-apocalypse*. Minneapolis: Minnesota University Press.
Buell, L. 2003. *From apocalypse to way of life*. London and New York: Routledge.
Castree, N. 2014. The Anthropocene and geography III: Future directons. *Geography Compass* 8:464–76.
Clark, N. 2011. *Inhuman nature: Sociable life on a dynamic planet*. London: Sage.
———. 2014. Geo-politics and the disaster of the Anthropocene. *The Sociological Review* 62:19–37.
Colebrook, C. 2012. Not symbiosis, not now: Why anthropogenic climate change is not really human. *The Oxford Literary Review* 34:185–209.
The Colony. 2013. J. Renfroe, director. Toronto, ON; Montreal, QC; and Vancouver, BC: Alcina Pictures, Item 7, and Mad Samurai Productions.
Contagion. 2011. S. Soderbergh, director. Burbank, CA: Warner Bros.
The Day After Tomorrow. 2004. R. Emmerich, director. Los Angeles, CA: Twentieth Century Fox.
De Landa, M. 1997. *A thousand years of nonlinear history*. New York: Zone.
District 9. 2009. N. Blomkamp, director. Wellington, NZ: WingNut Films.
Doolittle, W. 2004. Gardens are us, we are nature: Transcending antiquity and modernity. *Geographical Review* 94:391–404.
Ellis, E., J. Kaplan, D. Fuller, S. Vavrus, K. Klein Goldewijk, and P. Verburg. 2013. Used planet: A global history. *Proceedings of the National Academy of Sciences* 110 (20): 7978–85.
Ellsworth, E., and J. Kruse, eds. 2013. *Making the geologic now: Responses to material conditions of contemporary life*. Brooklyn, NY: Punctum.
Falling Skies. 2011. R. Rodat, creator. Universal City, CA: DreamWorks Television, TNT Originals, Invasion Productions.
Grosz, E. 2012. Geopower. *Environment and Planning D: Society and Space* 30:973–75.
Hageman, A. 2013. Cinema and ideology: Do ecocritics dream of a clockwork green? In *Ecocinema theory and practice*, ed. S. Rust, S. Monani, and S. Cubbitt, 63–86. London and New York: Routledge.
Hall, J. 2009. *Apocalypse: From antiquity to the empire of modernity*. Cambridge, MA: Polity.
Haraway, D. 2008. *When species meet*. Minneapolis: University of Minnesota Press.
Houellebecq, M. 2005. *The possibility of an island*. London: Weidenfeld.
Hulme, M. 2008. The conquering of climate: Discourses of fear and their dissolution. *The Geographical Journal* 174:5–16.
Ingram, D. 2004. *Green screen: Environmentalism and Hollywood cinema*. Exeter, UK: University of Exeter.
Ivakhiv, A. 2013. *Ecologies of the moving image: Cinema, affect, nature*. Waterloo, Canada: Wilfred Laurier University Press.
Katz, C. 1995. Under the falling sky: Apocalyptic environmentalism and the production of nature. In *Marxism in the postmodern age*, ed. A. Callari, S. Cullenberg, and C. Biewener, 276–82. New York: Guilford.
Kelemen, P. 2012. Interview. In The thinking image, R. Koehler. http://cinema-scope.com/cinema-scope-magazine/interview-the-thinking-image-fred-kelemen-on-bela-tarr-and-the-turin-horse (last accessed 2 May 2014).
Latour, B. 2010. An attempt at a "compositionist manifesto." *New Literary History* 41:471–90.
———. 2013a. Facing Gaia: A new enquiry into natural religion. Gifford Lectures, University of Edinburgh, Edinburgh, UK. http://www.ed.ac.uk/schools-departments/humanities-soc-sci/news-events/lectures/gifford-lectures/archive/series-2012-2013/bruno-latour (last accessed 12 September 2013).
———. 2013b. *An inquiry into modes of existence*. Cambridge, MA: Harvard University Press.
Levene, M. 2013. Climate blues: Or how awareness of the human end might re-instill ethical purpose to the writing of history. *Environmental Humanities* 2:147–67.
The Life of Pi. 2013. A. Lee, director. Los Angeles, CA: Fox 2000 Pictures.
MacDonald, S. 2004. Toward an eco-cinema. *ISLE* 11:107–32.
Malm, A., and A. Hornborg. 2014. The geology of mankind? A critique of the Anthropocene narrative. *The Anthropocene Review* 1:62–69.
Melancholia. 2011. L. von Trier, director. Copenhagen, DK: Zentropa.
Mitchell, T. 2011. *Carbon democracy: Political power in the age of oil*. London: Verso.
Morton, T. 2013. *Hyperobjects: Philosophy and ecology after the end of the world*. Minneapolis: Minnesota University Press.
Oxford Martin Commission. 2013. *Now for the long term: The report of the Oxford Martin Commission for future generations*. Oxford, UK: University of Oxford.

The Prefab People. 1982. B. Tarr, director. Budapest, HU: Balázs Béla Stúdió, Mafilm, Magyar Televízió Fiatal Müvészek Stúdiója, Társulás Stúdió.

Rancière, J. 2004. *The politics of aesthetics*. London: Continuum.

———. 2013. *Béla Tarr, The time after*. Minneapolis, MN: Univocal.

Robbins, P., and S. Moore. 2013. Ecological anxiety disorder: Diagnosing the politics of the Anthropocene. *Cultural Geographies* 20:3–19.

Rust, S., S. Monani, and S. Cubbitt, eds. 2013. *Ecocinema theory and practice*. London and New York: Routledge.

Sátántangó. 1994. B. Tarr, director. Hungary; Pottsdam, Germany; and Las Vegas, NV: Mozgókép Innovációs Társulás és Alapítvány, Von Vietinghoff Filmproduktion, Vega Film.

Skrimshire, S. 2014. Climate change and apocalyptic faith. *WIREs Climate Change* 5:233–46.

Steffen, W., Å. Persson, L. Deutsch, J. Zalasiewicz, M. Williams, K. Richardson, C. Crumley, et al. 2011. The Anthropocene: From global change to planetary stewardship. *AMBIO* 40:739–61.

Swyngedouw, E. 2010. Apocalypse forever?: Post-political populism and the spectre of climate change. *Theory, Culture & Society* 27:213–32.

———. 2013. Apocalypse now! Fear and doomsday pleasures. *Capitalism Nature Socialism* 24:9–18.

Szerszynski, B. 2012. The end of the end of nature: The Anthropocene and the fate of the human. *The Oxford Literary Review* 34:165–84.

Tarr, B. 2011a. Interview. In A conversation with Bela Tarr, P. Sbrizzi. http://www.hammertonail.com/interviews/a-conversation-with-bela-tarr-the-turin-horse (last accessed 21 April 2014).

———. 2011b. *The Turin Horse*: Interview with Bela Tarr. http://www.electricsheepmagazine.co.uk/features/2012/06/04/the-turin-horse-interview-with-bela-tarr (last accessed 21 April 2014).

———. 2012. Interview. In An interview with Béla Tarr, E. Kohn. http://www.indiewire.com/article/bela-tarr-explains-why-the-turin-horse-is-his-final-film (last accessed 21 April 2014).

The Turin Horse. 2011. B. Tarr, director. Budapest, HU: TT Filmmûhely.

Yusoff, K. 2013. Geologic life: Prehistory, climate, futures in the Anthropocene. *Environment and Planning D: Society and Space* 31:779–95.

Zalasiewicz, J., M. Williams, A. Haywood, and M. Ellis. 2011. The Anthropocene: A new epoch of geological time? *Philosophical Transactions of the Royal Society A: Mathematical, Physical and Engineering Sciences* 369: 835–41.

Imaginaries of Hope: The Utopianism of Degrowth

Giorgos Kallis* and Hug March†

*ICREA, ICTA, and Department of Geography, Universitat Autònoma de Barcelona, Barcelona, Spain
†Internet Interdisciplinary Institute, Universitat Oberta de Catalunya, Barcelona, Spain

This article analyzes *degrowth*, a project of radical socioecological transformation calling for decolonizing the social imaginary from capitalism's pursuit of endless growth. Degrowth is an advanced reincarnation of the radical environmentalism of the 1970s and speaks to pertinent debates within geography. This article benefits from Ursula Le Guin's fantasy world to advance the theory of degrowth and respond to criticisms that degrowth offers an unappealing imaginary, which is retrogressive, Malthusian, and politically simplistic. We argue instead that degrowth is on purpose subversive; it brings the past into the future and into the production of the present; it makes a novel case for limits without denying that scarcity is socially produced; and it embraces conflict as its constitutive element. We discuss the politics of scale of the incipient degrowth movement, which we find theoretically wanting, yet creative in practice.

本文分析"去成长"——一个呼吁对资本主义追求无止尽成长的社会想像进行去殖民的激进社会生态变革计画。去成长是1970年代激进环境主义的进阶式再度化身，并与地理学中的相关辩论进行对话。本文受益于娥苏拉．勒瑰恩（Ursula Le Guin）的梦想世界，藉此推进去成长的理论，并回应有关去成长提供了退化、马尔萨斯主义、且政治上简化而不具吸引力的想像之批评。我们则主张，去成长反而具有颠覆的目的；它将过去带往未来、并进入了当下的生产；它创造了限制的崭新案例，同时承认匮乏是社会生产的；它更欣然接受冲突作为其构成的元素。我们探讨最初的去成长运动的尺度政治，并发现该议题在理论上有所阙如，但在实践上却相当具有创意。

Este artículo analiza el decrecimiento, un proyecto de radical transformación socioecológica que clama por descolonizar el imaginario social del objetivo de crecimiento sin fin del capitalismo. El decrecimiento es una reencarnación avanzada del ambientalismo radical de los años 1970 y busca como interlocutor los debates pertinentes dentro de la geografía. Este artículo se apoya en el mundo fantástico de Ursula Le Guin para plantear la teoría del decrecimiento y responder a críticas de que el decrecimiento ofrece un imaginario poco atractivo, que es regresivo, maltusiano y políticamente simplista. Por el contrario, argüimos que el decrecimiento es deliberadamente subversivo; que lleva el pasado al futuro y a la producción del presente; hace un caso novedoso por los límites sin negar que la escasez es producida socialmente; y que abraza al conflicto como su elemento constitutivo. Discutimos la política de escala del incipiente movimiento del decrecimiento, al que hallamos teóricamente deficiente, aunque creativo en la práctica.

Utopias imagine a preferable system to the status quo, freeing us from the determinisms of history (Jameson 1975). Geographers have called for the revitalization of utopianism "to think the possibility of real alternatives ... and galvanize socioecological changes" (Harvey 2000, 156, 195), encouraging a "foregrounding and naming [of] different ... futures" (Swyngedouw 2010, 228). A movement of activists and scientists in France, southern Europe, and beyond has given the name *décroissance* (degrowth) to its alternative to capitalist socioecological relations. Degrowth draws from the postdevelopment and antiutilitarianism literatures, Georgescu-Roegen's entropic limits to growth, and the continental post-Marxist ecologies of the 1970s with the likes of Gorz, Illich, and Castoriadis (D'Alisa, Demaria, and Kallis 2014). It marks a rebirth of a radical environmentalism against the apolitical consensus on sustainable development. Degrowth should interest geographers, because it is increasingly invoked by prominent scientists (Anderson and Bows-Larkin 2013) and radical intellectuals (Klein 2013) in the context of climate change; because it inspires grassroots practices, movements, and politics (Baykan 2007); and because its theory speaks to pertinent debates in the field, such as scarcity and limits, or the politics and scales of a post-capitalist transition.

The most prominent advocate of degrowth is Serge Latouche (2009), an economic anthropologist, who brought the critique to development from his fieldwork in Africa back home to France. For Latouche, degrowth is a project of decolonizing the imaginary from growth. His utopia involves eight interdependent transitions, "the eight Rs of degrowth" (33): reevaluate (shift values); reconceptualize (e.g., wealth vs. poverty or scarcity vs. abundance); restructure production beyond capitalism; redistribute between North and South and within countries; relocalize the economy; and reduce, recycle, and reuse resources. Latouche envisions autonomous communities with restricted trade, organized in confederations of autonomous municipalities and bioregions. The transition is to be facilitated by policy reforms, such as taxes on consumption, advertising and finance, investment in peasant agriculture, reduction of working hours, and a basic citizen's income.

Critics take issue with the term *degrowth* because it is oblivious to the devastation of recession (Foster 2011). Degrowth frightens people unnecessarily, they argue: in a desirable socioecological transition, only bad things, such as military expenditures or dirty industries, should degrow, whereas others, such as renewables, education, or organic agriculture should grow (Schwartzman 2012; Chomsky 2014). Others are critical of a romanticization of poverty and of oppressive non-Western societies (Navarro 2013), Malthusianism, in conceiving limits to growth rather than to capital (Χοβαρδάς 2014), a fetishization of the local (Romano 2012) and an incomplete politics of transition that glosses over dispossession and conflict under capitalism (Foster 2011; Saed 2012; Navarro 2013).

Although there is ground for such criticism, this article follows a more sympathetic reading of Latouche and engages with a burgeoning literature beyond him (e.g., Schneider, Kallis, and Martinez-Alier 2010; D'Alisa, Demaria, and Kallis 2014) to elaborate a nuanced theory of degrowth. An analytical limitation to this task is that degrowth, like any utopia, is nowhere to be seen. There is theory and there are small experiments broadly inspired by degrowth, but there is no spatialized "degrowth world" in its full plentitude. This article overcomes this limitation by studying a unique case of a territorialized degrowth world in Ursula Le Guin's (1974) fantasy novel *The Dispossessed*. The novel was written in a similar intellectual environment of crisis and limits to growth as today's. In the words of its author, the book aspired to put a "pig on the tracks . . . in a one-way future consisting only of growth" (Le Guin 1982, 4). We are not alone in finding *The Dispossessed* as one of the most complete and thought-provoking expressions of a noncapitalist, ecological world with all its "seductions and snares" (Gorz 1994, 81). But following Le Guin's plea, we do not treat the novel as the expression of an idea but as "an experimental variation on our own empirical universe" (Jameson 1975, 4). We rethink the possibilities and pitfalls of degrowth through the lens of *The Dispossessed*, hinting at a way geographers could use science fiction.

We first argue that degrowth is frightening because its purpose is subversive. Second, we sustain that degrowth's call is not for a return to a past that never existed, but for a simultaneous production of the present by the past and the future. Subsequently we develop a case for limits without scarcity. Fourth, we argue that conflict, external and internal, should be constitutive of any transitional project. We then reflect on the articulation of scale in degrowth. We conclude that degrowth signals a new socioecological imaginary that expresses features of Harvey's (2000) spatiotemporal utopias.

A Subversive Utopia

> Because I've written science fiction I am always accused of writing about the future. I've never ever tried to write about the future. We have nothing but the present tense.
>
> —Ursula Le Guin (2014)

The Dispossessed tells the personal journey of Shevek, a physicist who, in pursuit of a unified theory of time, crosses borders from the planet of Anarres to Urras and back. Anarres is a small, dry, and barren planet supporting a frugal standard of living. Revolutionaries fleeing en masse the capitalist state of Io in Urras colonized Anarres seven generations ago. There is no private property, state, money, police, or military there. Mutual aid and sharing have replaced competition and "profiteering."

The Dispossessed subverts a dominant imaginary, by positioning utopia in a territory, Anarres, which from our gaze, as well as those of its contemporaries in Urras, looks like a postapocalyptic, dystopian catastrophe. There are no animals and only a few plants in Anarres. There is dust everywhere. Food consists of two simple vegetarian meals a day. Luxuries, other than simple ornaments, are absent, and citizens do communal work every few years in the mines and in reforesting a desert. Yet the Anarresti, as we see them through the eyes of Shevek, do not mind this physical hardship and are relatively happy. This does not mean that utopia can only flourish within a catastrophe but that material abundance is not a universal requirement for well-being.

In a similar vein, degrowth is thrown as a "missile word" to subvert the dream of continuously growing material wealth (Demaria et al. 2013). The purpose of using a negation for a positive project is not to frighten but to overcome a fear. For degrowthers it is precisely the fear of a future without growth that has to be confronted if the discussion for a future outside capitalism is to open up (Latouche 2009). Calling degrowth "the society of frugal abundance," Latouche (2012) performs a similar subversion to that of Anarres, decoupling utopia from a one-way future of material abundance.

The growth–degrowth or abundance–frugality pairs serve as dialectical oppositions, which "by way of negation ... grasp the moment of truth in each term" (Jameson 2004, 48). Le Guin's story is precisely about opposites: Anarres is the moon of Urras, Urras the moon of Anarres. Anarres is what Urras is not and Urras is what Anarres is not: dispossessed–possessed, barren–lush, horizontal–hierarchical. It is the (romanticized) imagination of Anarres that spurs revolutionary passions in Urras. And for an Anarresti like Shevek, it is not in Anarres itself but in traveling between Anarres and Urras, that utopia is found as he reappreciates the value of Anarres. His trip reanimates also social change in an Anarres stalled by defining itself as the opposite of Urras. The utopian affirmation is not to be found in the negation itself but in the synthesis provoked by the negation. Degrowth is not an affirmative imaginary that signifies the opposite of growth; it is an imaginary that by confronting growth opens up new imaginaries, spaces, and key words (D'Alisa, Demaria, and Kallis 2014).

Critics of degrowth miss this subversive element, stuck in an idea of utopia as a prefixed positive destination, such as Schwartzman's (2012) "solar communism," rather than a horizon chartered through confrontation with hegemonic worldviews. Precisely because the desired socioecological change is first and foremost qualitative rather than quantitative, and because this qualitative change has nothing in common with the dominant imaginary of what comes to be understood and measured as "growth," the prefix *de* is apt: a *de* for decolonization, a *de* for 'liberation from.' Of course, certain things will grow under degrowth; children, syndicates, or sharing do "grow" in Anarres. There will be growth with a lowercase g, in an overall process of degrowth with an uppercase D. Degrowth is therefore not a call for "less of the same"; that is, recession and declining gross domestic product in a capitalist economy where everything else stays the same (D'Alisa, Demaria, and Kallis 2014). It is a call for an altogether new, qualitatively different world that will evolve through confrontation with the existing one.

A Utopia Where Past and Future Produce the Present

Under enlightenment, modernism, and in deterministic versions of Marxism, the new world to come is a sequential outcome of the old one. In *The Dispossessed*, instead, the two are simultaneously present (Somay 2005). Anarres coexists with Urras. It is its past, a society frozen in the time of the exodus, and its future, as Shevek's coming from Anarres inspires revolts in Urras. For Le Guin, past and future are not linked causally. Instead the future, or a variety of futures, "can only be realised through human action taken in the present" and is therefore "*latent* in the past" (Ferns 2005, 256; Davis 2005). Unlike other science fiction, which imagines technological progressions from the capitalist present, Anarres embodies elements from the "so-called primitive societies" crushed by colonialism and capitalism (Davis 2005, 17). Past and present, real and fictional, encounter each other in Anarres.

Likewise, what degrowth envisages is not a return to the past, but a mobilization of elements of the past. What Χοβαρδάς (2014) reads in degrowth as a criticism of enlightenment and Navarro (2013) as a romanticization of feudalism is instead a rejection of a sharp distinction of "a before" and "an after," this temporal distinction that takes a spatial–geographical expression into an "us advanced in the West" and "they backwards in the rest" (Graeber 2004). Reflecting Latouche's roots in critical anthropology, degrowth reads the capitalist present as full of latent elements from a noncapitalist past, such as the gift economies of barter markets or the commons of urban gardens; it is these that carry the seeds for a different future (D'Alisa, Demaria, and Kallis 2014).

A Utopia of Limits Without Scarcity

> All that city. ... You just couldn't see an end to it. ... It wasn't what I saw that stopped me, Max. It was what I didn't see. ... In all that sprawling city, there was everything except an end. ... Take a piano. The keys begin, the keys end. You know there are 88 of them ... They are not infinite, you are infinite. On those 88 keys the

> music that you can make is infinite. . . . But you get me up on that gangway and roll out a keyboard with millions of keys, and . . . there's no end to them, that keyboard is infinite. But if that keyboard is infinite there's no music you can play. (From the movie *The Legend of 1900*)

Latouche's depoliticizing invocations of growing footprints, resource peaks, and "pedagogical disasters" are highly problematic as they might justify the techno-authoritarian and market solutions that degrowth opposes (Romano 2012). Alongside this environmental determinism, however, Latouche brings a strong anthropological understanding of scarcity "not as an intrinsic property of technical means . . . but as a relation between means and ends" (Sahlins 1972, 5). Latouche sees a precursor of degrowth in Sahlin's (1972) thesis that the "original affluent" societies of hunter-gatherers did not experience scarcity not because they had a lot but because "want not, lack not" (11). Permanent mobility and the absence of (the need for) accumulation went hand in hand with a lack of a sense of scarcity; stone agers' ends stayed within their limited technical means.

Critical geographers, following Harvey (1974), also understand scarcity as socially produced. They have, however, not always distinguished between two radically different strategies to confront scarcity: namely, the limitations of ends versus the expansion of means. Whereas the first dissolves the very sense of scarcity, the latter reproduces it at higher levels. Limitations feature prominently in degrowth manifestos, proposed "as a social choice and not imposed as an external imperative for environmental or other reasons" (Schneider, Kallis, and Martinez-Alier 2010, 513). Instead of Malthus, degrowthers find inspiration in the "neo-Malthusian" anarcho-feminists of Emma Goldman, who advocated conscious procreation to stop capitalism from exploiting female bodies to produce soldiers and cheap labor (D'Alisa, Demaria, and Kallis 2014). Such self-limitations do not submit to objective limits but act to produce particular (noncapitalist, egalitarian) socionatures. Illich (1974), for example, was against fossil fuels, not because of environmental or climatic limits but because "socialism can only arrive by bicycle" (1). A society of high energy use and advanced technologies, even a "solar communism" à la Schwartzman (2012), would need experts to manage them and by necessity will be undemocratic and nonegalitarian.

Likewise, for Castoriadis (2005), ecology is not the love of nature but the need for self-limitation, which brings true freedom. The pianist in *The Legend of 1900* lives all his life in an Atlantic cruise ship refusing to cash out on his fame. By limiting himself, he is liberated from the unbearable choice offered by the city outside. Unlimited wants are the foundation of modern economics and the sine qua non of capitalism. Scarcity is permanent if wants are unlimited (Skidelsky and Skidelsky 2012). Capitalism produces relative scarcities by enclosures, by positional inequalities, and by the promise of unlimited choice. In framing scarcity as a universal, production-related problem, capitalism is legitimized as the system best suited to expand the means of production. Only a society that "has had enough" can refrain from accumulation and liberate itself from capitalism.

A society that paradoxically has enough is Anarres. Anarres's signification is its liberating self-limitation and not as Jameson (1975, 12) argues a "sociopolitical hypothesis about the inseparability of utopia and scarcity." Indeed, revolution is as likely in materially rich Urras as counterrevolution in materially poor Anarres (Stillman 2005). It is self-limitation that distinguishes the Anarresti, manifested in their choice to move to a limited planet with harsh conditions and in their continued commitment to work only a few hours a day or to leave their mineral resources to Urras without reclaiming them for their own "development." This commitment to leave human and natural resources idle protects Anarres from a trajectory that would make it capitalist like Urras.

Self-limitation in Anarres does not translate into a sense of scarcity. It is in materially wealthy Urras that people live in perceived scarcity, whereas the materially poor residents of Anarres experience abundance. Initially marveling at the plenty of Urras, Shevek finally concludes that "there is nothing, nothing on Urras that we Anarresti need" (Le Guin 1974, 279). And he notes that by Anarres's standards, "even the poor in Urras are not so very poor" (275). Poverty in Urras, which so surprises Shevek, is not material. It is not a matter of absolute levels of resources or consumption; it is social, the result of enclosures and inequality.

The question then is how a society can limit itself. Sahlins's work might be read as a hypothesis of an inextricable link between sharing in common and self-limitation. Indeed, sharing the (reclaimed) commons in common is a major signification of degrowth. Proposals include work sharing, cohousing and communes, car and bike sharing, regaining the collective control of water or energy, or reclaiming and sharing public spaces (Latouche 2009; D'Alisa, Demaria, and Kallis 2014). In Anarres, as in degrowth communes

(Cattaneo and Gavaldà 2010; Carlson 2012), everyone shares equal access to the same commons: dormitories, kitchens, transport, and work.

Sharing might lead to self-limitation for three reasons: first, because sharing and consuming all surplus of production, as in stone-age societies, precludes accumulation and the investment-driven growth dynamic; second, because the abolishment of private property, as, for example, by the voluntary "dispossessed" Anarresti, ensures that everyone has access to the same basic resources necessary for survival and social reproduction; and third, and more important, because sharing in common establishes equality and equality dissolves the relative comparisons that breed a personal sense of scarcity and unsatisfied want. As Shevek tells us, it is not that the Anarresti are not poor because "nobody goes hungry." They are not poor because "nobody goes hungry while another eats" (Le Guin 1974, 229). The end of enclosures therefore and the sharing of the commons brings the end of scarcities and of accumulation, making living within limits possible.

The thesis that scarcity is social and not absolute does not suggest a relativistic denial of material reality. There will always be periods of "plenty" or "shortage," periods of temporary excess or deficit of socially determined aspirations over socioecologically determined resources. The argument here is not that hunter-gatherers or the Anarresti never experience hunger. Indeed, when an unprecedented drought hits Anarres, the planet suffers food shortages and famine. The difference is that this did not translate to a generalized sense of scarcity or a push for growth; it was a temporary disaster of shortage, which the society suffered through in common. Temporary lack is not generalized scarcity.

A Utopia of Conflict

Latouche has been criticized for his imaginary of localized self-sufficient communities. Romano (2012) argues that despite Latouche's reclaimers, there is an in-built risk of closure, authoritarianism, and xenophobia in any locality that pretends to be self-reliant and not dependent on the rest. A central problematic of any utopian project is indeed this tension between openness and closure, or change and stability. "The materialization of anything requires, at least for a time, closure around a particular set of institutional arrangements and a particular spatial form" (Harvey 2000, 188). Yet closure carries a risk of authoritarianism for securing the particular form. Le Guin's fiction is part of a utopian genre that transcends the openness–closure dichotomy by recognizing "that societies and spatialities are shaped by continuous processes of struggle" (Harvey 2000, 190). In these novels, "[t]he reader is not, therefore, introduced to a stable world already made and discovered, but is taken through the dialectics of making a new socio-ecological world" (190). Utopia is not an end-state of stability and perfection but a state of struggle and conflict (Le Guin 1989).

Unlike many other literary or political utopias, history and politics have not ended in Anarres. The closure necessary for internal order and protection from a hostile outside has started undermining the revolution. The teachings of Odo, the female leader of the first revolutionaries, are morphing into a religion, social control sustained through the ever-vigilant eye of peers. A hidden hierarchy emerges around the planet's central Production and Distribution Committee, which uses its powers to allocate people to jobs, oversee contacts with the outside world, and centralize control during the famine. It is the freedom of the individual and of a collective to take an initiative against the community, however, that allows conflict to be exposed. Shevek and friends set up a printing syndicate to challenge the Committee opening communication with Urras. In the novel the wall that protects the spaceport of Anarres, and that Shevek first crosses, symbolizes the ultimate closure of Anarres, one from which most other literary and real-existing utopias suffer (Somay 2005). Shevek opens this unnegotiable closure. Opening, however, comes with a risk: Shevek might revitalize the revolution in Anarres or expand it to Urras, but he also risks bringing its end, corrupting Odonian ethic by exposure to Urrasti values. The novel does not tell us what happens, because for Le Guin there is no end, only a process of renewal of an initial commitment that comes with inescapable risks.

A process-based understanding of utopia can help us rethink the politics of degrowth. Problematic here is Latouche's (2009) denial of conflict, his expectation of "a virtuous cycle of quiet contraction" (33), and a "concrete utopia of degrowth" that appears as an idyllic end state. Latouche defends his position on the grounds that true revolution means institutional and cultural change, which can only come through reforms, not violent takeovers of power that would drift into bloodshed. This leads, however, to a paradoxical proposal of an electoral program for degrowth, when he himself recognizes the capturing of parliamentary politics by oligarchic interests. Also contradictory is his advocacy for slow cultural and behavioral change in the face of extreme environmental changes, such as climate change, that ask for urgent action (Romano 2012). Latouche (2012) is right that degrowth is an imaginary that challenges the very spirit of capitalism but has no convincing answer to Foster's (2011) criticism that a degrowth transition cannot emerge without conflict within a capitalist system that cannot do without growth.

Indeed, a new generation of degrowth scholars considers conflict as constituent of a degrowth transition. Studying degrowth projects in Barcelona, D'Alisa, Demaria, and Cattaneo (2013) distinguish between civil and "uncivil" actors, the latter defined as those who refuse to be "governmentalized" and be part of a civil society that constitutes the social capital that fuels growth. Examples of "uncivil" actions in Barcelona include the reclaiming of public space by urban gardeners and *Indignados* protesters, the occupation of private property by the vibrant squat movement of the city (see also Cattaneo and Gavaldà 2010), or a much-publicized act of fiscal disobedience against private banks by degrowth activist Enric Duran (D'Alisa, Demaria, and Cattaneo 2013). Disobedience—conflict without violence—forms part of the repertoire of degrowth strategies (D'Alisa, Demaria, and Kallis 2014). Another strategy is the founding of prefigurative, nowtopian projects such as eco-communes in and out of the city, back-to-the-land communities, or various cooperative projects dealing with reproduction (food cooperatives; cohousing; child care, education, and health cooperatives). Against Romano's (2012) concerns, such exoduses from the capitalist system remain politicized, recognizing the inevitability of conflict if they are to expand beyond activist havens.

Anarres is a colony of exodus founded on conflict, a civil insurrection in Urras. The novel reminds us that if an exodus is to remain relevant it has to maintain interaction with the world from which it escaped, changing it and itself. Changing itself means dealing with internal conflict. As in Anarres, hidden hierarchies are omnipresent in degrowth nowtopias, where considerable time and resources are spent in exposing and managing interpersonal conflict through advanced techniques of facilitated deliberation (Cattaneo and Gavaldà 2010; Carlson 2012). Like Anarres, the solution to conflicts is procedural, not pregiven; it is the freedom allowed to individuals to contest the direction of a common project that keeps lurking hierarchies in tentative check. The "to close or to open" or "to isolate or to engage" dilemmas and the risks of irrelevance versus appropriation are constantly faced, when recruiting new members or deciding whether to engage with institutions, political parties, or the market economy. What one observes in practice is a plurality of strategies (Demaria et al. 2013). Our experience, for example, with Barcelona's degrowth movement suggests that whereas some activists refuse any engagement with the state and electoral politics (other than through protest and disobedience), others engage with radical left parties, seeing in their nowtopian degrowth projects embodiments of new values that could hegemonize society and be taken up by a radical government to reform the state. Similarly, whereas some activists aspire to an abolition of money or private property, others endorse a transitory politics of engagement with the formal economy and hybrid forms of property and money (e.g., squats where rent is eventually paid to local authorities, food cooperatives that pay farmers in Euros but at "socially fair" prices, and communities that use time banks, own currencies, peer-to-peer virtual currencies, or Euros, depending on purpose). What most projects share is a rejection of hermetic closures, tolerance to members' right to hold a different stance than the collective, and a constant case-by-case deliberation on the terms of external engagement.

Questions of Scale

A radical socioecological transformation cannot be conceived in nonspatial terms. Harvey (2000, 177) raises a point that should resonate with degrowthers: "how [utopia] gets framed spatially and how it produces space become critical facets of its tangible realization." As Brenner (2009, 32) argues, any (transformative) sociospatial process implies the "crystallization of multiple, intertwined geographical dimensions." Among those dimensions "scale" is critical. The process of rescaling—that is, the reconfiguration of relations among and across scales and the production of new scales of action—might open up the possibilities to produce inclusive, ecological, and democratic socionatures (Swyngedouw 2004).

Rescaling is obliterated in the existing theory of degrowth. Latouche reproduces uncritically the local–national–global distinctions produced by capitalism, positioning degrowth as a project of a fetishized localization (Romano 2012) against the forces of globalization; a project that is to be pursued (paradoxically) through political action at the national level. True,

Latouche refers to bioregions but without specifying how these could become the basis of political or economic organization. Unfortunately, *The Dispossessed* does not help much either, as Le Guin deemphasizes geography on purpose, to avoid imagined communities that segment and separate (Stillman 2005). As opposed to Urras, with its nations, capitals, and borders, Anarres is borderless. Communities are scattered in the territory, connected by communication and transport networks that permit the free circulation of resources and people. Interestingly, no one resides permanently in a community, as the Committee regularly posts citizens to jobs around the planet. Anarres is flattened by this constant mobility of its residents and resources; the only community is Anarres itself.

We have to turn to the actually existing, albeit partial, degrowth nowtopias to illuminate Harvey's call to understand how utopias get framed spatially and engage with scale. One such example is the *Cooperativa Integral Catalana* (CIC), a legally registered network with 600 members and 2,000 participants in Barcelona offering an integrative structure for producers and consumers of organic food and artisanal products, residents of eco-communes and squats, cooperative enterprises, and regional networks of exchange (*Ecoxarxes*, or eco-networks; Carlson 2012). CIC could be seen as an attempt "to overthrow the structures (both physical and institutional) that the free market has itself produced as relatively permanent features of our world" (Harvey 2000, 186). By challenging extant capitalist "spaces of dependence" that tie social–political and socioeconomic relations to certain scales and by rescaling relations of care, production, and social reproduction, CIC produces new "spaces of engagement" (Cox 1998), to be spread by self-replication and networking. This can be observed, for instance, in the health and education cooperatives of CIC that emerge as a (partial) alternative both to public services severely touched by austerity measures and to the growing fully private sector. Elsewhere, the *Ecoxarxes* not only subvert the extant scalar arrangements of food provision by connecting consumers to producers directly through alternative networks but also challenge the European Union scale of circulation and accumulation of capital by issuing their own social currencies (the ECOs; Carlson 2012).

Notwithstanding the discursive importance of the local scale for the cooperative model, CIC does not abide by a reactive retrenchment and closure to an idealized "locality" or a new form of neoliberal communitarianism. Instead, it shows a proactive will of transforming extant and producing new scalar relations, while engaging in political struggles such as the protests for protecting desirable services at existing scales. This embodies what De Angelis (2012) calls for; that is, a social movement of territorialized commoning combined with a political movement of protecting the state "deals" that secure collective resources for such commoning (e.g., public land, unemployment benefits, free health care, or a basic income). In this sense we argue that projects like CIC overcome the quandary of choosing a single scale of action (either–or) by critically engaging simultaneously with different ones (both–and; Harvey 2000), synthesizing new scalar relations. The spatiality of CIC, thus, as an example of really existing degrowth conforms to what Brenner (2009, 30) calls crystallization "at the interface between inherited sociospatial configurations and emergent spatial strategies intended to transform the latter."

Conclusion

This article presented a socioecological future named degrowth. The radical imaginary of Ursula Le Guin's fantasy world helped us spatialize Latouche's abstract-level theory, rethink a case for limits without environmental determinism, and argue that a degrowth transition will involve conflict, and will be necessarily open-ended. Degrowth, in the way it is advanced here, signifies a utopia that resembles in many ways a utopianism advocated by Harvey (2000): it imagines a future of process and conflict, not a blissful end state; it subverts the hegemonic desires on which capitalism rests; it brings the past into the present and into the production of the future; and it aspires to the production of egalitarian socionatures. Although the production of scale is theoretically wanting in the degrowth literature, actual projects create spaces of engagement that allow for new scalar linkages and geometries.

Crucially degrowth offers an ecological imaginary that differs fundamentally from a Malthusian "limits to growth" vision, without falling into eco-cornucopian dreams, capitalist (e.g., "green growth") or anticapitalist (e.g., "solar communism"). The proposition developed here is that scarcity is reproduced, not overcome, by the development of the means of production. Only a collective self-limitation, premised on sharing the commons, dissolves scarcity and opens up the possibility for a society that is not capitalist. The intriguing idea behind degrowth is that we do not need to "develop" to get enough, because we already have, and in a sense always had, enough. What we need is to struggle for the institutions that will allow us to live with enough.

Acknowledgments

The authors thank Dídac Costa for his time and insightful comments on the *Cooperativa Integral Catalana*. Karen Bakker, David Saurí, Ramon Ribera-Fumaz, Louis Lemkow, Kaysara Khatun, and several colleagues from "Research & Degrowth" commented on earlier versions of the article. The ideas on limits and scarcity draw from conversations with Giacomo D'Alisa.

Funding

Giorgos Kallis acknowledges grant 289374 under the EU Marie Curie project ENTITLE (European Network of Political Ecology). Hug March received support from the Spanish Ministry of Economy and Competitiveness under JCI-2011-10709.

References

Anderson, K., and A. Bows-Larkin. 2013. Avoiding dangerous climate change demands de-growth strategies from wealthier nations. http://kevinanderson.info/blog/avoiding-dangerous-climate-change-demands-de-growth-strategies-from-wealthier-nations/ (last accessed 23 July 2014).

Baykan, B. G. 2007. From limits to growth to degrowth within French Green politics. *Environmental Politics* 16 (3): 513–17.

Brenner, N. 2009. A thousand leaves: Notes on the geographies of uneven spatial development. In *Leviathan undone? Towards a political economy of scale*, ed. R. Keil and R. Mahon, 27–49. Vancouver, Canada: UBC Press.

Carlson, S. 2012. *Degrowth in action: How the Cooperativa Integral Catalana enacts a degrowth vision*. Unpublished master's thesis, Lund University, Lund, Sweden. http://www.lunduniversity.lu.se/o.o.i.s?id=24965&postid=3045186 (last accessed 24 July 2014).

Castoriadis, C. 2005. *Une société à la dérive* [A society adrift]. Paris: Ed. Seuil.

Cattaneo, C., and M. Gavaldà. 2010. The experience of rurban squats in Collserola, Barcelona: What kind of degrowth? *Journal of Cleaner Production* 18 (6): 581–89.

Chomsky, N. 2014. *The greening of Noam Chomsky: A conversation*. Canadian Dimension. http://canadiandimension.com/articles/5874/ (last accessed 24 July 2014).

Cox, K. R. 1998. Spaces of dependence, spaces of engagement and the politics of scale, or: Looking for local politics. *Political Geography* 17 (1): 1–23.

D'Alisa, G., F. Demaria, and C. Cattaneo. 2013. Civil and uncivil actors for a degrowth society. *Journal of Civil Society* 9 (2): 212–24.

D'Alisa, G., F. Demaria, and G. Kallis, eds. 2014. *Degrowth: A vocabulary for a new era*. London and New York: Routledge.

Davis, L. 2005. The dynamic and revolutionary utopia of Ursula K. Le Guin. In *The new utopian politics of Ursula K. Le Guin's* The Dispossessed, ed. L. Davis and P. G. Stillman, 3–36. Oxford, UK: Lexington Books.

De Angelis, M. 2012. Crisis, movements and commons. *Borderlands* 11 (2): 1–22.

Demaria, F., F. Schneider, F. Sekulova, and J. Martinez-Alier. 2013. What is degrowth? From an activist slogan to a social movement. *Environmental Values* 22 (2): 191–215.

Ferns, C. 2005. Future conditional or future perfect? The Dispossessed and permanent revolution. In *The new utopian politics of Ursula K. Le Guin's* The Dispossessed, ed. L. Davis and P. G. Stillman, 249–64. Oxford, UK: Lexington Books.

Foster, J. B. 2011. Capitalism and degrowth: An impossibility theorem. *Monthly Review* 62 (8): 26–33.

Gorz, A. 1994. *Capitalism, socialism, ecology*. London: Verso.

Graeber, D. 2004. *Fragments of an anarchist anthropology*. Chicago: Prickly Press.

Harvey, D. 1974. Population, resources and the ideology of science. *Economic Geography* 50 (3): 256–77.

———. 2000. *Spaces of hope*. Berkeley: University of California Press.

Illich, I. 1974. *Energy and equity*. Vol. 6. New York: Harper & Row.

Jameson, F. 1975. World-reduction in Le Guin: The emergence of utopian narrative. *Science Fiction Studies* 2 (3): 7.

———. 2004. The politics of utopia. *New Left Review* 25:35–54.

Klein, N. 2013. How science is telling us all to revolt. *New Statesman* 29 October 2013. http://www.newstatesman.com/2013/10/science-says-revolt (last accessed 24 July 2014).

Latouche, S. 2009. *Farewell to growth*. Cambridge, UK: Polity.

———. 2012. *La sociedad de la abundancia frugal* [The society of frugal abundance]. Barcelona, Spain: Icaria.

Le Guin, U. 1974. *The dispossessed*. New York: Avon.

———. 1982. *A non-Euclidean view of California as a cold place to be*. Lecture, University of California, San Diego, CA.

———. 1989. *Dancing at the edge of the world: Thoughts on words, women, places*. New York: Grove.

———. 2014. Plenary lecture at Anthropocene: Arts of Living on a Damaged Planet Conference, University of California, Santa Cruz, 8–12 May.

Navarro, V. 2013. Ivan Illich, Serge Latouche, el decrecimiento y el movimiento ecologista [Ivan Illich, Serge Latouche, degrowth and the ecologist movement]. Público.es. http://blogs.publico.es/dominiopublico/7733/ivan-illich-serge-latouche-el-decrecimiento-y-el-movimiento-ecologista/ (last accessed 24 July 2014).

Romano, O. 2012. How to rebuild democracy, re-thinking degrowth. *Futures* 44 (6): 582–89.

Saed. 2012. Introduction to the degrowth symposium. *Capitalism Nature Socialism* 23 (1): 26–29.

Sahlins, M. 1972. *Stone age economics*. Chicago: Aldine.

Schneider, F., G. Kallis, and J. Martinez-Alier. 2010. Crisis or opportunity? Economic degrowth for social equity and ecological sustainability. *Journal of Cleaner Production* 18 (6): 511–18.

Schwartzman, D. 2012. A critique of degrowth and its politics. *Capitalism Nature Socialism* 23 (1): 119–25.

Skidelsky, R., and E. Skidelsky. 2012. *How much is enough? Money and the good life*. London: Penguin.

Somay, B. 2005. From ambiguity to self-reflexivity: Revolutionizing fantasy space. In *The new utopian politics of Ursula K. Le Guin's* The Dispossessed, ed. L. Davis and P. G. Stillman, 223–48. Oxford, UK: Lexington Books.

Stillman, P. G. 2005. The Dispossessed as ecological political theory. In *The new utopian politics of Ursula K. Le Guin's* The Dispossessed, ed. L. Davis and P. G. Stillman, 55–76. Oxford, UK: Lexington Books.

Swyngedouw, E. 2004. Scaled geographies: Nature, place, and the politics of scale. In *Scale and geographic inquiry: Nature, society, and method*, ed. E. Sheppard and B. McMaster, 129–53. Oxford, UK: Blackwell.

———. 2010. Apocalypse forever? Post-political populism and the spectre of climate change. *Theory, Culture and Society* 27 (2–3): 213–32.

Χοβαρδάς, T. 2014. *Ανάπτυξη, αποανάπτυξη και κρίση: μια οικοαριστερή προσέγγιση* [Growth, degrowth and crisis: An ecoleft approach]. In *Οικολογία και αριστερά*, ed. T. Χοβαρδάς, 11–35. Athens: Poulantzas Institute.

On the Possibilities of a Charming Anthropocene

Holly Jean Buck

Department of Development Sociology, Cornell University

The Anthropocene—the geological epoch in which human activities are signaled in Earth's geological records—often appears as an age to be met with grim resignation. Anxiety-driven narratives about this era can translate into very material landscapes of surveillance, tightened borders, farmland acquisitions, and so on, landscapes where speculation shapes lived realities. This article proposes that instead of joining the chorus of dark predictions, or rejecting the flawed concept altogether, geographers are well positioned to experiment with articulating a different Anthropocene. Fragments of a beautiful Anthropocene are already under design: agroecology, green roofs and buildings, distributed renewable energy systems. Yet to weave together a vision compelling enough to provoke cultural and political change, other elements are necessary: a reawakened sense of wonder, an ethic of care, and aesthetic and cultural production around these. This article proposes enchantment as a concept to evoke these elements and discusses the merits and dangers of imagining an enchanted Anthropocene. It looks at emergent alternative framings for thinking about a human-shaped earth and examples of related practices—rewilding, biophilic cities, planetary gardening, smart landscapes—which could make for a more habitable and welcoming epoch.

人类世——这个人类活动在地球的地质纪录上产生信号的地质纪元——经常呈现作为无情地听天由命的时代。此一世纪由焦虑所驱动的叙事，可以转译成监控、强化边界、农地获取等相当物质化的地景，而在此般地景之中，猜疑行塑了生活的现实。本文主张，与其加入悲观的预测行列，或是全然反对具有瑕疵的概念，地理学者位于相当好的位置，对于接合一个不同的人类世进行试验。美好的人类世片断正在着手设计中：农业生态学、绿屋顶与绿建筑、分布式可再生能源系统。但编织一个足以令人信服的愿景以引发文化及政治变革，则同时需要其他元素：重新甦醒的惊奇感受、照护伦理，以及与之相关的美学和文化生产。本文提出魅化做为唤起这些元素的概念，并探讨想像一个魅化的人类世的益处及危险。本文检视想像由人类所形塑的地球的浮现中之另类框架， 以及相关的实践案例——再野化、 亲生物的城市、地球的园艺、智慧地景——这些实践能够创造出更宜居且更欢迎的纪元。

El Antropoceno—la época geológica durante la cual las actividades humanas son señaladas como parte de los registros geológicos de la Tierra—aparece a menudo como una edad abordable con sombría resignación. Las narrativas gestadas dentro de la ansiedad acerca de esta era pueden traducirse en paisajes de vigilancia muy materializados, fronteras endurecidas, adquisiciones de tierras de labranza, y demás, en fin, paisajes donde la especulación configura las realidades vitales. Este artículo propone que en vez de unirnos al coro de predicciones tenebrosas, o de rechazar de plano el concepto plagado de defectos, los geógrafos nos hallamos en una buena posición para experimentar en la articulación de un Antropoceno diferente. Ya se hallan en proceso de diseño fragmentos de un hermoso Antropoceno: la agroecología, techos y edificaciones verdes, sistemas distribuidos de energía renovable. Pero para entretejer una visión lo suficientemente cautivadora, que provoque cambio cultural y político, son necesarios otros elementos: un renacer del sentido del asombro, una ética de la preocupación y, alrededor de estas cosas, una producción estética y cultural. Este artículo propone el encantamiento como un concepto que evoque estos elementos y discuta los méritos y peligros que sobrevendrían de imaginar un Antropoceno encantado. Se mira a esquemas alternativos emergentes para pensar en una tierra humanamente configurada y en ejemplos de prácticas relacionadas—e-naturismo, ciudades biofílicas, jardinería planetaria, paisajes inteligentes—que pueden hacer de esta una época más habitable y acogedora.

The Anthropocene, supposed as our new geological home, is more than a single metaphor or narrative. It is those: a "gloomy metaphoric insistence that people are like forces of geology" (Robbins 2013, 316), or a denouement (Szerszynski 2012, 168), a climax in a tale of becoming. It is more useful, though, to see the Anthropocene as a collection of multiple, related stories, each calling up the reference of another—*People who liked this also read*—the whole narrative assemblage adding up to something more than its pieces. Stories in this Anthropocene anthology are uncanny. Land grabs for palm plantations and chemically treated cornfields, stripped and hydrofracked landscapes, whitened skies from solar radiation management: All of these interreferential horror stories take on new gravity as part of the Anthropocene, or Misanthropocene (Patel 2013), package.

On the flip side of graphic stories are dull tales of detachment, statistics about human appropriation of primary ecological production; for example, where humans become a collectively bland actant. Yet scientific and graphic descriptions can bleed into one another: Witness maps with viscerally red hot spots or descriptions and new vocabulary like Chesworth (2010) employs: "Since the Neolithic, agriculture has become an increasingly powerful forcing factor on processes at the Earth's surface. It attacks the vulnerable skin of the landscape and routinely increases physical and chemical change by one or two orders of magnitude over natural values" (35). Forcing factor, vulnerable skin; "agrobleme," agricultural scar; "anthrobleme," human scar (Chesworth 2010, 35): A graphic and scientific mire emerges. These horror stories are simultaneously *disenchantment* stories, in the Weberian sense. The Anthropocene is told as a sublime yet simultaneously rationalized era.

Critical scholars have identified many limitations of this imagined new era, in both its concept and its telling (Moore 2013; Malm and Hornborg 2014). It is gloomy, it is environmentally deterministic, it flattens, it obscures, and it collapses humans into one species. Chakrabarty (2009, 216) asks:

> Why should one include the poor of the world—whose carbon footprint is small anyway—by use of such all-inclusive terms as *species* or *mankind* when the blame for the current crisis should be squarely laid at the door of the rich nations in the first place and of the richer classes in the poorer ones?

These criticisms and probings are important and well deserved. This article, however, posits that coopting or retelling the Anthropocene might be more useful than arguing against or dismissing it. Already gaining popular resonance and reception, the term's flight beyond geochemistry journals indicates that it provides some function for people. Only after considering what work this word *Anthropocene* could do toward futures we might want should the notion be rejected. The signifier provides a linguistic jolt and further loosens the human–nature binary. It switches the thinker's temporal sense into the geological, offering, as Yusoff (2013) puts it, "a shift in the human timescale from biological life-course to that of epoch and species-life" (784). Offering a name for this unfamiliar time is an important step in recognizing and confronting it. Furthermore, as suggested by Dalby (2013), focusing on the present as the latest geological period—the next phase—suggests a continuity with the past that does not represent "the end times" but, rather, a call to "shape the future in ways other than those suggested by the Pentagon's planners" (191). Hence, here is a question to begin with: If the Anthropocene was not an anthology of scary tales, drawn from an awkward bricolage of science and preternatural fears, what else could it be?

Imaginative Forcings

Whose imaginations do Anthropocene stories originate from? First and most obviously, earth systems scientists give voice and visage to the concept. The humanity-as-earth-moving-agent also places "us" humans in a spectacular position. The current imagined storyline is perhaps the legacy of the Boomer generation, who grew up in an era of polarizing conflict and epic storylines—and who are thus enabled to continue carrying (for a few more years) what Latour (2013, 88) has called "Atlas's malediction," the "weight of the Globe, this strange Western obsession, the true 'White Man's burden.'"

Whose Anthropocene is it not? Who benefits from and is disadvantaged by this version of the Anthropocene? Moore (2013) has suggested that this is the Anthropocene of capital, the Capitalocene. Szerszynski (2012) suggests that "*Homo consumens*, that other-than-human assemblage of humans, technology, fossil fuels and capitalist relations," could be a contestant for "the onomatophore of the Anthropocene" (175). On one hand, it seems true that we are living in the imagination of capital; we can look out the nearest window and see its traces and logic inscribed on just about any landscape. Capital stalks the whole earth, as Smith ([1994] 2008) describes it; "no part of the earth's surface, the atmosphere, the oceans, the geological substratum or the biological superstratum are immune from transformation by capital" (79). On the other hand, we are also living in our imagination of its force, of capital as enchanted and enchanting sublime mover. Our imaginations are necessary to truly animate this force, and we trade away some power in the process. Hence, Bennett (2001) aims "to deny capitalism quite the degree of efficacy and totalizing power that its critics and defenders sometimes attribute to it," asking "Why should one bother to criticize what is inevitable or challenge what is omnipotent?" (115).

As for who benefits from this version of the Anthropocene, the sense of inevitability of geological machinations, as well as the constraints posited in an era of resource scarcity, help land and commodity speculators drum up investment. Stories of inevitability and constraint also aid extraction companies, whose high-capital projects need to construct a somewhat-certain near-term future of high prices to be worth pursuing. The infrastructure involved in these projects is weighty; it hangs around and inflicts some degree of lock-in. So do military technologies, and dark Hobbesian versions of the Anthropocene invite securitization. The "global farms race," for example, alludes to a new kind of securitization and competition but with referent to a Cold War mentality. With regard to either securitization or technological lock-in, accepting humans as rapacious earth eaters leaves scholars adrift in strange discourse coalitions with actors who have a dismal and dangerous politics. The point here is that Anthropocene storylines have fiscal, ecological, psychological, and other practical effects. These imagined futures shape present and future human and nonhuman ecologies, and geographers are well poised to examine them.

The Uses and Abuses of Enchantment

There are motions toward retelling the Anthropocene. The moment or movement variously known as green modernism, postenvironmentalism, or ecopragmatism (Brand 2009) posits that humans have been changing ecosystems for millennia, ecosystems are not static entities, and humanity's reshaping of the environment must be accepted. Steffen, Crutzen, and McNeill (2007, 618) posited a reflexive "third stage" to the Anthropocene, beginning now, when humanity might or might not rise to meet the challenge of being a self-conscious, active agent in its own life support system. Some of the green modernist approaches risk being bound up with the rational trade-offs and disenchantment that comes from quantifying the nature that we want to preserve. For example, Marris's (2011) vision of a "global, half-wild rambunctious garden" begins to shift toward enchantment, invoking energy with a well-chosen adjective, but her articulation gets caught up to some degree in calculating costs. Rather than critique the retelling of others, however, this project is exploratory: How could a better retelling happen?

The suggestion here is that one component of a compelling retelling is enchantment. *Enchantment* is understood here as something akin to Bennett's (2001) notion of a state of wonder, a mood centered around sensuous experience: "To be enchanted is to be struck and shaken by the extraordinary that lives amid the familiar and the everyday" (4). For, as Bennett asks, "What's to love about an alienated existence on a dead planet?" (4). Yet that is what the Anthropocene anthology offers, intimating that by "dominating" the planet, humans have effectively disenchanted it and are also alienated from it: Anthropocene as final blow to an enchanted prior state. There are many thought traps bound up in this idea. "How could we be capable of disenchanting the world, when every day our laboratories and our factories populate the world with hundreds of hybrids stranger than those of the day before?" asks Latour (1993, 115). Bennett (2001) asks, "Why must nature be the exclusive source of enchantment? Can't—don't—numerous human artifacts also fascinate and inspire?" (91).

The process of enchantment is understood here to have two parts: language and practice (or ritual, or performance). Enchantment is drawn from the roots of "enchant" and "charming," *incantare*, to sing, reflecting its linguistic or discursive process. As mentioned, speculative futures are created through word and image, whether this is in advertisements, risk prospectuses, or anti-immigration speeches. This first part, the linguistic, has some obvious dangers. Enchantment in the postmodern context speaks of simulation, of illusions, of cathedrals of capital, of the risks of losing reason or discernment. This provokes this question: Who are the enchanters? Capitalists are not the sole enchanters in the Anthropocene: Anyone can enchant an object, a habitat, a landscape, although people are not generally taught processes by which to do this, and there are not necessarily equal opportunities for these enchantments to blossom into widespread material changes. To be clear, I'm not suggesting going in an illusory direction with this idea. Nor am I advocating a new enchantment or romance with nature-as-object. A better project than reenchanting nature is to enchant humans-in-nature, which is about relationships. Hence, what is needed are practices where relationships can emerge, nonmediated and intimate. In general, the Anthropocene appears to us in mediated forms; one can sense it remotely, track its development, watch its representations evolve in print and on the Internet—but one is not immediately in it, working with it, part of it. The body is a forced temporal migrant within the Anthropocene, but the mind remains outside it, observing. Hence, the invitation here is for Anthropocene as practice, not Anthropocene as a container or setting for experiences. We need to not just retell the Anthropocene but redo it.

Of course, a disenchanted Anthropocene is also constructed through practices, but these happen largely in various states of alienation. The Anthropocene anthology offers the ultimate alienation: You did this and you didn't even know. Both consumers and producers are distant from their actions—and in popular representation, the more alienated a phenomenon is, the more Anthropocene it gets. Gathering firewood seems quite Holocene in its immediate labor. When it comes to Arctic mining or tar sands oil, the product and the processes and practices of getting it are mediated by railroads and pipelines, by the water used in processing, by spot prices and financial instruments, by experts in refining, and so on; this is hypermediation. This level of extraction, of geologic shaping, requires extreme specialization of labor to execute. These are Anthropocene practices available to a select few—those who are trained for it and who are often most immune to the effects of the practice's final waste products. These practices offer their own enchantments—that of the winning financial trade, the new hydrocarbon discovery. The embodied practices I want to focus on here, however, are characterized by two different things. They are (1) immediate, as in nonmediated, and (2) intimate—they open relationships with nonhuman nature.

The proposition here is that relational practices enable enchantment, and this is part of socioecological transformation. The sense of wonder evoked in encounters can lead to an ethic of care and tenderness—or it can lead to revulsion and perhaps action. These are not mutually exclusive directions. As Gibson-Graham (2011) asks: "While we might feel love for other earth creatures and want to accept a responsibility to care for them, might we also extend our love to parasites, or inorganic matter, or to the unpredictability of technical innovation?" (7). Being in relationship, or having enchanted experiences of humans-in-nature, can encompass several affective logics: mourning, comedy, a sense of the uncanny.

Enchantment is no substitute for structural, institutional, and political changes on various scales. Enchantment can be transient, and opportunities for engaging in some of the enchanting practices I mention later are subject to power dynamics. Rather, I think enchantment can enable the passion, care, revulsion, action, networks, sense of place, relationships, and so on that help bring about these socioecological transformations, offering greater momentum for mobilization than pure critique. It is transportive. Catastrophic narratives about the Anthropocene are less likely to motivate action on their own, and the science on climate change communication is applicable here. Evidence indicates that "fear framing" or risk-focused appeals to motivate public support of climate change policies do not work as well as positive, proenvironmental citizenship approaches or approaches emphasizing gains from taking action (Spence and Pidgeon 2010; Bain et al. 2012). As Moser and Dilling (2011) noted, "An excessive focus on negative impacts (i.e., a severe 'diagnosis') without effective emphasis on solutions (a feasible 'treatment') typically results in turning audiences off rather than engaging them more actively" (165). Publics suffering from apocalypse fatigue, or who believe that a grim future is inevitable, might have fewer incentives to do the hard work of socioecological transformation—whereas the immediacy of an enchanted, living, strange planet demands attention.

The second half of this article offers an invitation to stand within an alternative, charming Anthropocene and imagines its characteristics, tensions, and opportunities. In what follows, I offer four openings for how we might think differently about the Anthropocene. The aim is to continue to challenge some prevalent narratives of the Anthropocene by illustrating how things could be different, through referring to specific practices that offer relationship and enchantment.

Futuristically Ancient: Rewilding

First opening toward a charming Anthropocene: Humans have long shaped environments in a variety of ways, and understanding this helps imagine future practices in landscape creation and care. Evidence from paleoecology and environmental history continues to shake narratives of the "ecologically noble savage," the fall from grace that happened with the Industrial Revolution, or the human as necessarily and inherently destructive geological force. Since the late Pleistocene, land use change from hunting, foraging, land clearing, and agriculture has been profound in some regions (Ellis et al. 2013). More than 20 percent of temperate woodlands were "significantly used by 1000 BC and most other biomes by AD 1000," as predicted by a land use model (Ellis et al. 2013, 7980). Anderson's (2005) *Tending the Wild* illustrates how complex management practices on the part of native peoples shaped productive landscapes in what is now California: "Categorizing indigenous peoples as either hunter-gatherer or agriculturalist obscures the ancient roles of wildland managers and limits their use of nature to the two extremes of human intervention" (125). As Brand (2009) appraises Mann's *1491*, "Before the great dying [of Native Americans], the American continent was a managed landscape. . . . Afterward, it was an abandoned garden that the Europeans misinterpreted as wilderness" (239). If previous "wild" landscapes were in fact tended and managed, a future analog could be "rewilded" landscapes.

Rewilding is a framework and practice that holds some epic sway, as it alludes to both past and future: In an enchanted Anthropocene, humans are not reduced to simply removing species but reintroducing them. Proposed by conservation biologists Soulé and Noss (1998), rewilding started with species reintroduction but grew to mean rewilding whole ecosystems. The active "cores, corridors, and carnivores" approach was juxtaposed with biodiversity conservation, which focused on protecting diversity and particular species. Using both scientific and aesthetic justifications, Soulé and Noss argued that "by insuring the viability of large predators, we restore the subjective, emotional essence of 'the wild' or wilderness" (7). In 2005, a group of scientists sketched out an ambitious Pleistocene rewilding plan, which promoted the restoration of large wild vertebrates into North America—horses, Bactrian camels, lions, cheetahs, elephants, giant tortoises—in preference to an impeding landscape "dominated by rats and dandelions" (Donlan et al. 2005, 913). This would change the "underlying premise of conservation biology from managing extinction to actively restoring natural processes" (913). The authors argued that large vertebrate restoration is an ethical responsibility, as humans were at least partially responsible for their extinction. There are also ethical challenges, though, including the ethics of introducing predators, the ethics of introducing animals that might starve to death in the wild, and the critical question of whose land gets rewilded.

Assuming that rewilders could grow to navigate these concerns, the appeal of charming megafauna is obvious. Yet rewilding can be an enchanting practice beyond charming megafauna. Lorimer and Driessen (2013) studied Pleistocene rewilding in Oostvaardersplassen, a Dutch polder reclaimed in the 1950s from the sea. They suggested that the rewilded Heck cattle act as monsters that create "an unruly potential and an affective force" and that the project creates a time and space to engage experimentally with hybrid life forms (257). Monbiot's (2013) book *Feral: Searching for Enchantment on the Frontiers of Rewilding* describes how anarcho-primitivists have applied the concept to human life, imagining a rewilding of people and their cultures. He, too, suggests that rewilding is a reinvolvement with nonhumans and that "the rewilding of both land and sea could produce ecosystems, even in such depleted regions as Britain and northern Europe, as profuse and captivating as those that people now travel halfway around the world to see" (Monboit 2013, 9). Refreshingly candid about his position as a northern citizen who faces "ecological boredom," Monbiot (2013) writes that "our sublimated lives oblige us to invent challenges to replace the horrors of which we have been deprived" (6). Yet although there is a kind of middle-class comfort zone for these longings, dreams, and alliances, this does not mean that they must only dwell there. For one, rewilding visions enable strategic political performance. They can be "suited to the creation of opportunities for alliance with historically colonized places and people to produce what might best be described as *experimental conservation theatre*" (Robbins and Moore 2013, 13). This is not the only mode rewilding can operate in, however, as Lorimer and Driessen's (2013) reporting indicates. Rewilding might have the potential to be participatory practice and "social movement on a grand scale" (Fraser 2009, 250). In other words, rewilding efforts could become tactical performance, a genuine bottom-up social movement, or both.

Art and Craft: Building Biophilic Cities

This is the second proposition of a charming Anthropocene: Art and craft are innately human ways of shaping worlds, and a charming Anthropocene would incorporate these approaches into earth-shaping processes. Many of the words we use to describe human relations with the rest of nature—alteration, intervention, manipulation, artifice—have roots that are less sinister than their connotations: *manipulate*, by hand, human the tool-maker and craftsperson; *artifice* like art and design, *techne*. Loosening the connotations helps us imagine another version of what humans actually do with and in nature. Art and craft can allow for an enchanted, immersive state, and there is a relationship here with design as well, although contemporary design is often professionalized and performed for a client. There is also a relationship with aesthetics, which, as Yusoff (2010, 77) argues, is part of the practice of politics, as well as a space that configures the realm of possibility in politics.

The city is an illustrative site through which we can look at art, craft, and political aesthetics more concretely. The modernist urban planning of the twentieth century disenchanted cities; as Scott (2012) wrote, it "bears more than a family resemblance to scientific forestry and plantation agriculture," with its emphasis on visual order and the segregation of function (41). The contemporary moment, in terms of both culture and technology, however, makes a new conception of the city as integrated habitat possible. The various permutations of phrases involving urban-biodiversity-eco-design-politics point to this. "For the first time in history, an entire city can choose to become the functional urban equivalent of a natural ecosystem," enthused Despommier (2010, 2), who envisions vertical farms with hydroponics, aeroponics, drip irrigation, and advanced LED lighting as keystones of such ecosystems. Another approach is the biophilic city, which emerges from the idea that humans have an innate affiliation with and evolutionary need for contact with nature (Beatley 2011). Aesthetic and cultural elements include green roofing, community forests and orchards, edible landscaping, living courtyards, green utility corridors, pocket parks, vertical gardens, bird-friendly buildings, and so on, which make visible the ecosystems within, and blend art and craft on the part of citizens to form relationships. This is not a city–nature hybrid that mosaics together city and not-city elements; nor is it the vision of green urbanism or environmentally sensitive design, with their emphasis on better transit and building efficiency. Rather, biophilic cities emulate and incorporate natural forms, but also imply an expanded ethic, activities, attitudes, knowledge, institutions, and governance (Beatley 2011)—in short, conviviality. Perhaps a vision for a livable Anthropocene will crystallize around movements focused on the right to enchanting cities and transforming them through politics, art, and craft not into expensively designed green enclaves but into places where encounters happen.

Connection and Care: Planetary Gardening

The third proposal for a charming Anthropocene is a sense of connection that leads to communication and care, placing us in an Anthropocene resembling Berry's (2004, 39) "ecozoic," where "the first principle of this new era is to recognize that the universe is a communion of subjects, not a collection of objects." Connection, communication, and care have sometimes been considered gendered traits, and we can ask what a feminine Anthropocene would look like. Notably, ecofeminist discourses that essentialize connections between women's caring and ecological politics are wrought with ecomaternalism (MacGregor 2006). Furthermore, suggesting that the Anthropocene might be an era brought about by men ignores women's interactions with their environments. How women have shaped the earth over time is understudied. Scharff (2003) points out that although works like mega-dams and skyscrapers stand out, "bigness is no guarantee of ecological significance," and "mistaking size for significance confuses documenting the ways humans have left a mark on nature with Worster's far more ambitious goal of describing *interactions between people and all the other kinds of things on earth*" (10). A feminist retelling of the Anthropocene could begin with studying history to illuminate "how women's actions, desires and choices have shaped the world, including the things men have done" (Scharff 2003, 10). Feminist ecological citizenship, as theorized by MacGregor (2006), suggests a way to not romanticize but politicize the capacity to care for the earth: Care is a "form of work and moral orientation that has been feminized and privatized in Western societies" (7) and must be distributed fairly within and between societies to realize gender equality and sustainability.

The garden is a site through which we can examine connection and care in practice. It is a powerfully enchanting trope: the linguistic enchantment of the garden of love, the walled garden, the secret garden, and so on. The Anthropocene provokes the question of scale: As mentioned, large-scale industrial monocropped landscapes are a referent for Anthropocene horror tales. Planetary gardening imagines something quite different. "The garden is planetary, few can doubt this any longer," wrote Clément (2013), but then the question becomes, "'How does one become the gardener of such a garden?" (266). Perhaps there is a shift from the garden as control to a site of relationship: Braun

(2008) noted that "gardening was earlier an object of scholarly interest for its inscription of ideology onto the landscape," which "has taken on quite different meanings today, as a way to understand how people live in 'passionate, intimate and material relationships with soil, and the grass, plants and trees that take root there' (Hitchings, 2003)" (667). To take the garden beyond its walls, and to a planetary scale, agroforestry and advances in agroecological food production could produce edible landscapes. Popular books celebrate a grassroots movement of "agricultural creatives," where land stewards, food distributors, and "foodshed design teams" reawaken wonder and taste, offering a refute to the supremely disenchanted site of fluorescent-lit supermarket aisles (Cobb 2011, 6). At the same time, these agricultural creatives enchant production. This is not entirely new, especially the romantic strain—witness Thoreau's determination to "know beans" and cultivate a "long acquaintance" with the plants (cited in Marx 1964, 256). Yet a cultivation revolution would have important political and livelihood implications for actual farmers in many places. Agriculture is being reconceptualized as part of a movement to fight corporate interests and loss of control. It is also no longer seen as a strictly rural activity. This everywhere-garden goes beyond the pastoral ideal, loosening those binaries of rural–urban, civilization–wilderness, and simplicity–sophistication, thus offering a means of connection and care for many.

Convergences and Distributed Systems: Smart Landscapes

There is a fourth suggestion for a charming Anthropocene: Whereas disenchanted Anthropocene stories are tales of hierarchical planning and control (or utter chaos), a charming Anthropocene will build on the peer-to-peer, distributed, open-source, rhizomatic notes of our time. Despite the dominance of hierarchical systems in many arenas, increasing attention and enthusiasm is being routed toward peer-to-peer networks, the "sharing economy," and "disruptive" models of distributing goods and services. Distributed food, energy, and information systems allow for more direct and intimate experiences. They can be worked on, tweaked, and customized. Connection could thus be not merely affective but built into the infrastructure of new systems. For example, an outdoor electric meter lacks intimacy, but with rooftop solar panels or neighborhood wind turbines, there is a relationship to develop there: with the weather, with the form. A sunny or windy day has a new importance. Distributed systems can also imply a greater sensitivity, which Latour (2013) sees as "the real meaning of what it is to live in the Anthropocene"—this ability to feel consequences. Distribution can help with what Latour calls "explicitation": "Everything that earlier was merely 'given' becomes 'explicit.' Air, water, land, all of those were present before in the background: now they are explicitated because we slowly come to realize that *they* might disappear—and *we* with them" (Latour 2007, 3). Hierarchical systems increase distance; distributed ones can increase sensitivity and attachment, perhaps opening opportunities for the care described earlier.

Does this desire to invent a name for this new geologic epoch belie an inability to invent a new cultural or economic epoch? "Information Age" seems weakly inadequate as a name for our time. Yet we would not know we were in the Anthropocene without environmental data collection. Environmental informatics will figure into rewilding, smart city creation, smart agriculture, smart grids, and other relational practices. The convergence of the Anthropocene and the Information Age can be seen as offering us new infrastructure projects, and in some ways it makes sense to think of retelling these epochs together, as part of the same endeavor. The basic ecological infrastructure of the planet is under strain and needs care, whether that means cultivation, leaving places alone, rewilding, or crafting and reworking urban ecologies. At the same time, we are building an information infrastructure that interacts with the material world in new ways: sensed cities, mobile devices for DNA barcoding, geotagging, and so on. Big data enthusiasts Mayer-Schonberger and Cukier (2013) declared that with regard to data, "we are in the midst of a great infrastructure project that in some ways rivals those of the past, from Roman aqueducts to the Enlightenment's encyclopedia" (96). This is partly immaterial but not entirely so. The suggestion here is that environmental informatics properly structured can help create awareness of ecological function, and that awareness can aid in enchantment. Smart landscapes are vague and new, and some smart infrastructure is designed to function silently and invisibly, like automatic precision irrigation. But other aspects of it, like wildlife monitoring, or the camera in a local library that streams a hawk nesting in the nearby parking lot live, can increase wonder by raising awareness of the creatures cohabitating nearby, making them familiar parts of the landscape to relate to and perhaps creating demand for urban design that is more wildlife friendly.

Conclusion

Unless we build participatory, experiential infrastructure that offers room for enchantment, a data-driven future of surveillance, disciplinary architecture, and algorithmic decision making seems grim. Power, again, is the crux of the problem: Enchantment is influenced by who is doing the enchanting, designing, making, and relationship building. A relationship built between a subject and the world, through a practice, is a different kind of enchantment than one provoked by external design. Who has the power to experience and proffer enchantments? Brand's (1998) famous line from the *Whole Earth Catalog* is often excerpted in green modernism debates as both mandate and example of hubris—"We *are* as gods and might as well get good at it"—but his continuation is less often quoted. Against "remotely done power and glory" by big business and government obscures, he states that "a realm of intimate, personal power is developing—power of the individual to conduct his own education, find his own inspiration, shape his own environment, and share his adventure with whoever is interested." Infrastructure, cities, wilderness corridors, or gardens that are remotely designed would merely offer new flavors of the same old experiences. The participatory element is key to the whole project. Rewilded, biophilically designed, gardened, or smart landscapes shaped by the people who live in them could offer stronger and more lasting enchantments. As Gibson-Graham (2011) pointed out, there is an ethical project "of actively *connecting* with the more than human, rather than simply *seeing* connection" (5).

Retelling and practicing the Anthropocene asks even more of us than to speak or to act. It asks us to imagine another sort of human, a different character than the rapacious antagonist of the horror stories, who has shaped environments in a variety of ways throughout history. Human traits like tending, altruism, creativity, art and craftsmanship, and cooperation need to reclaim their status as basic human nature, although the competing economic and geopolitical actors of the (mis)Anthropocene minimize them. Imagining another human thus invites us to imagine another human involvement in nature, one that is not managerial or technocratic. The price, though, is giving up stories about calculability or control, as well as stories of despair and tragic guilt, which have a sublime fascination and enchantment of their own. As Baudrillard observes, the "tonality of disenchantment" is *itself* enchanted ([1981] 1994, 162). In some sense, we are trading one sense of enchantment for another, more ambitious sort, as a different human involvement in nature demands new roles, responsibilities, and practices. The stakes, however, are too high not to experiment. We know about sea level rise and ocean acidification and the changing nitrogen cycle, about planetary boundaries and potential tipping points. Enchanting practices are no stand-in for large-scale political change, but as companion to proactive critique, they can help create the critical mass of engagement and care to give humans and nonhumans a habitable Anthropocene.

References

Anderson, M. 2005. *Tending the wild: Native American knowledge and the management of California's natural resources*. Berkeley: University of California Press.

Bain, P., M. Hornsey, R. Bongiorno, and C. Jeffries. 2012. Promoting pro-environmental action in climate change deniers. *Nature Climate Change* 2:600–03.

Baudrillard, J. [1981] 1994. *Simulacra and simulation*. Ann Arbor: University of Michigan Press.

Beatley, T. 2011. *Biophilic cities: Integrating nature into urban design and planning*. Washington, DC: Island.

Bennett, J. 2001. *The enchantment of modern life: Attachments, crossings, and ethics*. Princeton, NJ: Princeton University Press.

Berry, T. 2004. Thomas Berry, interview. In *Listening to the land: Conversations about nature, culture, and eros*, ed. D. Jensen, 35–43. White River Junction, VT: Chelsea Green.

Brand, S. 1998. We are as gods. *Whole Earth Catalog*. http://www.wholeearth.com/issue/1340/article/189/we.are.as.gods (last accessed 30 October 2013).

———. 2009. *Whole Earth discipline*. New York: Viking.

Braun, B. 2008. Environmental issues: Inventive life. *Progress in Human Geography* 32:667.

Chakrabarty, D. 2009. The climate of history: Four theses. *Critical Inquiry* 35:197–222.

Chesworth, W. 2010. Womb, belly and landscape in the Anthropocene. In *Landscapes and societies*, ed. I. P. Martini and W. Chesworth, 25–39. New York: Springer.

Clément, G. 2013. The emergent alternative. In *Architectural theories of the environment: Posthuman territory*, ed. A. L. Harrison, 258–77. London and New York: Routledge.

Cobb, T. D. 2011. *Reclaiming our food*. North Adams, MA: Storey.

Dalby, S. 2013. Biopolitics and climate security in the Anthropocene. *Geoforum* 49:184–92.

Despommier, D. 2010. *The vertical farm*. New York: St. Martin's.

Donlan, J., H. W. Greene, J. Berger, C. E. Bock, Jane H. Bock, D. A. Burney, J. A. Estes, et al. 2005. Re-wilding North America. *Nature* 436:913–14.

Ellis, E., J. Kaplan, D. Fuller, S. Vavrus, K. Goldewijk, and P. Verburg. 2013. Used planet: A global history. *PNAS* 110 (20): 7978–85.

Fraser, C. 2009. *Rewilding the world: Dispatches from the conservation revolution*. New York: Metropolitan.

Gibson-Graham, J. K. 2011. A feminist project of belonging for the Anthropocene. *Gender, Place & Culture* 18 (1): 1–21.

Latour, B. 1993. *We have never been modern*. Cambridge, MA: Harvard University Press.

———. 2007. A plea for earthly sciences. Keynote lecture, British Sociological Association, London.

———. 2013. The Anthropocene and the destruction of nature. Paper presented at the Gifford Lecture Series, University of Edinburgh, 25 February.

Lorimer, J., and C. Driessen. 2013. Bovine biopolitics and the promise of monsters in the rewilding of Heck cattle. *Geoforum* 48:249–59.

MacGregor, S. 2006. *Beyond mothering Earth: Ecological citizenship and the politics of care*. Vancouver, Canada: UBC Press.

Malm, A., and A. Hornborg. 2014. The geology of mankind? A critique of the Anthropocene narrative. *The Anthropocene Review* 1:62–69.

Marris, E. 2011. *Rambunctious garden: Saving nature in a post-wild world*. New York: Bloomsbury.

Marx, L. 1964. *The machine in the garden: Technology and the pastoral ideal in America*. Oxford, UK: Oxford University Press.

Mayer-Schonberger, V., and K. Cukier. 2013. *Big data: A revolution that will transform how we work, live, and think*. New York: Houghton Mifflin.

Monbiot, G. 2013. *Feral: Searching for enchantment on the frontiers of rewilding*. London: Allen Lane.

Moore, J. 2013. Anthropocene or Capitolocene? http://jasonwmoore.wordpress.com/2013/05/13/anthropocene-or-capitalocene/ (last accessed 30 October 2014).

Moser, S., and R. Dilling. 2011. Communicating climate change: Closing the science-action gap. In *The Oxford handbook of climate change and society*, ed. J. Dryzek, R. Norgaard, and D. Scholsberg, 161–74. Oxford, UK: Oxford University Press.

Patel, R. 2013. Misanthropocene? *Earth Island Journal Spring*. http://www.earthisland.org/journal/index.php/eij/article/misanthropocene/ (last accessed 30 October 2013).

Robbins, P. 2013. Choosing metaphors for the Anthropocene: Cultural and political ecologies. In *The Wiley-Blackwell companion to cultural geography*, ed. N. C. Johnson, R. H. Schein, and J. Winders, 307–19. New York: Wiley.

Robbins, P., and S. Moore. 2013. Ecological anxiety disorder: Diagnosing the politics of the Anthropocene. *Cultural Geographies* 20 (1): 3–19.

Scharff, V. J. 2003. *Seeing nature through gender*. Lawrence: Kansas University Press.

Scott, J. 2012. *Two cheers for anarchism*. Princeton, NJ: Princeton University Press.

Smith, N. [1994] 2008. *Uneven development: Nature, capital, and the production of space*. Athens: University of Georgia Press.

Soulé, M., and R. Noss. 1998. Rewilding and biodiversity: Complementary goals for continental conservation. *Wild Earth* Fall: 18–28.

Spence, A., and N. Pidgeon. 2010. Framing and communicating climate change: The effects of distance and outcome frame manipulations. *Global Environmental Change* 20 (4): 656–67.

Steffen, W., P. Crutzen, and J. McNeill. 2007. The Anthropocene: Are humans now overwhelming the great forces of nature? *Ambio* 36 (8): 614–21.

Szerszynski, B. 2012. The end of the end of nature: The Anthropocene and the fate of the human. *The Oxford Literary Review* 34 (2): 165–84.

Yusoff, K. 2010. Biopolitical economies and the political aesthetics of climate change. *Theory Culture Society* 27 (2–3): 73–99.

———. 2013. Geologic life: Prehistory, climate, futures in the Anthropocene. *Environment and Planning D: Society and Space* 31 (5): 779–95.

Banking Spatially on the Future: Capital Switching, Infrastructure, and the Ecological Fix

Noel Castree* and Brett Christophers†

*Department of Geography & Sustainable Communities, University of Wollongong, and Geography, SEED, Manchester University
†Institute for Housing and Urban Research, Uppsala University

Since the onset of the global economic crisis, financiers and the institutions regulating their behavior have been subject to far-reaching criticism. At the same time, leading geo-scientists have been insisting that future environmental change might be far more profound than previously anticipated. Finance capital has long been a crucial mechanism for melting present solidities into air to create different futures. This article asks what the prospects are for the switching of credit money into green infrastructures—a switching increasingly recognized as necessary for climate change mitigation and (especially) adaptation. Most research into geographies of finance has ignored ecological questions and few contemporary society–nature researchers examine major fixed-capital investments. Unlike those geographers who criticize capitalism without offering feasible alternatives, we take a pragmatic view underpinned by democratic socioenvironmental values and attempt to identify leverage points for meaningful change. This programmatic article identifies reasons and examples to be cautiously hopeful that liquidity can be fixed in less ecologically harmful future infrastructures, thereby addressing crucial extraeconomic challenges for the century ahead.

自从全球经济危机爆发以来，金融家、以及管理其行为的机构，受到了广泛的批判。于此同时，引领潮流的地理科学家，亦坚称未来的环境变迁，有可能较过去所预测的更为深刻。长久以来，金融资本一直是消解当前实存之物、 使之灰飞烟灭， 以创不同的未来之关键机制。 本文探问将信用货币转移至绿色(环保) 基础建设的展望——此一转换，逐渐被视为减缓与 (特别是) 调适气候变迁的必要措施。诸多探究金融地理的研究， 忽略了生态问题， 而鲜少有当代的社会—自然研究者， 检视主要的固定资本投资。不同于批评资本主义、却未能提出任何实际替代方案的地理学者，我们採取了由民主社会环境价值所支撑的务实视角，并企图指认有意义的变化的转捩点。此一计划性的文章，指认原因与案例，以谨慎地期望流动资产可被固着于生态上较不具破坏性的未来建设，以此处理未来一世纪中，经济之外的关键挑战。

Desde el comienzo de la crisis económica global, los financistas y las instituciones que regulan sus actividades han estado sometidos a críticas sostenidas. Al mismo tiempo, los geocientíficos de avanzada siguen insistiendo en que el anunciado cambio ambiental podría ser mucho más profundo de lo anticipado. Durante mucho tiempo, para crear futuros diferentes, el capital financiero ha funcionado como mecanismo crucial para derretir las solideces del presente. Este artículo se pregunta cuáles son las posibilidades de que el dinero de crédito pueda transformarse en infraestructuras verdes—un cambio crecientemente reputado como necesario para la mitigación del cambio climático y para la adaptación (especialmente). La mayor parte de la investigación en geografías de las finanzas ha ignorado cuestiones ecológicas y pocos son los investigadores contemporáneos interesados en la relación sociedad–naturaleza que se ocupan en examinar en su contexto inversiones mayores en capital fijo. A diferencia de geógrafos que critican el capitalismo sin ofrecer alternativas factibles, nosotros adoptamos una visión pragmática apuntalada en valores socioambientales democráticos e intentamos identificar puntos de apoyo para el cambio significativo. Este artículo programático identifica razones y ejemplos que generen esperanza cautelosa porque la liquidez pueda afincarse en futuras infraestructuras menos dañinas ecológicamente, abordando así retos extraeconómicos cruciales para el siglo en curso.

Given the inherent unknowability of the future, foreseeing its socioecological parameters and possibilities is a challenging task at the best of times. Notwithstanding all of the future uncertainties with we which we must deal, however, two things are increasingly clear. The first is that anthropogenic climate change is occurring and to a degree likely to prove game changing for most countries worldwide. The second is that such change will necessitate significant, proactive socioeconomic reconfigurations that both respect and yet also help to write the new rules of the ecological game. One such vector of reconfiguration, and the one considered in this article, pertains to the built environment—that is, those long-lasting, expensive-to-construct physical infrastructures that both enable and constrain the quotidian activities involved in reproducing social life.

In what specific senses does anthropogenic climate change call for such reconfiguration? For several years analysts and policymakers have talked incessantly about mitigation and adaptation strategies. With respect to infrastructural changes, Pacala and Socolow (2004) have argued influentially that a rapid reconfiguration of the existing built environment—prioritizing "efficient buildings"—can contribute substantially to limiting future atmospheric warming. Given the relative failure of most national governments to pursue mitigation strategies aggressively, however, adaptation is now at least as vital thinking ahead. The recent (fifth) assessment report of the Intergovernmental Panel on Climate Change (2014) indicates with high certainty that an average atmospheric temperature rise of as much as 4°C by century's end is possible. In this light, it is hard to disagree with Sayre's (2010) contention that "a rapid and comprehensive reconfiguration of the built environment is imperative if we are to . . . adapt to global warming" (95). Sayre then goes on to identify what we agree is a pivotal question: that of how, and by whom, a comprehensive remaking of built environments might be financed.

Drawing on Harvey's (1982) seminal work on the distinctive political economy of major fixed-capital investment, Sayre (2010) reminds us that the built environment "depends heavily on financial instruments and institutions that permit large-scale borrowing and long-term amortization" (102). Recognizing such economic distinctiveness, our article's contribution to the imagination of socioecological futures is to explore how, in principle, the financing of major new infrastructure might work and to consider how likely such financing is to eventuate soon. Notwithstanding the tarnished image financiers have because of the global financial crisis and its fallout (plus various high-profile scandals), their decisions will be crucial in determining the road humanity will travel in the decades to come. There is a need for analysts and policymakers not only to understand these decisions but to shape them to realize important noneconomic goals. In effect, the financial sector is an unelected government whose power is such that it needs to be carefully governed through a set of endogenous and exogenous norms, rules, and institutions. With respect to infrastructure, one might say that (some) financiers are the ultimate geographers, literally graphing the face of the Earth as they translate liquidity into enduring fixed assets essential to our shared future. To use another metaphor, in designing the arteries through which future capital will flow, these financiers stand to determine the future health not only of the economy but of the bodies social and ecological too.

Normatively, and contrary to much geographical research into human–environment relationships, our explicit aim is to offer an even-handed appraisal of the possibilities capitalism offers for enabling humanity—or significant parts of humanity—to cope with what is likely to be an ecologically challenging future. Consider here differing uses of the word *fix*, which completes our article's title. For many critical geographers (and we consider ourselves such), the word usually has negative, even pejorative, connotations over and above its explanatory meanings. In the context of our current political economy, a fix is typically seen as capitalism trying to negotiate its inherent crisis tendencies to reproduce itself in perennially iniquitous forms. Thus, for Swyngedouw (2010, 222), capitalist enthusiasm for eco-technologies like smart energy meters is essentially about "producing a socio-ecological fix to make sure nothing really changes. Stabilizing the climate seems to be a condition for capitalist life as we know it to continue." We do not diminish such concerns. But here we use the word *fix* differently to denote the possibility of an ecologically and socially progressive reconfiguration of existing built environments. Such a fix can never be problem-free, and what counts as progressive is relative to be sure. Moreover, naïve optimism should never substitute for reasoned and realistic hope that certain specific varieties of capitalism can deliver wider socioecological benefits for more than just a few. But equally, as Gibson-Graham has argued consistently (e.g., Gibson-Graham 2008), such hope is absent when critics see capitalism as little more than a malign and impregnable force that disadvantages the majority of humanity. In sum, this is a programmatic article intended to highlight real-world issues pertaining to future infrastructure financing and to inspire thought among geographers about how they choose to examine these issues looking ahead.

Anthropogenic Climate Change and Capital Switching

In 2012 the British government, in the shape of the Department for Environment, Food & Rural Affairs, published the first of its new five-yearly Climate Change Risk Assessment reports (Defra 2012). Among other things, the document was notable for highlighting perceived "priorities for adaptation" under five headings, one of which was Buildings & Infrastructure. This identification of the need for built environment renewal reflects increasingly widespread consensus, not only in the United Kingdom but internationally, regarding the centrality of infrastructure upgrade to what Defra calls "climate resilience" (flooding and overheating of buildings being seen as the most significant risks in the British case). But there is also, as already intimated, a perceived role in climate change mitigation. Defra's conviction that building a "new low carbon infrastructure" that can assist in "the transition to a low carbon economy" is underpinned by the fact that, in Sayre's (2010) words, the existing built environment "mediates economic production, exchange and consumption in ways that both presuppose and reinforce high rates of greenhouse gas emissions" (95).

Given the growing acceptance that our infrastructures need to be substantially redesigned and renewed, planners and other experts are turning their attention to the critical question of what more resilient and energy-efficient ones might look like. This is not the place to consider this question in any detail. Suffice to say that the proposals that are now beginning to emerge share two main attributes. First, there is a strong emphasis on coastal regions, where the impact of rising sea levels will be most pronounced. Second, there is an emphasis on cities, reflecting the fact that the world's built environment is primarily urban. Stone's (2012) *The City and the Coming Climate* is a good example: It makes concrete proposals for new energy, residential, and transport infrastructures under the "coming" climate and associated physical geography.

Alongside such envisionings of future built environments, we are also seeing tentative appraisals of the potential costs thereof. Such costs, needless to say, will be colossal, and their quantification at this juncture is challenging. For example, a recent report (Hallegatte et al. 2013), assuming a mean sea level rise of 0.2 to 0.4 m by 2050 and seeking to quantify the impact of flooding on the world's 136 largest coastal cities, estimated necessary infrastructure adaptation costs at a cumulative US$50 billion per annum. The authors stress, however, that although such figures might sound high, they are "far below [their] estimate of aggregate damage losses per year in the absence of adaptation" (Hallegatte et al. 2013, 805). This qualifier echoes the now famous conclusion of the *Stern Review* (Stern 2006) that the relative economic costs of inaction vastly outweigh those of acting meaningfully now.

All of this tells us that the redesigning and rebuilding of infrastructures will require a long-lasting diversion of financial investment from existing uses. Global society, in other words, will have to effect a massive *capital switch*. Harvey (1978) conceptualized this as the process whereby investment is rerouted from one circuit of capital to another. He was most interested in the case where the switch is from the primary circuit of productive capital (investment in the wage-labor–based production of goods and services for sale on the market) to the secondary circuit of investment in the (re)construction of built environments (e.g., roads and factories). This begins to explain the appeal of the capital switch optic for us here. It refers to large-scale, temporally concentrated diversions of investment that serve to alter systematically the historically contingent forms that capitalism assumes. It is just such diversions that are pivotal to global attempts to cope with future climate change.

Such switches have certainly happened before—indeed, Harvey regards them as part of capitalism's metaphorical DNA. As such, the general dynamics of the ecological switch envisioned in this article would not be unprecedented, even if its magnitude might be. Such a switch would involve protecting existing natural capital (e.g., large areas of forest, today reframed as providers of ecosystem services), creating green infrastructure en masse, and fabricating built environments designed to cope with higher sea levels, warmer temperatures, and so on. Yet, as theorized by Harvey and examined by scholars using his theoretical lens, capital switching is ultimately a function of short-term capitalist profit-seeking. It is, at heart, a response to an immediate crisis of economic capital (specifically, a crisis of overaccumulation and thus of surplus capital in search of a home). At best, it also provides a temporary fix to the crisis tendencies necessitating it. This being so, how can we foresee a switch necessitated in large part for ecological reasons and with the long-term viability of biophysical goods and services as a key motivation? If not from an acute dearth of current opportunities for profitable investment in the productive circuit, and thus for ongoing economic growth, where will the impetus to switching come from (and over what timescale)? Additionally, how enduring will the infrastructural products of such a switch, should it and they eventuate, prove to be?

We do not pretend to know the answers. On several accounts, though, the theory and historical reality of capital switching suggest that the fix envisioned here is not unimaginable. First, if Harvey is right, capital switching occurs necessarily—it happens regardless of the extra-economic reasons that may be used to encourage, or justify, such a switch. Second, when extra-economic factors come to have serious economic consequences, they become internalized within the logic of capitalism. To quote Sayre (2010) once more on infrastructure, "By a fortuitous paradox, these [fixed capital] investments are threatened with devaluation whether or not we act to stabilize atmospheric GHG [green house gas] concentrations; in highly uneven, unpredictable, and potentially abrupt ways, global warming will make our current built environment increasingly untenable and uneconomical" (95). This is a reminder that capital functions in a world that never was, nor will ever be, designed to meet the changing needs of this mode of production. Third, financial institutions have long been accustomed to translating short-term money-making opportunities into the sort of long-term commitments needed to build and maintain something like a nuclear power facility. Given the right incentives by governments, financiers are well capable of crafting investment vehicles to help decarbonize the world's current infrastructural assets.

Fourth, and relatedly, although global capitalism remains broadly neoliberal, national governments retain the power to radically alter the regulatory conditions that define rational and permissible financial activities. They also, of course, can be direct participants in a capital switch by choosing to make large public investments. Even "night watchman" states have some more than nominal sense of the public interest, not least because civil society actors force them to look beyond the immediate needs of business. In the present case, an increasingly large and vocal community of geoscientists is urging governments to take global environmental change far more seriously than heretofore. Alarmed by the so-far weak attempts to decarbonize the world economy, these geoscientists are sounding a drumbeat of alarm intended to be heard by leading politicians (see, e.g., Nobel Laureates 2011). Their concerns are echoed by leading nongovernmental organizations, foundations, and charities and are part of a wider questioning of contemporary capitalism that precedes the recent financial meltdown. The fact that none of this is at all new—indeed, it is all too familiar—helps to create the material preconditions for the sort of policy shift that will be one "form of appearance" of the causal necessity Harvey's switching idea identifies. In short, if capital switching can never be a purely economic affair, there are general reasons to believe causal necessity and contingent causality might fortuitously combine to reconfigure capitalism's operating "hardware" along more ecofriendly lines. These reasons might not be sufficient to win any argument on the subject, but neither should they be dismissed.

Of course, should such a reconfiguration occur, two important distributional questions arise. First, because the scope and scale of contemporary capitalism is unprecedented, the infrastructural transformation being envisaged here could not simply be regional in scale. It would have to be global, raising this question: How feasible is this? Second, as several commentators have shown (e.g., Graham and Marvin 2001), many contemporary infrastructures offer declining "public goods" functions as they are increasingly targeted to privileged social groups. In this regard it is clearly not hard to imagine a bleak future in which new ecologically resilient infrastructures serve—and generate economic growth for—only those few who can pay for them. This therefore begs a further question: Can a different future eventuate?

Again, there are general reasons to believe that this question can be answered in the affirmative. First, unless capitalism were to deglobalize through force of circumstance, its future trajectory virtually requires the coincident alteration of infrastructures across continental boundaries. The time–space compression that is one of capitalism's hallmarks typically forges built environments conducive to commodity circulation on the largest possible scale at any one historical moment. Barriers to circulation are typically overcome because they are both opportunities for and drags on productive investment. Second, although in many places future infrastructures will likely not bring benefits to a great many citizens (see the next section for more on this), in many other cases a grassroots Polanyian countermovement against such exclusion is conceivable. In still other cases, nonneoliberal governments might have strong public interest agendas; we should remember, too, that capital ultimately suffers (indeed, faces the crises that trigger switches) if too few people consume the things it commodifies.

Let us now move away from these general considerations toward a more focused examination of how ripe finance capital is to effect the switch we have been discussing. The central investment issues, as we see them, are threefold and intimately connected. The remainder of the article proceeds with this generic outlining of future financing possibilities as background:

1. *Source of funds*. Here the question is a nominally simple one: Who would pay? Although funding could, of course, come from many different sources, the key generic distinction is between private and public finance. Would the rebuilding of our built environments be financed by private, commercial interests or by the state (and thus, ultimately, various taxpayers)? Although some of the new infrastructure would be privately owned and revenue generative, much of it—transportation infrastructures (without tolls or other user fees), coastal defenses, and so forth—could not and would not be. In the latter case, it is more difficult—although clearly not impossible—to envision a significant role for private capital.
2. *Nature of funds*. Would the capital being invested be cash or credit? That is to say, would it be derived from actually existing financial assets without matching liabilities, or would the funding entity be required to borrow to make the investment? An example of the former would be a pension fund or sovereign wealth fund (SWF). An example of the latter, at least in most Western countries, would be the state—as we are only too well aware, few Western governments today have the luxury of spending without borrowing first. Borrowing is likely to be essential to one degree or another.
3. *Form of investment of funds*. Where the infrastructure investor and owner are not one and the same, would capital be invested as debt or as equity? To put it another way, to what types of returns would the investor be entitled—interest and principal repayments (debt) or payments tied to the market value of the investment (equity)? The latter is only conceivable in instances where ownership of the infrastructure is wholly or partly private and thus would be likely to be relatively limited. This has significant implications, with debt and equity investments typically characterized by different risk profiles and attractive to investors with different risk appetites.

Barriers to Realization

For all our desire to present a balanced outlook on the need and possibility of effecting an ecological fix underpinned by new societal infrastructures, we are mindful of the necessity to acknowledge the many reasons to be skeptical. In this section we enumerate some of the more important of these. To be clear, our concern here is not with technical barriers but with the obstacles that existing configurations of finance capital pose and with ominous ongoing dynamics in and around current (financial) political economy. We identify five related sets of challenges in this sphere:

1. Given that much of the new infrastructure will not be designed to be revenue generative, and will be publicly owned and controlled, it is difficult to imagine that it can be financed without a significant contribution from public sources. Yet we stand at a moment in history where, certainly in the Global North, public finances are under—or, at least, are seen and asserted to be under—exceptional pressure. The timing, in other words, is anything but propitious: In a milieu where the political and public appetite for taking on more long-term public debt is virtually nonexistent and, moreover, where the appetite for higher taxes (the other "unmentionable" mechanism for raising funds) is equally limited, it is hard to see public financing as solving the problem *in toto* or even in large part. Witness, for instance, the situation in the United States, where plans for a national infrastructure bank using federal funds remain stalled despite backing from President Barack Obama and numerous influential senators.
2. There is also something of a philosophical—as opposed to a political–economic—barrier to the notion of public financing of major infrastructure renewal. As O'Neill (2013) has shown, a major trend of the late twentieth and early twenty-first centuries has been toward privatization of the ownership, financing, and operation of infrastructure. In the resulting ideological-discursive environment, where "state responsibility for infrastructure and public acceptance for its finance and delivery" are "no longer intrinsic to Western capitalist ideology" (O'Neill 2013, 3) in the way they once were, the challenges confronting future public financing of built environment reconstruction loom larger still.

3. If the necessary means to finance reconstruction publicly are neither readily available nor perhaps readily justifiable, a different set of obstacles threatens the viability of private financing. Perhaps the biggest concerns the question of whether investment in such reconstruction would be deemed sufficiently attractive (i.e., profitable). The fact that much of the new infrastructure would not be revenue generative is certainly part of the problem, although private sources could, in theory, fund public, non-revenue-generative infrastructure if, for example, invested as debt rather than equity. Yet two hurdles would still remain. First, as Clark et al. (2012) have noted, short-termism continues to dominate institutional investment, even among investors who can (as infrastructure requires) invest over intergenerational time spans. Second, private investment of all types remains, for the most part, profit-maximizing above other considerations, whereas investment in public infrastructures with a climate change mitigation and adaptation remit would, by its very nature, need to be constituted as low risk and low reward. This tension is already writ large in the fact that private, ecologically oriented infrastructure financing has yet to substantively materialize. As Zadek (2013) notes, if a significant switch is not happening now—"mechanisms like the Green Climate Fund are struggling to get off the ground, and have yet to offer a vision for unlocking capital at scale," as major institutional investors continue to invest heavily in nonrenewable fossil fuel industries, remaining, collectively, "resolutely brown and dirty"—even with "the low cost of capital and the need to stimulate the global economy," it is arguably hard to see when it will.
4. The lack of substantive progress on the private financing front is probably also explicable in part by the fourth important obstacle requiring recognition: namely, active resistance (broadly defined). Sayre (2010) regards one form thereof as particularly important: resistance from existing owners of current built environments to the devaluation that would necessarily accompany such capital's redesign, disuse, and rebuilding. (Although, as Sayre also remarks, this capital will be devalued regardless, by climate change, if not by proactive social decision making; but who says the owners of capital are rational?). Resistance also emanates from other influential sources. One is the oil and gas sector, with its entrenched power. Another, equally powerful, is the financial services sector. As Zadek (2013) explains, a prerequisite for major long-term private financing of built environment reconstruction is meaningful reform in and reregulation of financial markets. Such reforms are not even close to happening, though, stymied as regulators are by "a fierce headwind of lobbying by financial firms" deeply invested in existing institutional configurations of risk and reward.
5. Last but not least, there is a pivotal geographical dimension to the financial challenge that lies ahead and one closely bound up in complex ways with the distributional issues identified in the previous section. Put simply, there is marked geographical unevenness internationally both in the need to upgrade infrastructures and in the financial wherewithal to do so, rendering the geopolitical and geofinancial consensus and collectivity that is likely to be a necessary condition of success all the more difficult to achieve. Climate change, as Davis (2010, 37) writes, "will produce dramatically unequal impacts across regions," in the process "inflicting the greatest damage upon poor countries with the fewest resources for meaningful adaptation" and in a way that may "undermine ... pro-active solidarity." Put this concern alongside the more pointed fear that any climate change–related financial transfer from Global North to South "would be so firmly ring-fenced with conditionalities that it would auction away the sovereignty of African nations at the altar of 'Green Capitalism'" (Tandon 2011, 141), and the daunting difficulties of achieving a global capital-switch-cum-ecological-fix are clear.

Grounds for Cautious Optimism

With all of this said, it is not in our view sufficient—conceptually, empirically, or politically—to leave things there: to bemoan, effectively, that capital is capital and the future thus constrained by the inertia built into the current historical–geographic conjuncture. Amidst the undoubted grounds for skepticism, we see three main reasons for believing that finance capital can potentially facilitate the type of investment and rebuilding that is needed in the near future. These particular reasons can be added to the four very general ones described earlier in the article.

First, and in many ways least significantly, there are signs of potentially promising developments on the ground. In its recent report on cities and what it calls "infrastructure transitions," the United Nations' International Resource Panel (IRP 2013) included thirty case studies of "innovative approaches to sustainable infrastructure change across a broad range of urban contexts." Wholesale infrastructure renewal is not just a theory, in other words. One city often mentioned in this context, by way of example, is Chicago. The Chicago Infrastructure Trust (CIT) was established in 2012 to incentivize both debt- and equity-based private financing of a range of infrastructure projects. Chicago is also the leader of the international Sustainable Infrastructure Finance Network, which is a collective initiative, also launched in early 2012, by the so-called C40 group of cities.

Although there does not appear to be anything particularly revolutionary about the financing models used in the aforementioned cases—Citibank and JP Morgan are both actively involved in the CIT—it is vital to appreciate, second, that different types of financial entities, with the potential to embrace different financing rationales, do exist and are, in many cases, extremely well capitalized. "Insurance companies, pension funds, sovereign wealth funds, endowments, foundations and family offices," note Clark et al. (2012, 104), "all have the ability to invest over inter-generational spans" and in "long-term, illiquid assets"—in, most notably, infrastructure assets. Such investors—which Oulton (2012), in the context of the funding of a low-carbon future, describes collectively as constituting "patient capital"—were reported as having had $27 trillion in assets under management in 2011 (Clark et al. 2012, 104).

SWFs appear to represent an especially encouraging investor type. For one thing, they occupy a "unique position" among investors insofar as they are "not constrained by liabilities" or "subject to increasing solvency requirements." Furthermore, they typically have "greater discretion over tactical and strategic asset allocation," which explains why they are often described as "unconstrained investors" (Clark, Dixon, and Monk 2013, 8–9). Of special interest in relation to our own hypothesized capital switch and the alternative rewards profile that financing thereof would be likely to exhibit—lower financial returns but higher social-collectivist, nonmonetary ones—is the moralist SWF subcategory identified by Clark, Dixon, and Monk (2013). Such funds include, inter alia, Australia's Future Fund and Norway's Government Pension Fund-Global. The former has an obligation to ensure that its investments are sustainable in the sense of not harming the prospects of future generations; the latter represents "an expression of Norway's commitment to global justice" (64, 68). Both would seem, at least from the outside, possible participants in a broadly based ecological-fix financing initiative.

Third, finally, and perhaps most significantly of all, it is possible to identify precedents for socially progressive, large-scale, infrastructure-oriented capital switches. In other words, it has—sort of—been done before, even if not on the same scale, and even if not with specifically environmental ends. One such precedent is discussed briefly now; we think others are equally pertinent, but space constraints preclude their elaboration.[1] Our aim here is not to advance particular financing models to be replicated but to highlight a historical–geographical achievement—of political will as much as financial innovation—in the face of socioeconomic need from which we can hopefully learn.

This particular example is somewhat ironic in this context, given the environmental implications it ultimately had, but it is instructive nonetheless. It is a case very much of market failure. When the mass-market provision of electricity was rolled out in the United States in the 1910s and 1920s, rural areas were largely neglected: The economics of building out infrastructure to dwellings and farms not clustered together could not be made to work from the service providers' perspective, meaning that only those end users able to advance the necessary financing themselves were initially connected. The result was that by 1930 only approximately 10 percent of U.S. farms had electricity, whereas rural electrification levels had reached in excess of 50 percent in countries including Czechoslovakia, France, and Germany (Brown 1980).

The Great Depression put ambitions for rural electrification on hold, but from 1934 renewed impetus was provided by two men in particular: President Franklin D. Roosevelt (who had been an outspoken advocate of rural electrification as Governor of New York) and the engineer Morris Cooke, both of whom recognized the importance of electrification to rural economic and social life. The latter was the author of a hugely influential report that estimated the cost of comprehensive rural electrification and made proposals for how it could be financed—a report that

Cooke later referred to as "the detonating force which started rural electrification" (Brown 1980, 42).

In 1935, the Rural Electrification Administration (REA) was established. In time, the specific financing model used would come to be seen as critical to the program's success. The REA was, effectively, a funding agency. It did not procure and install infrastructure itself—instead, it provided loans to those who did, among whom were both private companies and, most significantly, cooperatives, which were formed (on a strictly not-for-profit basis) by rural residents to build distribution systems and provide their own electrical services. The REA's loans, toward which Congress in 1936 made available a maximum of $410 million over ten years, featured three key (connected) components: federal subsidization; low (since subsidized) interest rates; and repayment schedules designed to match, in temporal terms, the time frame (decades rather than years or months) over which the "returns" on investment would be realized by borrowers.

What was the outcome of the REA? As early as 1939, it had 417 cooperatives serving 268,000 rural households. After this initial burst of borrowing and investment, the pace of uptake slackened off during wartime before then picking up again from 1945. By 1953, "over 2.5 million farms were connected to REA and for all practical purposes, all American farmers had electricity, whether through REA or otherwise" (Schurr 1990, 234; and Brown 1980, x), bringing the story up to the end of the 1970s, reported a default rate on REA loans since the agency's genesis of less than 1 percent.

Conclusion: From Bad Debt to Healthy Credit, From Present Liquidity to Future Solidities

It must be admitted that now is a particularly challenging time to be arguing for the positive potential of finance capital in making alternative socioecological futures. It goes against the grain of the bulk of writing on environmental questions in critical human–environment geography. Moreover, debt has gotten a particularly bad name in the past five years in governmental and public spheres.

But finance capital, including credit creation, will be crucial to enabling us to adapt to a changing biophysical world and to mitigating some of that change. In particular, and in line with the focus of this article, built environments will need to be substantially reformatted and constituted afresh. Unless capitalism is replaced by another mode of production altogether (and, on some accounts, even if it is), this cannot be done without putting finance capital to work. Granted, such capital is in many of its forms centrally bound up with capitalism's most exploitative and ecologically harmful circuits. We refuse to accept, however, that it must always be so or at least to the degree some critics suggest. We acknowledge that there are significant obstacles to an ecologically and socially progressive mobilization in the service of the capital switch envisioned here, yet we see signs of hope, not least in historical (and contemporary) examples of finance being put to extraeconomic ends.

The often sweeping condemnation of finance capital, and debt in particular, therefore needs revisiting. In recent years, few books have done more to give debt its bad name than Graeber's (2011) eponymous history. As such, we conclude by citing from a review of this weighty monograph that, although recognizing its many merits, calls for research and thinking about "healthy credit" alongside "bad debt." It is a sentiment, in the context of socioecological futures, with which we fully concur:

> This hefty treatise on bad debt shows the need for further work on healthy credit. . . . [We can denounce] capitalist "gambling." Yet all investment in future facilities involves an element of uncertainty and a claim on resources. Governments can conjure the needed money out of thin air if those investments turn out to meet a genuine and effective social need. Capitalists are often not very good at spotting and sponsoring necessary social innovations, especially those which require large-scale and complementary infrastructure. Public enterprise—carried out on a broad and varied canvas—is the vital missing ingredient in a world dogged by indebtedness, weak demand, climate catastrophe, poverty, crumbling infrastructure and counter-productive austerity. (Blackburn 2013, 150)

In this light, we might hope that leading governments that help finance capital realize its profound potential to remake the arteries through which capital flows and that are the lifeblood of the biological and social reproduction of most of contemporary humanity. Indeed, we need far-sighted state action because, in most parts of the world, we lack collective agents capable of forcing the hand of (often unwilling or hamstrung) governments. We might hope, too, that more social scientists, including many human geographers, can be generative of transformative ideas, evidence, and proposals geared to creating a future we would like our successors to inherit. To help change the world for the better we might, as O'Brien (2013) recently opined, have to change our own modus operandi. A critical geography of finance capital and infrastructural transformation clearly cannot and should not be a Polyanna, but neither can it afford to remain analytically skeptical or normatively gloomy about the prospects for a greener and more socially just capitalism.

Note

1. An important contemporary example of a substantial capital switch geared to financing infrastructure with a social orientation (broadly defined), for instance, is that of European social housing, where in many countries innovative financing methods (e.g., the government acting as loan guarantor) have partially arrested the decline in social housing construction widely associated with the state's withdrawal from direct provision and with housing associations' growing reliance on otherwise expensive private financing.

References

Blackburn, R. 2013. Finance for anarchists. *New Left Review* 79:141–50.

Brown, D. 1980. *Electricity for rural America: The fight for REA*. Westport, CT: Greenwood.

Clark, G., A. Dixon, and A. Monk. 2013. *Sovereign wealth funds: Legitimacy, governance, and global power*. Princeton, NJ: Princeton University Press.

Clark, G., A. Monk, R. Orr, and W. Scott. 2012. The new era of infrastructure investing. *Pensions* 17 (2): 103–11.

Davis, M. 2010. Who will build the ark? *New Left Review* 61:29–46.

Defra. 2012. UK climate change risk assessment: Government report. https://www.gov.uk/government/uploads/system/uploads/attachment_data/file/69487/pb13698-climate-risk-assessment.pdf (last accessed 30 April 2014).

Gibson-Graham, J.-K. 2008. Diverse economies. *Progress in Human Geography* 32:613–32.

Graeber, D. 2011. *Debt: The first 5,000 years*. New York: Melville House.

Graham, S., and S. Marvin. 2001. *Splintering urbanism: Networked infrastructures, technological mobilities and the urban condition*. London and New York: Routledge.

Hallegatte, S., C. Green, R. Nicholls, and J. Corfee-Morlot. 2013. Future flood losses in major coastal cities. *Nature Climate Change* 3:802–06.

Harvey, D. 1978. The urban process under capitalism: A framework for analysis. *International Journal of Urban and Regional Research* 2:101–31.

———. 1982. *The limits to capital*. Oxford, UK: Blackwell.

Intergovernmental Panel on Climate Change. 2014. *Climate change 2014: Synthesis report*. Nairobi: UNEP.

Nobel Laureates. 2011. The Stockholm memorandum: Tipping the scales towards sustainability. *Ambio* 40 (7): 781–85.

O'Brien, K. 2013. Global environmental change III: Closing the gap between knowledge and action. *Progress in Human Geography* 37 (4): 587–96.

O'Neill, P. 2013. The financialisation of infrastructure: The role of categorisation and property relations. *Cambridge Journal of Regions, Economy and Society* 6 (3): 441–57.

Oulton, W. 2012. Funding a low-carbon future. http://www.environmental-finance.com/content/analysis/funding-a-low-carbon-future.html (last accessed 30 April 2014).

Pacala, S., and R. Socolow. 2004. Stabilization wedges: Solving the climate problem for the next 50 years with current technologies. *Science* 305:968–72.

Sayre, N. 2010. Climate change, scale, and devaluation: The challenge of our built environment. *Washington and Lee Journal of Energy, Climate, and the Environment* 1:93–105.

Schurr, S. 1990. *Electricity in the American economy: Agent of technological progress*. Westport, CT: Greenwood.

Stern, N. 2006. *Sterm review on the economics of climate change*. London: HM Treasury.

Stone, B. 2012. *The city and the coming climate: Climate change in the places we live*. Cambridge, UK: Cambridge University Press.

Swyngedouw, E. 2010. Apocalypse forever? Post-political populism and the spectre of climate change. *Theory, Culture & Society* 27:213–32.

Tandon, Y. 2011. Kleptocratic capitalism, climate finance, and the green economy in Africa. *Capitalism Nature Socialism* 22 (4): 136–44.

United Nations' International Resource Panel. 2013. City-level decoupling: Urban resource flows and the governance of infrastructure transitions. http://www.unep.org/resourcepanel/portals/24102/pdfs/Cities-Full_Report.pdf (last accessed 30 April 2014).

Zadek, S. 2013. Greening financial reform. http://www.project-syndicate.org/commentary/integrating-the-green-growth-imperative-and-financial-market-reform-by-simon-zadek (last accessed 30 April 2014).

Biomimetic Futures: Life, Death, and the Enclosure of a More-Than-Human Intellect

Elizabeth R. Johnson* and Jesse Goldstein†

*Department of Geography, University of Exeter
†Department of Sociology, Virginia Commonwealth University

The growing field of biomimicry promises to supplant modern industry's energy-intensive models of engineering with a mode of production more sensitively attuned to nonhuman life and matter. This article considers the revolutionary potentials created by biomimicry's more-than-human collectives and their limitations. Although biomimicry gestures toward a radical reontologization of and repoliticization of production, we argue that it remains subject to entrenched onto-political habits of social relations still dominated by capitalism and made part of a "terra economica" in which all is potentially put to profitable use and otherwise left to waste. With reference to Marx's notions of general industriousness and the general intellect, we find that this universalizing tendency renders myriad biological capacities and ways of knowing invisible. Drawing a comparison with the reworkings of life and knowledge explored in Shiebinger's work on nineteenth-century abortifacients, we show how biomimicry's more recent ontological remakings reproduce some forms of knowledge—and life—at the expense of others. Reflecting on biomimicry's inadvertent erasure of nonindustrial ways of knowing, we advance the notion of a pluripotent intellect as a framework that seeks to take responsibility for the cocuration of forms of life and forms of knowledge. We turn to Jackson's Land Institute as a grounded alternative for constructing more-than-human techno-social collaboratives.

成长中的生物模拟领域，承诺以对非人类生活与物更为敏感的生产模式，取代现代工业能源密集的工程模式。本文考量生物模拟的"超人类"集体所创造出的革命性潜能及其限制。我们主张，儘管生物模拟示意迈向对生产进行激进的再本体化与再政治化，生物模拟仍然受制于社会关係的本体—政治习性，该习性受到资本主义所支配，并构成了部分的"全球经济境况"（terra economica)，其中所有事物皆可潜在作为利润生产之用， 否则便会被当作垃圾遗弃。 我们参考马克思的"普遍勤奋"与"一般智力"之概念， 发现此般普世化的趋势，让各式各样的生物能力与认知方式变得隐而不见。我们比较席宾格对十九世纪堕胎剂之研究中所探讨的生命与知识的再製， 展现生物模拟更为晚近的本体论再製， 如何复製了部分的知识形式及生命，并以损及其他知识形式为代价。我们透过反思生物模拟对非工业的理解方式的不经意抹除， 推进多能分化智力之概念， 作为寻求为生命与知识形式的共同监护（cocuration）进行负责的框架。我们转而求助杰克逊的土地机构，作为建构"超人类"的科技—社会协作的扎根式另类取径。

El campo de la biomimesis, en creciente desarrollo, promete suplantar los modelos de ingeniería intensivos en energía que maneja la industria moderna con un modo de producción más sensiblemente sintonizado con la vida no humana y con la materia. Este artículo considera los potenciales revolucionarios creados por los colectivos más que humanos de la biomimesis, y sus limitaciones. Aunque la biomimesis apunta hacia una radical reontologización y repolitización de la producción, nosotros sostenemos que esta sigue sujeta a los arraigados hábitos onto-políticos de las relaciones sociales todavía dominadas por el capitalismo e hizo parte de una "terra económica" en la que todo se pone potencialmente en uso rentable o es dejado de lado como desperdicio. En referencia a las nociones de industriosidad generalizada e intelecto general de Marx, encontramos que esta tendencia universalizadora genera miríadas de capacidades biológicas y formas de conocer invisibles. Derivando una comparación con los nuevos esquemas de vida y conocimiento explorados en el trabajo de Shiebinger sobre los abortivos del siglo XIX, mostramos cómo las más recientes re-creaciones ontológicas de la biomimesis reproducen algunas formas de conocimiento— y de vida—a expensas de otros. Reflexionando en la involuntaria enmendadura de la biomimesis de los modos de conocer no industriales, proponemos la noción de un intelecto pluripotente a título de marco que busca asumir la responsabilidad por la co-curación de formas de vida y formas de conocimiento. Tornamos hacia el Land Institute [Instituto de Tierras] de Jackson como alternativa fundamentada para construir colaborativos tecno-sociales más que humanos.

Expanding our collective consciousness beyond Enlightenment ideologies and liberal humanism seems to be an increasingly central task of the humanities and social sciences, particularly given current pressures of ecological change. Latour's (2009) renovated political ecology, the "thing theory" of Bennett (2010) and Connolly (2013), the "multi-species knots" of Haraway (2008), and the "agental realism" of Barad (2007) offer important new frameworks that are more sensitively attuned to nonhuman life and matter than the anthropocentric accounts of Enlightenment origins that remain predominant today. These propositions offer more nuanced tools for socioecological diagnostics and open up space for political imaginaries that extend beyond the human.

The field of biomimetic innovation seems to provide a practical expression of these theoretical propositions. Encompassing a diverse range of efforts to apply insights from the biological sciences to military and industrial applications, biomimicry is as much an ideal as it is a paradigm for research and development. Biomimeticists cast nature as a participatory "mentor" of engineering, an inventive companion capable of generating solutions more sophisticated than our clunky, industrial society can imagine. By working with nature in this way, biomimicry elevates what Connolly (2013) has called *pluripotentiality*—the generative capacities of living and nonliving processes—as the driving force of technological innovation. In doing so, biomimicry promises to beget a new form of production that decenters and devalues human ingenuity, leaving open the possibility of a new industrialism that is more attuned to nature's needs.

Such a transformation in the political economy of knowledge production would seem to be cause for celebration and, indeed, mainstream accounts of biomimicry view the paradigm with joyful approbation (see Benyus 1997; Hawkins, Lovins, and Lovins 1999; Forbes 2006; Allen 2010; Harman 2013). Here, however, we want to encourage caution and critique. Although biomimicry gestures toward a radical reontologization and repoliticization of production, in practice it must contend with the entrenched onto-political habits of social relations still dominated by capitalism. Instead of regarding capitalism as a set of overbearing structures, we see instead an unstable performance, repeated with difference; regularly renewed and diligently maintained but far from inevitable (Gidwani and Reddy 2011). Or as Williams (1978) would write, a set of weakly determined relationships that nonetheless pattern our social, political, and ecological lives. Biomimicry offers an example of how capital's onto-politics habituate creative producers to channel emerging assemblages of human and nonhuman potentialities back into a paradigm of industrial progress, renewing and even accelerating the ongoing "treadmill of production" without destabilizing its unsustainable trajectory.

In this article, we consider how biomimicry draws nonhuman participants into an ongoing process of enclosure, which we understand first and foremost to be the creation of capital's ontological conditions of production. The project of creating an enclosed nature entails the production of a world apprehended as *terra economica*—an undifferentiated landscape in which all is potentially put to profitable use and otherwise left to waste.

Biomimicry's transformations of nature's productive capacities parallel the social transformations under industrial capitalism that Marx described. Lauding the emergence of a technoscientific society made possible by capitalism and united by a "general intellect," Marx promoted a highly selective vision of knowledge and value production. We locate our critique of biomimicry around a similar universalizing gesture. We argue that the most troubling element of biomimicry's emergence is not the privatization of specific forms of life (although this too creates numerous issues) but the way that life and all of its pluripotent potentials are incorporated into *terra economica*. In other words, biomimicry reduces, channels, and flattens a more-than-human pluripotent intellect into a general intellect focused on commercial and industrial advance. This follows in the wake of earlier (and ongoing) forms of colonial conquest that have entailed the erasure of local knowledges, ontologies, and sciences, all lost in the pursuit of a generally industrious, universally rational mode of (profitable) progress. Drawing a comparison with the reworkings of life and knowledge explored in Shiebinger's work on nineteenth-century abortifacients, we show how these more recent ontological remakings validate and reproduce knowledge—and forms of life—created in some human–nonhuman collectives while laying others to waste (Povenelli 2014).

If we are to acknowledge, as Bennett (2010) suggests, the "vibrancy" of nonhuman matter to generate "healthy and enabling instrumentalizations" (12), we might need to look beyond the creation of a more-than-human ontological framework to recognize how dangerous combinations of capitalism and colonialism reduce our sensitivity to life's pluripotency, as well as the inescapable and uncomfortable responsibility of curating life and death. In thinking through our more-than-human (and more-than-capital) potentialities, perhaps concepts such as Marx's general intellect need to give way to a conceptual language capable of proliferating more-than-human affiliations and transspecies subjectivities along multiple forms of knowing and being. Reflecting on biomimicry's inadvertent erasure of nonindustrial ways of knowing, we ask what such a pluripotent intellect and attendant biomimetic practices might look like, offering one provisional example in the work of Jackson's Land Institute.

Where the Bedbugs Bite: Biomimicry and the Making of More-Than-Human Production

Throughout the early part of the twenty-first century, cities across the United States have been plagued by the return of bedbugs. Although the widespread use of synthetic pesticides in the twentieth century had all but eradicated the pests, an evolved resistance to chemical deterrents has led to their resurgence and an urban public health crisis. Existing options for bedbug control range from the ineffective to the prohibitively expensive, prompting a team of biomimetic scientists at the University of California, Irvine (UC Irvine), to explore an old trick used by Balkan women: kidney bean leaves. Spread along a bedside overnight, the leaves trap bedbugs as they look to feed and have been known to eradicate the pests without chemical pesticides. UC Irvine scientists have studied the precise mechanisms for the bean leaves' success in harvesting bedbugs. Using scanning electron microscopy, they revealed that tiny barbs protruding from the underside of the leaves called trichomes hook into the legs of any traversing bedbugs, impaling them as they struggle to escape. Drawing on the trichomes as "a source of inspiration," they are working to "design and fabricat [e] biomimetic surfaces for bedbug trapping" (Szyndler et al. 2013). As part of the advancing field of biomimicry, the kidney beans are inspiring scientists and engineers to work with some natures—the bean leaves—to work against others—the bedbugs—as part of an effort to engender more healthful, enduring ecologies suitable for human life. Although the recipe for these synthetic trichomes has yet to be perfected, it promises a new, pesticide-free way of eradicating one of the most tenacious irritants of urban life.

Since the 1990s, biomimicry has made biological research an increasingly important element in technological innovation. Here, nonhuman life and human production meet anew: rather than raw material to be put to work, biomimicry draws on nonhuman life as a source of inspiration. It is a fractured field, driven by diverse aims.[1] Many advocates and designers view mimicking natural processes as a means to develop "green" technologies as well as a catalyst for thinking—and making—beyond anthropocentric approaches to production.[2] Although for some, the eco-friendly narrative that accompanies biomimetic innovation is little more than a new greenwashing ploy, biomimicry's most ardent believers insist that it is much more. Following Benyus, the appointed "guru" of biomimicry and author of *Biomimicry: Innovation Inspired by Nature* (Benyus 1997), designers, naturalists, investors, and biologists have come together to catalyze what they see as a paradigm shift in production, one that not only looks to natural systems to solve problems of climate change and ecological degradation but that also promises a new—nontoxic, nonexploitative—future mode of production modeled on nonhuman processes.

Discursive claims that have emerged alongside the field insist that biomimicry flips conventional scripts of human–nonhuman relations. Against ideologies of nature as a passive resource to be extracted and enslaved by *Homo industrialis*, biomimicry recasts the nonhuman world as a collection of active, collaborative participants, recognized for their "genius." For these advocates and practitioners, biomimicry is a praxis capable of leaving behind Francis Bacon's cruel vision of nature as something to be violently subdued and harnessed. Benyus's vision of a biomimicry-inspired "Parliament of Species," for example—reminiscent of Latour's (1993) "Parliament of Things"—gestures toward a would-be democratic body in which human and nonhuman agents (or Latour's "actants") can no longer be easily separated or hierarchized but together collaborate in the constitution of new worlds. The practice seems to offer a way toward what Haraway has referred to as a more-than-human cosmopolitanism: a framework for creating the world together while proliferating the ties that bind human and nonhuman life (Haraway 2008; see also Johnson 2010).

Terra Economica: Enclosure as Capital's Ontological Conditions of Production

Although biomimetic discourse promises to catalyze a future of expanded human–nonhuman relationships, actually existing biomimetic research and development might do more to constrain our perceptions of this more-than-human world. From within the extant political economy of techno-science, biomimeticists isolate parts of a comprehensively creative, wise, and innovative nature, harnessing individual kernels of wisdom that can be translated into intellectual property and mobilized for commercial application. Examples are manifold: The silk-making capacity of the golden orb weaver spider inspires the production of durable, bulletproof fabrics; the ventilation structures in a termite mound inspire greener building designs; lotus leaves inspire self-cleaning paint; and, as we saw earlier, the bedbug-snaring capacity of the kidney bean leaf inspires new pest-resistance commodities.

There is nothing inherently wrong with the biomimetic approach to industrial design and, in fact, many biomimetic projects offer inspiring ways to envision a much more benign form of industrial production that mitigates the materially intensive model of chemical engineering undergirding modern industry. But, born of the realities of industrial funding streams from government programs (including the Defense Advanced Research Projects Agency and the Advanced Research Projects Agency-E) through to the neoliberal university and the high returns that the financial industry expects from early-stage technologies, biomimicry remains beholden to the entrained pursuit of profitable, commercially (or militarily) viable technologies.[3] Accordingly, it contributes to the perpetuation of the very same techno-social infrastructure that it intends to transform in the first place. As we have written elsewhere, the reproduction of life is supplanted by the reproduction of capital (Goldstein and Johnson 2015).

As biomimicry draws nonhuman life into the process of production, its reontologizations of nature or embrace of its pluripotent possibilities are presently occurring through the ongoing expansion of capital's own ontology, which casts nature as a general condition of production, a whole earth available to become capital (see Cronon 2003; Smith 2007, 2008). Goldstein (2013) has termed this distinctly capitalist ontology of nature *terra economica*, tracing its emergence through the process of enclosure that transformed the early modern English countryside and providing the conditions of possibility for the transition to capitalist social property relations (Wood 2002).

It is important to consider the ways that enclosure entails discrete acts of dispossession and the attendant violence enacted on those made newly landless, separated from their means of production and reproduction and therefore left to the whims of the labor market. Here, however, we consider the ways that enclosure, as a more general project, entailed the production of a new nature, one that was no longer a world of diverse, noncapitalist productivities (what we call a pluripotent nature) but was instead reconceptualized as a site for potential capital accumulation. This new nature, or *terra economica*, was a world waiting to become capital and otherwise left to waste in its unimproved state. In this way, enclosure represented, and still represents, a reontologization of the more-than-human world, whose potentials (or the *potentia* of nature) are reduced and narrowed into the potential to become capital.

Accordingly, we consider biomimicry's predominant practices as one example of what many political economists refer to as "new enclosure." These new enclosures revive the originary land grabs of the seventeenth century, only instead of extending the process of privatization across territory, they are said to be intensive, targeting a diverse set of matter and practices. Elements like water, wildlife, fisheries, and forms of matter once considered incompatible with the notion of enclosure have led to novel forms of privatization (Shiva 2002; Correia 2005; St. Martin 2005; Robbins and Luginbuhl 2006). The establishment of regulatory markets and intellectual property patents has generated frontiers for profit generation around what McCarthy (2004) has called "impossible subjects of enclosure" (337). In other words, the process of enclosure allows for specific objects or relations to be incorporated within established patterns of valorization by apprehending objects formerly considered impossible to commodify as potential capital.

Along with emerging research on cell lines, nanoparticles, animal bodies, air molecules, and other forms of knowledge, biomimicry has become one frontier of new enclosure, generating novel forms of property that must be regulated through patents on intellectual processes as well as material products (Katz 1998; McCarthy 2004; Prudham 2007; Jasanoff 2012; Rajan 2012; Robertson 2012). To reiterate, this process is more than theft and dispossession. It represents a profound shift in how meaning is attributed to land and to life.[4] For example, agricultural crops (providing starch and sugar) are grown, harvested, and consumed as raw materials. By contrast, through the lens of biomimicry, plants are instead made evidence of nature's "genius," repositories of skills—for killing bedbugs in the preceding example or as "photosynthesis reactors" guiding solar cell research to take another example (Benyus 1997). In the latter case, Benyus goes so far as to explain that although plants have become "experts" at producing sugars out of sunlight, the biomimetic challenge is to apply the genius of photosynthesis to the production of more desirable energy products—such as hydrogen—that can be used to fuel industrial machines. Instead of being considered objects to be cultivated and consumed, plants and other nonhumans are put to work as the bearers of creative wealth. The evolutionary process—said to unfold within nature's own "laboratory" for "research and development" (Benyus 1997)—becomes a site bursting with potential for commercial revenue. This is part of an active effort on the part of biomimeticists like Benyus who strive to suture production practices that "create conditions conducive to life" together with conditions conducive for capital. The pluripotent potentials of the not-capital, nonhuman world appear, to them, as not-yet-capital.

The Eco-Social Promises of a Generally Industrious Nature?

A return to Marx's writings on industrialization will help locate a kernel of transformative optimism in the biomimetic project, while also providing grounds for a feminist critique (of Marx and of biomimicry). In his often-cited "Fragment on Machines," Marx (1993) suggests (albeit vaguely) that the rise of technological development radically transforms labor by prioritizing innovation and intellectual creativity over manual labor. For him, the systemic violence of enclosure and wage-labor relations remade human life by creating conditions for social unification. Marx saw therein the creative capacity of humans, which he termed the *general intellect*. This source of collective social power was made manifest in the advance of science and technology that accompanied industrial production. Marx uses the term *general industriousness* to describe the modernizing potential of the techno-social system made possible by this general intellect, in which production for money—or general social need—releases human productivity from the narrow constraints of local, particular needs.

Marx's excitement for the techno-social possibilities of general industriousness ground his hope for a noncapitalist future, characterized by a new kind of communality realized through "call[ing] to life all the powers of science and of nature, as of social combination and of social intercourse" (Marx 1993, 706). He thought this transformation would unify human production with technological apparatuses, gesturing to the dissolution of the very class relations that brought capitalism into being in the first place. For Marx, as well as more recent autonomous Marxist thinkers, these techno-social possibilities are one of the most promising residues of capitalism's advancing intensity (Negri 1991; Marx 1993; Casarino 2008).[5]

In many ways, biomimicry's narrative extends Marx's analysis to a more-than-human sociability. Biomimicry and, for that matter, the broader bioeconomy attribute to nature the same innovative capacities that Marx attributed to social labor. The nonhuman world is seen as a contributor to the general intellect of a more than human socionatural body and therefore a more-than-human general industriousness. Along these lines, biomimicry offers a further twist to the contradictory pathway toward revolutionary forms of production pursued by Marx and Marxists alike. It simultaneously furthers the process of capital accumulation while also encouraging a fundamentally transformed vision and practice of world making inclusive of nonhuman life. It might in fact be that biomimicry's more-than-human collaborators are the long-awaited "gravediggers" that capital itself will hail into being, threatening the ontological habits of a social logic that can only see the entirety of the world as an extension of itself.

What troubles here, however, is the ontological limitations imposed by the seemingly universal conception of a "general" condition of intellectual or industrial advance. Although Marx's conception of general industriousness provides a way of imagining that profound sociopolitical change will emerge from within the capitalist economy, it also shares with biomimicry a tendency to mark Western techno-science as a preferred means of making use of nature and its productive capacities. Biomimicry not only makes some lives—and not others, certainly not bedbugs—seem "naturally" worthy of reproduction and mimesis; it also endows the progress of technological development and the tendencies of the market with decisive power to laud some collaborations and forms of knowledge—those considered "productive," capable of incorporation into regimes of innovation, and "conducive to life"—erasing others. This runs parallel to Marx's tacit and somewhat uncritical endorsement of the "general needs" established by the community of money and the specific forms of technology and science aimed at meeting them. It is most evident in the numerous passages in which Marx describes nonmodern, noncapitalist forms of social production as lacking the dynamism of capitalism's general industriousness. In those passages, we find a premodern world filled with the idiocy of peasant life, nature idolatry, and encrusted forms of customary production that are predicated on maintaining existing social formations, as opposed to progressing down the unquestionably desirable path of modernity.

We dwell on these parallels between Marx and biomimicry to reveal the uncritical embrace of industrial progress coursing through otherwise radical or even revolutionary discourses of total social transformation. The problem with biomimicry—and with Marx—is not that a "general intellect" categorically supports the onto-habits of enclosure but that this generalized worldview tends to presume that all of life—human and nonhuman—is enrolled toward a singular emancipation through industrial advance. To consider what is lost in the rise and reproduction of Western techno-science, we now turn to other transitional histories well before the advent of biomimesis when life similarly entered new trajectories of production.

The General Ignorance: Erasing Knowledge Through Production

Feminist–Marxist critique has already done much of the work for us in condemning this Promethean tendency in Marx's work. Feminist historians and theorists, such as Merchant, Mies, and Federici, have documented the ways that Enlightenment thought and the project of modernity came at the expense not only of nature—which had been reduced to an enslaved machine—but also at the expense of many indigenous and feminized forms of knowledge that were either incompatible with or at the very least uncontrollable by the emerging scientific order (Mies 1986; Merchant 1990; Federici 2004). Federici's history of the European witch hunts, for example, demonstrates how women were demonized and often executed for displaying their autonomous relation to the world—human and nonhuman alike. Elsewhere, regimes of knowledge production were erased as Western medicine took hold as the only legitimate form of knowledge. Schiebinger's (2004) work on the history of women caregivers in the West Indies during the eighteenth and nineteenth centuries offers one notable example.

As Schiebinger writes, late eighteenth-century knowledge of how to induce abortion using several native and transplanted species was the product of a globalized form of medicine. Through loosely interconnected networks of Native Americans, African slave populations, and West and East Indians, knowledge of women's bodies and nonhuman materials supported women's autonomy and provided traction for forms of political and social resistance (Schiebinger 2004; Proctor and Schiebinger 2008). Enslaved women opposed the conditions of their lives by cultivating plants such as the peacock flower, an abortifacient that could help women manage the liveliness of their pregnant bodies. With the plants, women caregivers offered their communities a choice between bringing new life into the world or terminating unwanted pregnancies in a place and time where the conditions of life were considered not fit for living.

As a male-dominated medical profession began to systematize the study of pharmacology, these circuits of women's medicinal knowledge were no longer considered valid or safe. The rise of the university system and the professionalization of medicine forcibly excluded women from learning and barred nonprofessionals from practicing their knowledge. As a result, a once vibrant and well-circulated system of knowing—and the attendant relationships between human and plant species—was overturned for commercial and ideological reasons, changing how life was managed and by whom across the West Indies. Even the history of these practices and the social relevance of certain forms of plant life were largely lost. Male botanists and physicians exploring the West Indies failed to recognize the effectiveness of species like the peacock flower for women's reproductive health. As Schiebinger (2004) explains, "Funding priorities, global strategies, national policies, the structures of scientific institutions, trade patterns, and gender politics all pushed investigation toward certain parts of nature and away from others" (226).

This story of abortifacients in the eighteenth and nineteenth centuries parallels what we see happening today with biomimicry's revalorization of nonhuman life. The biomimetic imaginary legitimizes itself, in part at least, through the inadvertent erasure of any alternative techno-scientific approaches to engaging with more-than-human counterparts. By glorifying discrete portions of nature as a distinctly nonhuman source of inspiration, practices that have historically bound humans to nonhumans are ignored, as nature contra society is lauded for its ingenuity.

We see this production of ignorance taking shape in our example of kidney bean leaves and bedbugs. By glorifying the bean plants as a distinctly nonhuman source of inspiration, the UC Irvine biomimeticists exploring synthetic trichomes effectively erase the Balkan women who were initially part of this human–plant collective. Consider the implications of this erasure: As attempts to synthesize trichome technology are met with failure, the wider public remains without any public health efforts to promote the far simpler and more direct use of actual bean leaves to manage these pests. Indeed, what might be most troubling of all is the loss of an imaginary capable of bringing into being human–bean leaf collectives that might ameliorate an ongoing urban health crisis. There is no shortage of urban growing space, even in a city such as New York. Yet community kidney bean patches, bean sprout giveaways, and city-wide campaigns celebrating the history of these Balkan women as leaders in a public health revolution all seem somehow more fantastical than "simply" patenting the production of synthetic bean leaves and letting the market do its work.

Even more, the production of biomimetic trichomes suggests a way of "choosing life" through eco-friendly products without addressing the highly political question of "which life?" (Neyrat 2010) and, even more uncomfortably, "which deaths?" Each of these examples illustrates that cultivating some lives to inhibit others would seem, at times, a socially desirable necessity.

As feminist historians such as Federici and Schiebinger demonstrate, this erasure of knowing ecologies is part of an ongoing tendency in Western knowledge production. Historically, recognition of nonmodern innovators only takes place—if it takes place—when those populations are coded, as Bannerji (1998) writes, as "people of the body" or, in biomimicry's case, people of nature. Agency and intention on their part as creative and knowing participants in the more-than-human environment are often either dismissed or reinterpreted as local source material for the "real" and universalized (i.e., industrial) forms of innovation. The result is a denial of the particularity of Western knowledge, and the erasure of other, indigenous epistemes (Sundberg 2014).[6]

Given this set of problematics, how might we think more-than-human planetary eco-social realities—and future imaginaries—without resorting to either a universalizing discourse or a naïve embrace of Indigenous or "alternative" forms of knowledge production? How might we consider a more-than-human, many-bodied form of production that is neither generalized nor exclusively industrious?

Unimagined and Reimagined Futures

As we suggest with the bean leaf example, it is important to push back against narratives of industrial socionatural production and narrowing presuppositions of what we want and what nature does. When evolution is codified as research and development, we lose site of the pluripotency of (re)productive forces that exist, whose potentials exceed the narrowing ontology habituated through capitalist social relations. It might not be possible to repattern capital at an individual scale, but it does remain possible, at every scale, to think against and break with these ontologies and even to work through an alternative biomimicry that might develop conditions conducive not to a generalized life but to some chosen form of it.

One final example of a more-than-human collaboration offers an imaginary that moves beyond biomimicry's naïve sense of a future guided by nature's "will." Jackson's Land Institute is researching the complex dynamics of prairie ecosystems to develop a viable alternative to soil-depleting agricultural practices based on the cultivation of annual monocultures. Through the technologically mediated, yet ultimately age-old practice of selective breeding and hybridization, the Land Institute is attempting to develop perennial grains, legumes, and oilseed crops that can be cultivated as part of complex perennial polycultures. Jackson thus offers a different kind of biomimicry. Instead of looking for inspiration in specific plants or isolated characteristics of discrete nonhuman "innovators," Jackson and the Land Institute are inspired by the complex and resilient prairie ecosystems that once thrived on the Kansas plains. This includes a diversity of plants, animals, and ecosystemic interactions, as well as a diversity of local and traditional information embodied in the everyday practices and ingenuity of the people who have historically lived on the land.

This is a project fully aware of and responsive to the political economy in which it is situated. After years of research and development, the Land Institute hopes to release their first perennial grain, Kernza, within the decade. Rather than privatizing the seed as a commodity, they plan to make it—and their methods of production—available for farmers to experiment with and continue to refine. They know that their ultimate success will depend on further experimentation in different ecological and social contexts. This kind of collaborative, distributed project serves as an invitation to continually recombine with other forms of life, other ecologies, and other systems of knowledge production. As the Land Institute's Web site explains:

> We are often asked when we will be done. The honest answer is never. Our germplasm must constantly evolve to be useful in different agroecosystems all around the world. For us there is no "endgame." We openly share our research and germplasm with scientists all around the world. (The Land Institute 2013)

We point to the Land Institute as an example of postproductivist scientific inquiry and nonenclosing biomimicry—a collaborative process that seeks to learn from the more-than-human to coproduce a more beautiful, livable, and sustainable world. For Jackson, coming out of the appropriate technologies movement of the 1960s and 1970s, another less acquisitive science is possible. In fact, he believes that evolutionary processes are capable of generating conditions that are useless, unproductive, and inefficient for capital, yet still valuable for more-than-human communities. As he writes, "The fundamental question is not, 'what can we do?' but rather, 'what kind of world do we want and what will nature require of us?'" (Jackson 2006, 23).

Conclusions

Biomimicry is at the forefront of a quiet confrontation between the acquisitive techno-science of industrial fabrication and other forms of knowledge production, including indigenous and "folk" knowledges. Simply put, nature's innovative capacities look much different depending on what social-industrial-technical project its "wisdom" is imagined to be in dialogue with. Thus, the question is not only, as Katz (1998) asks, "whose nature and whose culture?" but also "whose science, whose technology?" and "which forms of life?" are made and unmade in the development of eco-social alternatives to our present condition. In posing these questions, we do not merely offer a critique of biomimicry as a form of greenwashing but flag a much greater problem: that of the necessity of our own position as cocurators of life and death on Earth. Biomimicry's future imaginary reduces our political imperative to merely "choosing" life over the "hubris" of human history. But, if life is neither unified nor benevolent, we remain faced with the question "which life?" In the example of the bedbugs, it is clear that we have already begun to choose: We might yet develop ways to live with and care for bedbugs, but these disagreeable parasites are presently uninvited participants in our eco-social present.

The history of scientific racism, which regularly deploys natural metaphors to legitimize systemic violence, should provide pause to consider just how fraught this terrain can be (see Magubane 1996). There are no easy ways to navigate these questions; an embrace of all life (and disavowal of all death) is simply an abrogation of our ethico-political responsibility to be, as Haraway (2008) writes, "in responsible relation to always asymmetrical living and dying, and nurturing and killing" (42). Mimicking and abiding nature and its processes will not save us from the constant and dirty politics of having to "choose death" even as we "choose life."

In our encounter with global ecological and economic systems whose expanded reproduction has come to threaten the very viability of our entire planetary ecology, it is becoming increasingly apparent to us that we might need to question whether the "real sources of wealth" are in fact aligned with a general industriousness that "know no bounds" (Marx 1993, 224). If we are to craft a different kind of eco-social future, we must consciously shift away from this celebratory—and decidedly humanist—account of innovation's potential and the historical arc of progress.

Beyond any vibrant metaphor of more-than-human industriousness, we will need to actively consider what technologies and what sciences we want to pursue. This is not to suggest that modern science and technology be categorically disregarded but that we must continually ask how and in what ways these forms of knowing can be rehabituated with other forms of inquiry. If projects such as biomimicry are to further a more-than-human, and more-than-capital eco-social future, they will only do so by recognizing both the recuperative powers of capital's logic and its tendency to evacuate the political from notions of progress. Such questions, positioned at the intersection of material and knowledge production emanating from multiple forms of life, might help us locate the possibilities of a new form of (re)production, one focused on holding as much as it is on making and, with it, the quickening of a new eco-social future.

Notes

1. Corporate and military interests have championed biomimetic science for diverse reasons, many having nothing to do with "greening" technological production. Here, however, we focus on the more popular arm of biomimicry, which views the field as a powerful means of remaking the relationship between nature and production.

2. Humans have fashioned behavior on nonhuman life for millennia. Historical examples include da Vinci and Lilienthal's (failed) flying machines, the Wright brothers' (successful) airplanes, and Georges de Mestral's development of Velcro. But scientific researchers, engineers, and designers have only recently begun to embrace biomimicry as a new industrial paradigm.
3. Elsewhere, the authors have explored the empirical conditions of biomimetic production, exploring the relationship among its historical emergence and venture capitalism, the U.S. military, and the neoliberalization of university research (see Johnson 2010).
4. This is not so different from the mid-twentieth-century warnings of Heidegger ([1954] 1977) and Horkheimer and Adorno ([1944] 2002), who feared that a logic of instrumental reason had begun to govern perceptions of life, labor, and land. The crucial distinction, however, is that biomimetic approaches reconstitute life's ontology as a productive participant, rather than a passive object.
5. We also find this expressed in recent texts on "accelerationism," which suggest a temporary embrace and acceleration of capitalist innovation as a means toward a postcapitalist future (Williams and Srnicek 2013).
6. Sundberg's critique of the Eurocentricity of some posthumanist geographers and their tendency to sanitize knowledge production into a singular and universal form implicates our own disciplinary work as well. Many scholars critiquing nature–society dichotomies in the Western philosophical tradition (Sundberg calls out her own work, as well as Bennett 2010; Braun and Whatmore 2010; and Wolfe 2010) have failed to present our own intellectual tradition as spatially and temporally particular—or, more to the point, provincial. Sundberg argues that this inadvertently replaces one universalizing tradition with another.

References

Allen, R. 2010. *Bulletproof feathers*. Chicago: University of Chicago Press.

Bannerji, H. 1998. Gender, race, class and socialism: An interview. *New Socialist* 3 (1): 20–25.

Barad, K. 2007. *Meeting the universe halfway*. Durham, NC: Duke University Press.

Bennett, J. 2010. Thing power. In *Political matter: Technoscience, democracy, and public life*, ed. B. Braun and S. Whatmore, 35–62. Minneapolis: University of Minnesota Press.

Benyus, J. M. 1997. *Biomimicry: Innovation inspired by nature*. New York: HarperCollins.

Braun, B., & Whatmore, S. J., eds. 2010. *Political matter: Technoscience, democracy, and public life*. Minneapolis: University of Minnesota Press.

Casarino, C. 2008. Time matters: Marx, Negri, Agamben, and the corporeal. In *In praise of the common*, C. Casarino and A. Negri, 219–46. Minneapolis: University of Minnesota Press.

Connolly, W. 2013. *The fragility of things*. Durham, NC: Duke University Press.

Correia, D. 2005. From agropastoralism to sustained yield forestry: Industrial restructuring, rural change, and the land-grant commons in northern New Mexico. *Capitalism Nature Socialism* 16 (1): 25–44.

Cronon, W. 2003. *Changes in the land: Indians, colonists, and the ecology of New England*. New York: Hill and Wang.

Federici, S. 2004. *Caliban and the witch*. New York: Autonomedia.

Forbes, P. 2006. *The gecko's foot*. New York: Harper.

Gidwani, V., and R. N. Reddy. 2011. The afterlives of "waste": Notes from India for a minor history of capitalist surplus. *Antipode* 43 (5): 1625–58.

Goldstein, J. 2013. Terra economica: Waste and the production of enclosed nature. *Antipode* 41 (2): 357–75.

Goldstein, J., and E. Johnson. 2015. Biomimicries for and against capitalism. *Theory, Culture, Society* 32 (1): 61–81.

Haraway, D. 2008. *When species meet*. Minneapolis: University of Minnesota Press.

Harman, J. 2013. *The shark's paintbrush*. Ashland, OR: White Cloud Press.

Hawkins, P., A. Lovins, and H. Lovins. 1999. *Natural capitalism*. London: Earthscan.

Heidegger, M. [1954] 1977. *The question concerning technology and other essays*. New York: Harper Perennial.

Horkheimer, M., and T. Adorno. [1944] 2002. *Dialectic of enlightenment: Philosophical fragments*. Stanford, CA: Stanford University Press.

Jackson, W. 2006. *Altars of unhewn stone*. Wooster, OH: Wooster Book Company.

Jasanoff, S. 2012. Taking life: Private rights in public nature. In *Lively capital: Biotechnologies, ethics, and governance in global markets*, ed. K. S. Rajan, 155–83. Durham, NC: Duke University Press.

Johnson, E. 2010. Reinventing biological life, reinventing "the human." *Ephemera: Theory and Politics in Organization* 10 (2): 177–93.

Katz, C. 1998. Whose nature, whose culture?: Private productions of space and the "preservation" of nature. In *Remaking reality: Nature at the millenium*, ed. B. Braun and N. Castree, 45–62. London and New York: Routledge.

The Land Institute. 2013. What we do and why. http://www.landinstitute.org/vnews/display.v/ART/2013/03/06/513a22329f471 (last accessed 29 October 2013).

Latour, B. 1993. *We have never been modern*. Cambridge, MA: Harvard University Press.

———. 2009. *Politics of nature*. Cambridge, MA: Harvard University Press.

Magubane, B. 1996. *The making of a racist state: British imperialism and the union of South Africa, 1875–1910*. Trenton, NJ: Africa World Press.

Marx, K. 1993. *Grundrisse: Foundations of the critique of political economy*. New York: Penguin Classics.

McCarthy, J. 2004. Privatizing conditions of production: Trade agreements as neoliberal environmental governance. *Geoforum* 35 (3): 327–41.

Merchant, C. 1990. *The death of nature: Women, ecology, and the scientific revolution*. New York: HarperCollins.

Mies, M. 1986. *Patriarchy and accumulation on a world scale: Women in the international division of labour*. Atlantic Highlands: Zed Books.

Negri, A. 1991. *Marx beyond Marx: Lessons on the Grundrisse*. New York: Autonomedia.

Neyrat, F. 2010. The birth of immunopolitics. *Parrhesia* 10:31–38.
Povenelli, E. 2014. The four figures of the Anthropocene. Paper presented at the Anthropocene Feminism Conference, University of Wisconsin–Milwaukee, Milwaukee.
Proctor, R., and L. Schiebinger. 2008. *Agnotology: The making and unmaking of ignorance*. Stanford, CA: Stanford University Press.
Prudham, S. 2007. The fictions of autonomous invention: Accumulation by dispossession, commodification and life patents in Canada. *Antipode* 39 (3): 406–29.
Rajan, K. S., ed. 2012. *Lively capital: Biotechnologies, ethics, and governance in global markets*. Durham, NC: Duke University Press Books.
Robbins, P., and A. Luginbuhl. 2005. The last enclosure: Resisting privatization of wildlife in the Western United States. *Capitalism Nature Socialism* 16 (1): 45–61.
Robertson, M. 2012. Measurement and alienation: Making a world of ecosystem services. *Transactions of the Institute of British Geographers* 37 (3): 386–401.
Schiebinger, L. 2004. *Plants and empire: Colonial bioprospecting in the Atlantic world*. Cambridge, MA: Harvard University Press.
Shiva, V. 2002. *Water wars: Privatization, pollution and profit*. Cambridge, MA: South End Press.
Smith, N. 2007. Nature as accumulation strategy. *Socialist Register* 43.
———. 2008. *Uneven development: Nature, capital, and the production of space*. Athens: University of Georgia Press.
St. Martin, K. 2005. Disrupting enclosure in New England fisheries. *Capitalism Nature Socialism* 16 (1): 63–80.
Sundberg, J. 2014. Decolonizing posthumanist geographies. *Cultural Geographies* 21:33–47.
Szyndler, M. W., K. F. Haynes, M. F. Potter, R. M. Corn, and C. Loudon. 2013. Entrapment of bed bugs by leaf trichomes inspires microfabrication of biomimetic surfaces. *Journal of the Royal Society Interface* 10 (83): 20130174.
Williams, A., and N. Srnicek. 2013. #Accelerate manifesto. *Critical Legal Thinking*. http://criticallegalthinking.com/2013/05/14/accelerate-manifesto-for-an-accelerationist-politics/ (last accessed 10 December 2014).
Williams, R. 1978. *Marxism and literature*. Oxford, UK: Oxford University Press.
Wolfe, C. 2010. *What is posthumanism?* Minneapolis: University of Minnesota Press.
Wood, E. 2002. *The origin of capitalism: A longer view*. New York: Verso.

Agro-Ecology and Food Sovereignty Movements in Chile: Sociospatial Practices for Alternative Peasant Futures

Beatriz Cid Aguayo* and Alex Latta†

*Department of Sociology and Anthropology, University of Concepción, Chile
†Department of Global Studies, Wilfrid Laurier University, Canada

The agro-ecology and food sovereignty movements of southern Chile promote alternatives to the hegemonic agro-export regime that dominates the landscape. We explore these mobilizations and the strategies they employ, with a particular focus on a network of peasant women "seed curators." The global agri-food complex relies on a flat and universalizing spatiality of land as resource and food as commodity, in which the character and fate of individual places is of little importance. This is paired with a hierarchical monopolization of knowledge, where producers become recipients rather than creators and custodians of agricultural inputs and know-how. In response, peasant movements have given birth to alternative spatial practices based on horizontal networks that join together interdependent producers and places. By sharing traditional and agro-ecological knowledge, cultivating alternate circuits of exchange, and building urban–rural partnerships, these movements seek to reshape the horizons of possibility both for peasant communities and for the broader agri-food system.

智利南部的农业生态与粮食主权运动，提倡在支配地景的霸权农产出口政体之外，寻求另赖可能。我们探讨这些动员及其所运用的策略，并特别关注一个名为"种子监护人"的女性农民网络。全球的农产品复合体，仰赖土地的扁平及普遍化空间性作为资源，以及粮食作为商品，其中个别地方的特徵与命运，并不具太大的重要性。而与其搭配的是知识的阶层式独佔，其中生产者成为接收者，而非创造者及农业投入与技术祕诀的看守人。农民运动，催生了根据将相互依存的生产者及地方连结在一起的水平网络的另类空间实践。透过分享传统及农业生态知识、培育交换的交替迴路，以及建立城乡伙伴关係，这些运动企图同时为农民社群与更广泛的农产品系统，重塑可能性的范畴。

Los movimientos de agroecología y soberanía alimentaria del sur de Chile promueven alternativas al régimen hegemónico agro-exportador que domina el paisaje regional. Exploramos estas movilizaciones y las estrategias que ellas emplean, enfocándonos particularmente sobre una red de mujeres campesinas "curadoras de semillas." El complejo global agro-alimentario depende de una espacialidad plana y universalizadora de la tierra como recurso y del alimento como mercadería, en la que el carácter y suerte de los lugares individuales revisten poca importancia. Lo anterior va de la mano con una monopolización jerárquica del conocimiento, donde los productores se convierten más en receptores que en creadores y custodios de los recursos agrícolas y del saber. Como respuesta, los movimientos campesinos han dado lugar a prácticas espaciales alternativas basadas en redes horizontales que conectan a los productores interdependientes con los lugares. Al compartir conocimiento tradicional y agro-ecológico, cultivando circuitos alternos de intercambio y construyendo compañías urbano-rurales, estos movimientos buscan reconfigurar los horizontes de posibilidad tanto para las comunidades campesinas como para el más amplio sistema agro-alimentario.

Chile figures centrally in the story of neoliberal globalization, both for being the first full-fledged neoliberal experiment and for the depth and durability of the changes that experiment set in train. Among the first and most important policy changes implemented after the country's 1973 military coup was the reversal of agrarian reforms carried out under previous governments and the pursuit of a new agricultural model focused on export. Over the ensuing twenty years, the Chilean countryside was dramatically reshaped by an intensification of agriculture and forestry, not carried out by the traditional landowning elite but rather by a new capitalist landowning class (Gómez and Echenique 1988). Meanwhile, small peasant farmers became rural proletarians, providing labor within the newly industrializing agricultural model (Echenique and Rolando 1991).

Our article locates alternate socioecological futures in the context of this transformed countryside. In the interstices of the now widespread agro-industrial model, we examine mobilizations for agro-ecological alternatives and food sovereignty in the Biobío region of southern Chile. These mobilizations offer heterogeneous political responses to the various impacts of the neoliberal model on the Chilean countryside, which include the industrialization of production, the homogenization of agricultural commodities, and the externalization of risk to local producers. We explore these mobilizations and the strategies they employ as a networked politics of place, embodied in everyday practices and embedded in a broader challenge to capitalist territorializations of rural space and rural–urban relationships.

Where the agro-industrial model relies on a flat and universalizing spatiality of land as resource and food as commodity, agro-ecological movements are giving birth to insurgent territorial configurations based on the particularities of local socionatural relations, along with principles of ethical consumption, urban–rural partnership, and solidaristic economic exchange. We bring together three perspectives to frame our analysis: (1) food sovereignty, as both an analytical frame and as a principle mobilized in the discourse of peasant activists (Sevilla Guzmán 2006; Wittman, Desmarais, and Wiebe 2010; Gutierrez Escobar 2011; Andrée et al. 2014); (2) postdevelopment and alter-globalization approaches that chart the way specific places generate practices, agencies, and possibilities that constitute alternatives to capitalist globalization (e.g., Gibson-Graham 2003, 2006; Escobar 2008; White and Williams 2012; Gibson-Graham, Cameron, and Healy 2013); and (3) a related literature that considers local mobilizations as forms of spatial politics interwoven with and enabled by broader sets of sociospatial relations and practices (e.g., Massey 2004; Tsing 2005; Jessop, Brenner, and Jones 2008; Neumann 2009).

After a brief discussion of methods, our arguments are organized in four parts. We begin by generating a theoretical basis for considering alternative economies and spatialities of agriculture and food. Next, we explore debates around peasant responses to the global agri-food regime. In the second half of the article we address the agrarian context in southern Chile, describing first the impacts of the agri-business complex and the response from food sovereignty movements, before turning to a deeper examination of the sociospatial practices that constitute agro-ecological alternatives.

Method

The case study is based on interviews carried out by the first author during 2012 and 2013, with peasant women who are part of a network of *curadoras de semillas* (seed "curators" or "healers," hereafter *curadoras*) in southern Chile. The interviews are part of a larger research project that also examines two other sites of alternative agriculture in the region, namely, community-based urban agriculture and export-oriented certified organic farming. Research with the *curadoras* was carried out as part of a long-standing collaboration with Chile's National Association of Rural and Indigenous Women (ANAMURI by the Spanish acronym), although not all members of the *curadora* network (nor all those interviewed) are part of ANAMURI. A snowball sampling approach was taken to the study, beginning with movement leaders and requesting referrals from interviewees. Twenty-three in-depth ethnographic interviews were conducted, eight with movement leaders and fifteen with *curadoras*. The average length of interviews was two hours, although in some cases more than one meeting took place. Mostly conducted in the homes of the research participants, interviews were recorded, transcribed, and coded using NVivo software (NVivo 10 2012). The analysis is also informed by participant observation at seed exchange events, as well as past experiences of several political and technical collaborations with ANAMURI between 2010 and 2013.

The majority of the interviews took place in Chile's south-central Biobío region, where the network of *curadoras* is concentrated. The Biobío contains the largest remaining concentration of peasant farms and the most significant examples of alternative agricultural practices and movements in the country. That the members of the network are predominantly women speaks to a notable gendering of agricultural roles, where men are more likely to participate as laborers in industrial agriculture or other sectors of the mainstream economy, and women have taken increasing leadership in developing the kinds of alternative agricultural and economic practices associated with the food sovereignty movement.

Despite the fact that ANAMURI attempts to mobilize both indigenous as well as nonindigenous women, in Chile's south it has had only partial success in achieving this intercultural objective. This is due in part to the troubled relationships between settler society and the Mapuche people, characterized by a history of land conquest and racism, but it is also related to a tension between feminist values that inform ANAMURI and the traditional values of complementarity that structure gender relations in many Mapuche communities. Only three of the interviewees in the study were Mapuche, two *curadoras* and one movement leader. The gendered and racial dimensions of the *curadora* network are certainly deserving of further analysis. Nevertheless, we put these issues to one side to focus here on the sociospatial practices of the *curadoras*.

Economic and Spatial Dimensions of Contested Socionatural Selations

Gibson-Graham (1996, 2006) asserts that critical scholarship has overemphasized the pervasiveness of capitalism, ignoring the persistence of alternative modes of economic organization in the interstices of an always uneven and incomplete pattern of subjection to a global capitalist order (see also Williams 2005; Escobar 2008; Healy 2009; Williams et al. 2012). She argues that by embracing economic heterogeneity, "noncapitalism lost its negativity, becoming a multitude of specific economic activities and relations, and capitalism simultaneously lost its abstract singularity" (Gibson-Graham 2006, xxxiv). In the wake of this shift, research has connected economic heterogeneity with modes of adaptation or resistance in response to economic hardship, encapsulated by Gibson-Graham (2006) in the notion of "community economies" (see also Whitson 2007; Round, Williams, and Rodgers 2008; Williams et al. 2012). It has also taken up diverse economic practices as part of a political project of purposefully imagining and enacting alternatives to a capitalist order (e.g., Gibson-Graham 2006; Escobar 2008; Healy and Graham 2008; White and Williams 2012; Gibson-Graham, Cameron, and Healy 2013).

As Healy (2009) argues, there are key spatial dimensions to the project of rethinking economies as heterogeneous fields of possibility—what Gibson-Graham calls "a geography of ubiquity." Healy suggests that this perspective offers a nonhierarchical or flattened politics of the global as an alternative to hierarchical scalar imaginaries where global forces are seen to penetrate and reconfigure local spaces. This alternative imaginary is a potentially empowering ontological shift, but we assert that the politics of space and scale persist and must be addressed as part of social and political practice (Moore 2008; Neumann 2009). In addition, it is crucial to note that the spatial imaginary of dominant economic narratives is also horizontal. For neoliberal ideologues, the global market is a flat and universalizing space, where order is spontaneously generated by competition among nations, regions, cities, enterprises, and individuals (e.g., Friedman 2005; see also Christopherson, Garretsen, and Martin 2008). For neo-Keynesian economists, the current global economy is unevenly scaled in favor of powerful nations and corporations, but appropriate policy measures could render it more egalitarian and hence more flat (Sheppard and Leitner 2010). Finally, in the practices of transnational corporations that organize global commodity chains, local cultural and ecological differences are inconvenient spatiotemporal factors to be overcome through the standardization of agricultural outputs (e.g., see Cid Aguayo 2007).

Given this context, we argue that Gibson-Graham's place-based orientation (Gibson-Graham 2003, 2006; Gibson-Graham, Cameron, and Healy 2013; see also Escobar's [2001] "localization strategies") demands closer attention to counterhegemonic spatial practices that reconstruct localities and regions as sites where identity and agency are generated in negotiation with wider global forces. Moreover, following Jessop, Brenner, and Jones (2008), we seek to connect the politics of scale with a broader set of "sociospatial relations" that also include territories, places, and networks. Taking up Gibson-Graham's notion of the economy as a field of difference, we can begin to trace the way that self-conscious movements for alternative food systems emerge against the grain of the sociospatial relations of capitalist agriculture.

Agro-Industry, Peasant Agriculture, and Food Sovereignty

In recent debates, the agrarian question has been developed in relation to the so-called corporate food regime (Friedmann 1995) and the Supermarket Revolution (McMichael 2005). This regime or revolution has fomented a reconcentration of agricultural lands and promotes monocropping for export, leading to increasingly precarious conditions for peasant economies and a renewed focus on global processes of "depeasantization" (Araghi 2000). At the same time, it has become apparent that the plight of the peasantry cannot be understood simply in terms of peasants' economic and territorial displacement. The agro-food complex has also colonized and reshaped relations of production, leading many peasants to lose control over productive processes and develop dependency relations with agro-industry, becoming quasi-proletarians on their own land.

There are important sociospatial dimensions to this reorganization of peasant production. Goodman, Sorj, and Wilkinson (1987) describe parallel processes of "appropriationism" and "substitutionism" (see also Goodman and Redcliff 1991). The former involves the transformation of agricultural cycles into agro-industrial processes, replacing localized knowledge, practices, and agricultural inputs with deterritorialized scientific knowledge, globally standardized techniques, and industrially derived inputs. In spatial terms, these changes constitute a territorial flattening of agro-ecological relations, which facilitates the extraction of ecological capital and human labor. Appropriationism is complemented by substitutionism, which further standardizes food as it moves through the value chain, converting agricultural outputs into industrial materials and eventually into durable goods. This process furthers the standardization initiated in the appropriationism phase, usually constituting the total erasure of locational and temporal dimensions of agricultural production.

Through these related processes, mid- to large-size farms are articulated with agro-industry through "contract agriculture," whereby producers commit to the use of standardized inputs and procedures in exchange for purchase agreements. This approach allows corporations to download production risks like climate inconsistency, disease, and pests. Producers who absorb these risks often end up trapped in a debt cycle driven by the annual costs of inputs (Glover and Kusterer 1990; Grossman 1998; Warning and Key 2002; Cid Aguayo 2007). In this way, risk is perhaps the only element of agricultural production that remains strongly localized within the agri-food system. Meanwhile, a composite of globally distributed place-based production ensures reliable corporate profitability.

Some peasants have responded to the globalization of agro-industrial production by participating in contract agriculture, a successful strategy for producers who can acquire more land and scale up production. Others have sold their land to become rural wage laborers or migrate to cities. Finally, some peasants have resisted market pressures to get big or sell out, instead seeking to revitalize local and regional economies of smallholder production, in the process reclaiming peasant identity as a new site of cultural, economic, and political agency. It is this final group that has taken up food sovereignty as a critical perspective and political agenda, especially in the Global South.

Food sovereignty stands in contrast to food security by focusing on modes of production instead of just affordability and reliability of access. As such, it puts the dominant agro-industrial model in question and converts food into a broader site of political engagement (see, e.g., Wittman, Desmarais, and Wiebe 2010; contributions to Andrée et al. 2014). It is notable that the concept emerged directly from peasant movements, especially the largest global peasant network, Vía Campesina (Wittman 2009; Ortega-Cerdà and Rivera-Ferre 2010; Gutiérrez Escobar 2011). In this vein, Wittman (2009) identifies food sovereignty as the basis for a reconstitution of agrarian citizenship. Ortega-Cerdà and Rivera-Ferre (2010) identify five dimensions of the food sovereignty agenda: (1) ensuring peasant access to agricultural inputs like land and seeds; (2) nurturing diversified, locally appropriate agricultural practices; (3) protecting and promoting local markets for the products of peasant agriculture; (4) ensuring all citizens' rights to have healthy, locally produced food; and (5) promoting peasants' right to full participation in setting agricultural policy.

The food sovereignty literature frequently makes explicit links to principles of agro-ecology, which seeks to foster a synergy between peasant knowledge and alternative agricultural science informed by ecological principles (Sevilla Guzmán 2006; Altieri 2009). We also find it crucial to emphasize the link between food sovereignty and movements to build social solidarity economies (e.g., see Gutiérrez Escobar 2011). Following Miller (2013, 529), the latter consist of "constructing concrete relations of collaborative production, transaction, consumption, and surplus reinvestment that can constitute a growing 'outside' to capitalist relations." As Miller argues, movements for social solidarity economies can be seen as one potentially powerful political embodiment of the community economies paradigm set out by Gibson-Graham.

Agrarian Change and the Food Sovereignty Movement in Southern Chile

The 1980s saw a consolidation of the export-oriented agri-businesses model across Chile, especially in the market niches of fresh fruits and vegetables, premium wines, wood pulp, and Atlantic salmon. This process was fueled by state support for research and development, the flexibilization of the rural labor market, and a dramatic opening to foreign investment. The sector's expansion continued with little change in direction after the 1990 return to democracy under successive governments of the center-left parties of the *Concertación* (Cid Aguayo 2011).

Only the most isolated regions less apt for commercial agriculture have escaped the process of rural transformation, becoming "refuge areas" for peasant economies (Clapp 1998). The Biobío region, located in south-central Chile, contains an important concentration of those refuge areas. According to the most recent Farm Census (Instituto Nacional de Estadísticas 2007), this region holds 22 percent of the nation's agricultural lands less than ten hectares in size (comprising 64.9 percent of the holdings in the region). These smallholders continue to face pressures associated with the agro-export economic model, especially in terms of an expanding forest industry. During the past twenty years, 1,330,163 hectares of farmland previously dedicated to crops such as wheat and sugar beets have been converted to forestry plantations (Instituto Nacional de Estadísticas 2007). For peasant farmers this expansion of plantation forestry has meant competition for water resources, contamination from agrochemicals, and the increased risk of forest fires. Some have opted to also plant trees or lease their land to forestry companies. For those who remain in agriculture, government policy has constituted another kind of pressure on traditional patterns of livelihood, with assistance programs that encourage the use of external inputs like improved seeds and fertilizers, and gauge farm viability in terms of producers' integration into agro-industrial commodity chains.

In response to these pressures, the Biobío region has seen the rise of significant peasant organizing around the principles of both agro-ecology and food sovereignty. The region is home to two of the leading national organizations promoting agro-ecology, the Centre for Education and Technology of the South (CET Sur) and another called CET Yumbel. The region also concentrates the largest number of agro-ecological producers in Chile, including approximately 1,000 certified organic farms—half of all certified organic farms in the country (personal communication with organic certification agent, 23 March 2012). The agro-ecological movement in the Biobío region can be divided into three groups: (1) medium and large farms with a certain level of capitalization, focused on certified organic production for mainly international markets; (2) a small middle- and upper-class movement around the philosophy of permaculture; and (3) peasant federations, urban agricultural organizations, and nongovernmental organizations (NGOs) focused on small-scale production for subsistence and local markets. This third group exhibits a strong affinity with the principles of food sovereignty, especially evidenced by the participation of female peasant producers in the National Association of Rural and Native Women (ANAMURI). ANAMURI was formed in the 1990s as an offshoot of existing male-dominated peasant organizations. It began its work with temporary agricultural employees as a response to the growth of a female agricultural proletariat but has reoriented its priorities toward the reinvention and revalorization of peasant identities and rural life. It has become the most important organization for peasant advocacy at the national level and is a member of Vía Campesina (Cid Aguayo 2011).

ANAMURI is the principle instigator of the network of *curadoras de semillas* that is the focus of this study. The *curadoras* protect and multiply a wide variety of seed in household gardens. In the Biobío region, the practice has its contemporary origins in a series of seed curator schools, implemented over the past decade by NGOs such as CET Sur and CET Yumbel. The *curadoras* work with seeds in two senses: both to physically preserve and reproduce local genetic diversity, and more metaphorically to "heal" the seeds as elements of a landscape under attack by agro-industrial practices. As one *curadora* describes the practice:

> [A *curadora* is] a healer of seeds . . . [being a *curadora*] is to rescue the seeds and store them . . . to defend the seeds. The *curadora* defends everything that is not genetically modified. (CS01-I, 21 May 2012).[1]

Although the self-conscious identity of the *curadoras* is partly a product of the political discourse of ANAMURI, it also has its roots in peasant and indigenous traditions. The same *curadora* goes on to note the following:

> Who are really accomplished *curadoras*, real custodians of seed, are our people, our Mapuche people. . . . [The practice represents] the continuation of our ancestors . . . [what is important is] nourishing yourself with legumes, nourishing yourself with that which comes from the earth, directly from the earth. . . . We started to plant, and care for the seeds, then we started the exchanges, and from there we started to be *curadoras* of seeds. (CS01-I, 21 May 2012)

The principles of food sovereignty are taken up in different ways by the women who participated in the study. On the one hand, many of the *curadoras* conceive of food sovereignty in terms of their ability to put healthy food on the table and the right to consume products free of pesticides or potentially transgenic properties associated with industrial agriculture.

> Sovereignty is the healthy nutrition that one harvests: the vegetables ... the legumes, having a little farm for oneself, planting. ... That's healthy nutrition for a person. (CS02, 20 September 2012)

> Food sovereignty in general terms ... is a right that a person has to feed herself as she should, healthily. ... One has to have the right to consume uncontaminated food, that's what's fundamental, right? (CS03, 13 July 2012)

On the other hand, among women who have more political formation and take leadership roles in ANAMURI there is a broader articulation both of rights claims and of the productive roles of peasant farmers in the rural economy, as visible in the following comments from movement leaders.

> Food sovereignty ... [implies] a set of rights for us, the *campesinos* and *campesinas*. It implies the right to water; it implies the right to land; it implies, let's say, the principal right to continue maintaining our creole and native seeds from which the chain of nutrition starts. (ML01, 25 September 2013)[2]

> So we talk about food sovereignty because as I have always said, and I'm going to repeat it on this occasion: the small female farmer is the one who produces the food for the community, and the small farmer is the one in charge—we the women—of the conservation, care, reproduction and protection of our seeds. (ML02-I, 28 June 2013)

This same leader clearly articulates the connection between food sovereignty and agro-ecology, emphasizing traditional peasant (and indigenous) knowledge as a basis for more ecologically sound modes of production.

> [We are] trying to rescue natural forms of production ... returning to the ancestral practices because, well, we know those practices from when we were children. ... My paternal grandmother was a very intelligent woman ... she organized corner boxes in the corrals, huge boxes, and there she collected all the manure every day, the leaves that fell, the ashes she collected, throwing everything in these piles of compost and mixing it up. (ML02-I, 28 June 2013)

We assert that the eco-political discourses and productive practices of these women, joined together in a network of seed saving and exchange, signal the emergence of socioecological alternatives to dominant agro-industrial practices. In the next section, we explore their networked place-based agency as an alternative spatial order operating in the interstices of the flattened capitalist landscape of Chile's neoliberal model of rural modernization.

Theorizing the Sociospatial Practices of Food Sovereignty in Chile's Biobío

In the most basic sense, the continued presence of independent peasant agriculturalists in the region constitutes a form of active resistance against the global agri-food system, as it conserves a heterogenous field of economic possibility across a diversity of local places—Gibson-Graham's "geography of ubiquity." This presence and resistance, however, constitute more than mere negation of the dominant model. Peasants persist because peasant economies are sustained and adapted over time as a set of alternative sociospatial practices. Of particular importance are local relationships of exchange among peasant households, whether through farmers' markets, barter, or reciprocity arrangements. More collectively organized practices also exist, such as direct exchange systems between peasant federations and unions of fishermen or urban industrial workers, as well as community-supported agriculture. The latter links agro-ecological producers with urban consumers that value local, nonstandard, and agro-ecological products. Such patterns of exchange are not necessarily outside the cash economy, but they politicize consumption and reconstruct food exchange as a space of trust, solidarity, and proximity, against the grain of the industrial food complex and globalized circuits of capital accumulation. The political significance of such relations comes into focus when considered alongside food sovereignty movement leaders' explicit rejection of export-oriented agriculture and peasant integration into commodity chains.

> We defined ourselves immediately as being an organization against that capitalist and patriarchal [agro-export] model ... truly, the agricultural policies that there were during the period of the *Concertación*, and now even worse, were destined precisely to reduce the role of peasants. ... They gave you credit; they bought your produce; and if the dollar went down you were out of luck, which is what happened for example with those *compañeros* [fellow peasants] in the VIII Region with the famous "fattening of the tulips."[3] (ML03, 25 June 2013)

In this way, the peasant economy is sustained as part of a conscious strategy in response to the pressures to integrate with the export economy and the production systems associated with it.

The network of *curadoras* embodies an even more evidently noncapitalist set of exchange relations, based on principles of social solidarity. Participants of the network regularly meet to exchange seeds in rituals called *trafkintu*, where each *curadora* shares small amounts of seed with her peers. The term *trafkintu* is borrowed from rituals of exchange practiced by the Mapuche indigenous people of the region and represents a key dimension of ANAMURI's vision for an intercultural network. The *trafkintu* can be read not only as a performance of solidarity between individuals and organizations but also as an expression of the embodied relationships between human and nonhuman agency that lie at the heart of seed-saving practice. Seed saving and exchange is an agro-ecological response to agro-business's control over the genetic material of agricultural crops, epitomized in the invention of "terminator" seeds, genetically modified to grow plants whose own seeds are sterile. In this context, what is especially interesting about the spatial dissemination of seed diversity throughout the sociopolitical networks of the *curadoras* is that it mitigates the risk of displacement or disappearance felt both by peasant farmers and by the plants on whose generative capacities they depend.

The sociospatial practices of the *curadoras* are of particular importance but should be understood in relation to a broader set of agro-ecological practices, particularly in the way such practices embody an alternative epistemological model. Agro-ecology combines aspects of local peasant knowledge with scientific knowledge gleaned largely by NGOs through transnational circuits of innovation and exchange around nonindustrial and organic agricultural technologies. Just as the *trafkintus* distribute genetic diversity across localities, agro-ecological knowledge is shared horizontally across transnational circuits of the social movement—an alter-globalizing model of dissemination directly contrary to the hierarchical concentration of knowledge production by agro-business. The horizontal extension of agro-ecological knowledge also avoids the homogenization of agricultural practice that is so characteristic to agro-industry.[4] As knowledge flows through networks both regionally and internationally, peasants experiment with and adapt agro-ecological technologies to the specificities of their local ecosystems. In this way, the epistemological practice of agro-ecology is horizontal but not flat. The result is a diverse topology of knowledge and practice, in which the heterogeneity of place-based livelihoods is a reflection not of inward-looking localism but rather of numerous interplace and interorganizational engagements and affinities.

Food sovereignty links the sociospatial practices of agro-ecology to another suite of discourses and practices related to food consumption. At the level of the peasant household, participation in intensified agro-ecological production is also about conserving and celebrating a relationship with diverse and healthy food in the face of the homogenization of diets via global commodity chains and supermarkets. One research participant referred to this as follows:

> If the industry for example decides "let's not make any more tomatoes, no more peppers, no more carrots," then what are we going to be left with? . . . We are accustomed to the *buena mesa* (plentiful table). I call it the *buena mesa* when we can serve ourselves everything, when nothing is lacking, all the vegetables, isn't that right? (CS05, 25 September 2013)

More importantly, the role of women in the agro-ecological movement provides bridges among garden, kitchen, and peasant identity. Just as heirloom seeds are rescued and diffused across the region, traditional recipes are also experiencing resurgence, transmitting the enjoined agencies of soil, climate, seeds, and peasant producers into place-based experiences of food. This cultural dimension to food sovereignty is fundamental, not only because it conserves local diversity in the embodied socioecological act of eating but also because it provides a basis for new relationships between rural producers and urban consumers. Along these lines, the *Arca del Gusto* (Ark of Taste) initiative, started in 2012, links Chilean peasants with the global slow food movement and exemplifies agro-ecological and food sovereignty practices in this dimension. With its origins in the Biobío region, but a reach that extends to other parts of the country, the *Arca del Gusto* brings together organizations like ANAMURI, producers, and academics to raise the profile of traditional peasant foods, fomenting culinary tourism and promoting local flavors through festivals and popular markets.

Conclusions: New Agri-Food Territories?

The agri-business complex survives and grows by averaging risk across a hierarchically scaled and vertically integrated network of individually vulnerable localities. In this sense, it rests on a certain territorial logic that generates a "reserve army" of places. Agro-ecological practices and related food sovereignty movements present a diametrically opposed territorialization of the local, seeking to increase the resilience of individual places through a horizontally integrated network of producers and consumers. In other words, these competing food systems rely on very specific spatial practices focused on the way particular places are connected into broader assemblages of socioecological relations.

Perhaps what remains to be emphasized is that places themselves do not preexist the relations into which they become enfolded but rather must be conceived as "sites of negotiation" (Massey 2004, 7; see also Jessop, Brenner, and Jones 2008). They are not passively integrated into the agri-business complex, nor do they map seamlessly onto an alternate agro-ecological landscape. Rather, place-based struggles in the name of peasant livelihoods and food sovereignty are always part of intraplace contention over the shape of rural economies. At the same time, it must be emphasized that if we accept Gibson-Graham's (1996, 2006) assertion about economic heterogeneity, such contention need not be conceived as zero-sum. Both within and across places in Chile's southern agricultural landscape, an agro-ecological network can and does coexist with a hegemonic agro-industrial complex.

In this light, what does it mean to describe the Biobío's food sovereignty movement in the language of alternate economies or alter-globalization? This question raises issues from three different angles. First, given that the peasant producers who participated in this study must work against the grain of government policies, dominant political–economic relations, and pervasive cultural practices around food, their survival alongside the dominant agro-industrial model is far from guaranteed. In the same vein, asserting the possibility of coexistence does not erase the question of relative importance: Viable alternatives should be capable of growing and spreading. Second, we recognize that other observers might dispute our assertion that peasant agricultural economies can be considered alternative even when they continue to operate through forms of market exchange. Such a critique might be especially troubling to our thesis if the money that circulates in local economies is significantly derived from incomes in the mainstream economy, a dependence that is likely to increase as producers deepen their connection with urban consumers. Finally, we should not lose sight of the fact that certain places and actors are more represented than others in the movements we describe. In particular, the low participation of Mapuche women in ANAMURI calls out for further inquiry. These are some of the challenges that form an agenda for future research, but they also mark the horizon of debates among civil society and community leaders in the agro-ecology and food sovereignty movements of Chile's Biobío region.

Funding

The research, and also the collaboration between the authors, was funded by Fondecyt Initiation grant 11110020.

Notes

1. We use CS to refer to the interviewees who are *curadoras de semillas*, members of the seed saving network. We use I after the main coding to indicate when the interviewee is indigenous. All original recordings were in Spanish, with translation by the authors.
2. We use ML to refer to the interviewees who are movement leaders.
3. The fattening of the tulips refers to a state-sponsored projected to encourage peasant farmers to participate in the commercial bulb industry, where they would purchase seed bulbs and care for them until they were large enough for commercial sale.
4. Here it is important to set aside the certified organics sector. Certification is of limited importance to peasant producers due both to cost and administrative and technical requirements. At the same time it is also rejected for political reasons, precisely because it represents a top-down flow of knowledge and a loss of farmer autonomy.

References

Altieri, M. 2009. Agroecología, pequeñas fincas y soberanía alimentaria [Agroecology, small farms and food sovereignty]. *Ecología Política* 38:25–35.

Andrée, P., J. Ayres, M. J. Bosia, and M. J. Massicotte, eds. 2014. *Globalization and food sovereignty: Global and local change in the new politics of food.* Toronto: University of Toronto Press.

Araghi, F. 2000. The great enclosure of our times: Peasants and the agrarian question at the end of the twentieth century. In *Hungry for profit: The agribusiness threat to farmers, food, and the environment*, ed. F. H. Buttel, F.

Magdoff, and J. B. Foster, 145–60. New York: Monthly Review Press.
Christopherson, S., H. Garretsen, and R. Martin. 2008. The world is not flat: Putting globalization in its place. *Cambridge Journal of Regions, Economy and Society* 1 (3): 343–49.
Cid Aguayo, B. E. 2007. Frozen fates: The risky rationalization of a Chilean frozen vegetable company. *Latin American Perspectives* 34 (6): 40–51.
———. 2011. Between conventionalization and civic agriculture: Emerging trends in the Chilean agroecological movement. *Food Systems and Community Development* 1 (3): 53–66.
Clapp, R. A. 1998. Regions of refuge and the agrarian question: Peasant agriculture and plantation of forestry in Chilean Araucanía. *World Development* 26 (4): 571–89.
Echenique, J., and N. Rolando. 1991. *Tierras de parceleros ¿Dónde están?* [Lands of smallholders, where are they?]. Santiago, Chile: Agraria.
Escobar, A. 2001. Culture sits in places: Reflections on globalism and subaltern strategies of localization. *Political Geography* 20 (2): 139–74.
———. 2008. *Territories of difference: Place, movements, life, redes*. Durham, NC: Duke University Press.
Friedman, T. 2005. *The world is flat: A brief history of the twenty-first century*. New York: Farrar, Straus and Giroux.
Friedmann, H. 1995. Food politics: New dangers, new possibilities. In *Food and agrarian orders in the world economy*, ed. P. McMichael, 15–33. Westport, CT: Greenwood.
Gibson-Graham, J. K. 1996. *The end of capitalism (as we knew it): A feminist critique of political economy*. Cambridge, UK: Blackwell.
———. 2003. An ethics of the local. *Rethinking Marxism* 15 (1): 49–74.
———. 2006. *A postcapitalist politics*. Minneapolis: University of Minnesota Press.
Gibson-Graham, J. K., J. Cameron, and T. Healy. 2013. *Take back the economy: An ethical guide for transforming our communities*. Minneapolis: University of Minnesota Press.
Glover, D., and K. Kusterer. 1990. *Small farmers, big business*. New York: St. Martin's.
Gómez, S., and J. Echenique. 1988. *La agricultura Chilena, las dos caras de la modernización* [Chilean agriculture, the two faces of modernization]. Quito, Ecuador: FLACSO.
Goodman, D., and M. Redclift. 1991. *Refashioning nature: Food, ecology and culture*. London and New York: Routledge.
Goodman, D., B. Sorj, and J. Wilkinson. 1987. *From farming to biotechnology: A theory of agro-industrial development*. New York: Blackwell.
Grossman, L. 1998. *The political ecology of bananas: Contract farming, peasants, and agrarian change in the eastern Caribbean*. Chapel Hill: University of North Carolina Press.
Gutiérrez Escobar, L. M. 2011. El proyecto de soberanía alimentaria: Construyendo otras economías para el buen vivir [The project of food sovereignty: Constructing other economies for living well]. *Revista Otra Economía* 5 (8): 59–72.
Healy, S. 2009. Alternative economies. In *The international encyclopedia of human geography*, ed. N. Thrift and R. Kitchin, 338–44. Oxford, UK: Elsevier.
Healy, S., and J. Graham. 2008. Building community economies: A postcapitalist project of sustainable development. In *Economic representations: Academic and everyday*, ed. D. Ruccio, 291–314. London and New York: Routledge.
Instituto Nacional de Estadísticas (National Statistics Institute). 2007. *Censo agropecuario* [Agricultural census]. Santiago: Government of Chile.
Jessop, B., N. Brenner, and M. Jones. 2008. Theorizing sociospatial relations. *Environment and Planning D: Society and Space* 26 (3): 389–401.
Massey, D. 2004. Geographies of responsibility. *Geografiska Annaler: Series B, Human Geography* 86 (1): 5–18.
McMichael, P. 2005. Global development and the corporate food regime. *Research in Rural Sociology and Development* 11:265–99.
Miller, E. 2013. Community economy: Ontology, ethics, and politics for radically democratic economic organizing. *Rethinking Marxism* 25 (4): 518–33.
Moore, A. 2008. Rethinking scale as a geographical category: From analysis to practice. *Progress in Human Geography* 32 (2): 203–25.
Neumann, R. P. 2009. Political ecology: Theorizing scale. *Progress in Human Geography* 33 (3): 398–406.
NVivo10. 2012. Melbourne: QSR International.
Ortega-Cerdà, M., and M. Rivera-Ferre. 2010. Indicadores internacionales de soberanía alimentaria: Nuevas herramientas para una nueva agricutura [International indicators of food sovereignty: New tools for a new agriculture]. *Revista Iberoamericana De Economía Ecológica* 14:53–77.
Round, J., C. C. Williams, and P. Rodgers. 2008. Everyday tactics and spaces of power: The role of informal economies in post-Soviet Ukraine. *Social & Cultural Geography* 9 (2): 171–85.
Sevilla Guzmán, E. 2006. Agroecología y agricultura ecológica: Hacia una "re" construcción de la soberanía alimantaria [Agroecology and ecological agriculture: Towards a "re" construction of food sovereignty]. *Agroecología* 1 (1): 7–18.
Sheppard, E., and H. Leitner. 2010. Quo vadis neoliberalism? The remaking of global capitalist governance after the Washington consensus. *Geoforum* 41 (2): 185–94.
Tsing, A. L. 2005. *Friction: An ethnography of global connection*. Princeton, NJ: Princeton University Press.
Warning, M., and N. Key. 2002. The social performance and distributional consequences of contract farming: An equilibrium analysis of the Arachide de Bouche program in Senegal. *World Development* 30 (2): 255–63.
White, R. J., and C. C. Williams. 2012. The pervasive nature of heterodox economic spaces at a time of neoliberal crisis: Towards a "postneoliberal" anarchist future. *Antipode* 44 (5): 1625–44.
Whitson, R. 2007. Hidden struggles: Spaces of power and resistance in informal work in urban Argentina. *Environment and Planning A* 39 (12): 2916–34.

Williams, C. C. 2005. *A commodified world? Mapping the limits of capitalism*. London: Zed.

Williams, C., S. Nadin, P. Rodgers, and J. Round. 2012. Rethinking the nature of community economies: Some lessons from post-Soviet Ukraine. *Community Development Journal* 47 (2): 216–31.

Wittman, H. 2009. Reworking the metabolic rift: La Vía Campesina, agrarian citizenship, and food sovereignty. *Journal of Peasant Studies* 36 (4): 805–26.

Wittman, H., A. A. Desmarais, and N. Wiebe, eds. 2010. *Food sovereignty: Reconnecting food, nature and community*. Halifax, NS: Fernwood and FoodFirst Books.

School Gardens as Sites for Forging Progressive Socioecological Futures

Sarah A. Moore,* Jeffrey Wilson,† Sarah Kelly-Richards,† and Sallie A. Marston†

*Department of Geography, University of Wisconsin, Madison
†School of Geography and Development, University of Arizona

In this article we approach school gardens as sites of socioecological change where experiential politics work through the establishment of sustainable and socially just practices. We argue that for some children in "struggling schools," school gardens become spaces where the alienating aspects of neoliberal school reform in the United States can be overcome by forging connections with classmates, university students, plants, and animals. In these intimate urban ecologies, affective and playful labor become the bases for knowledge production that exceeds the disciplinary functions of standardized testing, individual achievement, and accountability emphasized in neoliberal school reform. Our empirics derive from garden projects involving university interns and school children in two underresourced schools in poor neighborhoods in Tucson, Arizona.

我们在本文中，将学校花园视为社会生态变迁的场域，其中实验性的政治，透过建立可持续发展与社会正义的实践发挥作用。我们主张，对部分"挣扎中的学校"的儿童而言，学校花园透过打造与同学、大学生、植物及动物的连结，成为克服美国新自由主义学校重构的异化面向之空间。在这些亲密的城市生态中，情感与嬉戏般的劳动，成为知识生产的基础，该知识生产，超越了新自由主义学校重构所强调的标准化测验、个人成就及可究责性这些规训式的功能。我们的经验主义，衍生自亚历桑那州土桑市的贫困邻里中，两座资源匮乏的学校纳入大学实习与学校儿童的花园计画。

En este artículo observamos los jardines de las escuelas como sitios de cambio socioecológico donde operan políticas experienciales a través del establecimiento de prácticas sustentables y socialmente justas. Sostenemos que para algunos niños de "escuelas en dificultades," los jardines escolares se convierten en espacios donde los aspectos alienantes de la reforma neoliberal de la escuela en Estados Unidos pueden ser derrotados al forjar conexiones con los compañeros de clase, estudiantes universitarios, plantas y animales. En estas ecologías urbanas íntimas el trabajo afectivo y lúdico se convierte en la base de producción de conocimiento que sobrepasa las funciones disciplinarias de los exámenes estandarizados, el logro individual y la responsabilidad enfatizados por la reforma neoliberal de la escuela. Nuestros datos empíricos provienen de proyectos de jardinería que involucran pasantes universitarios y escolares en dos escuelas escasas de recursos de las barriadas pobres de Tucson, Arizona.

The resurgent popularity of school gardens in the United States might indeed be reason to think that they are a solution in search of a problem (cf. Guthman 2011, 6). In fact, throughout their history in the country, school gardens have been lauded as solutions to problems ranging from the loss of rural knowledges and ways of being with the urbanization of the country in the first decades of the twentieth century (Clapp 1901; P. K. Miller 1904; Jewell 1906; L. K. Miller 1908; Greene 1910), to childhood and community obesity in the first decades of the twenty-first century (Robinson-O'Brien, Story, and Heim 2009). School gardens, though, have also been critiqued for reproducing neoliberal subjects who look to self-help and social entrepreneurialism to solve problems of access to food and other local development issues (Guthman 2008a; Pudup 2008). Further concerns have been raised over how school gardens and other nutrition education projects aimed at children are discussed as solutions to the "obesity epidemic" (Guthman and DuPuis 2006; Colls and Evans 2009, 2010; Guthman 2009) and risk demonizing fatness and normalizing certain (white, thin) bodies.

In this article, we engage this debate by drawing on several years of experience with a service-learning course that places undergraduate and graduate student interns in school gardens in Tucson, Arizona. It is clear that school gardens are not a panacea for all the ills of modern education and socioecological relations. It is equally clear, however, that school gardens are diverse urban ecologies through which to begin to address a range of problems. As sites where experiential learning and affective labor intertwine, we argue, school gardens have the potential to expand knowledge acquisition beyond an individualistic, atomistic, neoliberal view of nature and social relations toward one based on solidarity with and mutual care for human and nonhuman others. We support these claims with data from an experimental research project conducted in two underresourced urban schools in Tucson.

In the next section, we frame our project in the context of broader urban agriculture and alternative food movements. We then contrast the knowledge production practices of neoliberal school reform with the subjectivities and knowledges that can emerge through affective and playful labor in school gardens. After describing our program, sites, and methods, we focus on one example of affective and playful labor at each of two research sites. We conclude by arguing that we can choose to foster these capacities by creating collectives that cross university–community divides.

School Gardens in Context

Although the literature on urban agriculture and alternative food projects (AFPs) is too vast to discuss fully here, it raises several critiques pertinent to evaluating school gardens' potential transformative effects on their participants. First, AFPs often rely on strategies and narratives of self-help and provisioning that might reinforce, rather than challenge, existing socioecological relations and their concomitant inequalities (Guthman 2008c). They can do this by providing stop-gap measures that accompany a general rolling back of the state under neoliberal regimes or by disciplining citizen subjects to participate in individualizing programs, rather than making demands on the state or engaging in radical practice. Second, and relatedly, AFPs can be exclusionary, middle- or upper-class, and largely white, both in membership and in goals (Guthman 2008a, Guthman 2008b; Alkon and Mares 2012). In contrast to these critical stances, some argue that urban gardens are better understood as actually existing commons that resist the neoliberal city and governance (Eizenberg 2012a, 2012b) and that many urban agriculture projects might entail other moments for a radical critique of the status quo (Galt, Gray, and Hurkey 2014). Recent analyses have argued that all such projects, in fact, include several forms of noncapitalist and capitalist relations, as well as varied modes of subject formation (McClintock 2014). There is, then, no reason to assume that any particular AFP must necessarily reproduce neoliberal subjects. They might do so in some ways, while creating socioecological relations that potentially exceed, escape, or challenge this neoliberal disciplining in others. These potentials have been understood usefully as "enchantment" (Bhatti et al. 2009) and as "visceral politics" (Hayes-Conroy and Hayes-Conroy 2008, 2010).

The same is true of school gardens. Whereas some argue (Allen and Guthman 2006; Pudup 2008) that such programs reproduce neoliberal subjects, others (Hayes-Conroy 2011) argue that the results of such projects are more indeterminate. Hayes-Conroy (2011), in particular, finds that although school gardens can promote neoliberal ideals, they do not preclude complex and contradictory daily, material interactions that might also undermine them. In this article, we discuss some of these material interactions and the knowledges and subjectivities they enable by focusing on affective and playful labor in school gardens.

Affective labor, which produces "knowledge, information, communication or an emotional response" rather than goods (Hardt and Negri 2004, 109), is a form of immaterial labor that dominates post-Fordist production systems (Hardt 1999). It is work that engages both rational intelligence and passions or feeling. Although affective labor's creative aspects are sometimes commodified or coopted for the extension of capitalist enterprises, it also comprises an excess that might provide the seeds for new ways of being and doing (Hardt 1999; Hardt and Negri 2004). Although the political potentials of affective labor have been debated within geography (Ruddick 2010), Singh (2013) finds that it is fundamental to understanding how "environmental subjects" are created in the process of caring for forests. These are subjects created not through the disciplining strategies of Foucault's (neoliberal) environmentalities but, rather, through a "biopower from below" that challenges neoliberal socioecological relations and that focuses on "affective relations, cooperation and communication" (Singh 2013, 197). The caring labor to which Singh refers resonates with feminist analyses of reproductive labor (Hardt 1999; Weeks 2007). In the case of school gardens, this affective labor happens alongside the material production of the garden as well as in concert with play. Playful labor (Katz 2011), during which children both reenact contemporary socioecological relations and enact alternative imaginations, constitutes an important aspect of the school gardens in which we work. As Katz (2011) describes, the "tactile knowing" involved in play and the "wild imaginings" associated with it have "revolutionary potential," as these "fantasies and reveries are reservoirs for thinking and making new ways of living" (56). The political potential of imagination is also highlighted by Graeber (2009), who argues that if we "stop thinking of the imagination as largely about the production of free-floating fantasy worlds, but rather as bound up in the process by which we make and maintain reality, then it makes perfect sense to see it as a material force in the world" (523).

As Katz (2011) emphasizes, however, not every child is equally valued or encouraged to imagine and enact different futures. In an analysis of the figure of the child as waste, she discusses how certain children are constructed as disposable and subsequently dispossessed of adequate educational opportunities. This is a salient point in contextualizing school gardens in terms of their relationship to a broader political economy that produces both segregated neighborhoods and schools. Public schools in the United States today are more segregated by class and race than at any time in the past forty years (Orfield, Kucsera, and Siegel-Hawley 2012). This spatial segregation aligns with significant differences in curriculum, building conditions, and other educational components between underresourced urban schools and better funded suburban schools (Kozol 2005; Lipman 2007, 2011a, 2011b; Niesz 2010; Katz 2011). Rather than attending to the larger scale political and economic processes influencing these differences, policy rhetoric in the United States describes urban schools as localized sites of "failure," as sites of disorder and underachievement in need of large-scale neoliberal education reform (Lipman 2009; Lizotte 2013). In these reform efforts, creative pursuits, group learning and problem solving, recreation, and holistic learning are devalued in favor of intensified classroom efforts to raise individual scores on standardized tests, creating "a violent reification of human consciousness and creativity" (De Lissovoy and McLaren 2003, 132).

Seeking a universal equivalent to measure children's educational attainment is especially problematic when one considers the evidence that most standardized tests are culturally biased in favor of middle-class white students (De Lissovoy and McLaren 2003). Given this evidence, some have argued that what is confronted in the reform movement is "really a large-scale *staging* of the failure of students of colour" (De Lissovoy and McLaren 2003, 135).

This "staging of failure" is arguably underway in neoliberal educational reform in Tucson, Arizona, where, as in states and cities across the country, there is a broad move to disinvest in education—at all levels, from the university to the preschool. In 2013, the Tucson Unified School District (TUSD) closed eleven schools, all in low-income or lower-middle-income urban neighborhoods. This move dovetailed with xenophobic attacks at the state level directed at TUSD. Arizona House Bill 2281 (HB 2281) was created to eliminate the highly successful Mexican American Studies (MAS) program at Tucson High School, part of TUSD. The bill states no course shall be "designed primarily for pupils of a particular ethnic group." It further argues that "public school pupils should be taught to treat and value each other as individuals," and prohibits "any classes that . . . advocate ethnic solidarity" (HB 2281). As described by one University of Arizona faculty member, HB 2281 "augers a broader effort to bar access to thinking relationality. . . . The threat is not usually so explicit. The spaces and times for relational thought usually just get swamped by the flood of resources—material, institutional, cultural—that flow in support of the specialized bourgeois knowledge production, the production of one-sided knowledge in service to capital accumulation" (Crosby et al. 2012, 145). In its attempts to individualize knowledge production in classrooms and limit relational thinking and solidarity, HB2281 is part of broader neoliberal assaults on Tucson's public schools. As such, it is an example of what Graeber (2009) calls bureaucratic violence: "If violence is a force capable of radically simplifying complex social situations, if bureaucracy is largely a method of imposing such a simplistic rubric systematically, then bureaucratic violence should logically, consist first and foremost of attacks on those who insist on alternative interpretations" (519).

In contrast to neoliberal school reform that perpetuates this bureaucratic violence through standardized testing, individuation, and reification of human consciousness, school gardens might present an opportunity for "thinking relationality" through the creation and circulation of what Haraway (2008) calls "encounter value" in affective and playful labor. It is her hope that we can use encounter value to understand how, in "making companions," we might exceed the simple reproduction of society and its exploitative capitalist logics (65). In what follows, we describe how, through our work in school gardens, we have tried to foster such encounters.

The University of Arizona School and Community Garden Project

Background

The University of Arizona School and Community Garden Project (UASCGP) was developed in cooperation with a federally funded project awarded to the Community Food Bank of Southern Arizona (Food Bank) to introduce healthy eating to school children in underresourced schools. The Food Bank supported these efforts between 2010 and 2012 through infrastructural projects funded by federal stimulus dollars in the form of the Communities Putting Prevention to Work Initiative (CPPW). Through its Community Resources Center, the Food Bank continues to provide technical support to the school and community gardens initiated during the grant period. At the time of the grant, a few dedicated individuals at several schools had been struggling to create and maintain alternative educational experiences involving school gardens and ecological initiatives for their students. To help maintain such programs, faculty at the University of Arizona worked with the Food Bank to establish a service-learning course that placed undergraduate and graduate student interns in local schools. When the course began in 2008, six student-interns worked in three schools. During the 2013–2014 school year, more than fifty students per semester enrolled in the course and were placed at sixteen sites, mostly in the Tucson area but spanning the interstate corridor north to Phoenix.

The service-learning course involves weekly classroom sessions where interns study various aspects of school gardens and urban agriculture, including the racial and class dynamics affecting participation in and the effectiveness of such projects. The students additionally learn basic gardening skills involving the cultivation of native and native-adapted plants through the Food Bank. Most of the students' time, though, is spent working in school and community gardens. Interns assigned to schools spend ten or more hours a week working with schoolchildren in the garden. Each intern accumulates more than 150 hours of contact time per semester and many become important members of the school and local community. As the program has grown, we have observed informally that interns' work has had positive impacts on the school gardens (some turned from abandoned green or brown space to productive gardens), the teachers, and the schoolchildren. We have also noted positive changes in the interns themselves.

Methods

These observations led us to begin more formal research during the 2012–2013 academic year into whether and how the garden spaces affected the socio-ecological imaginaries of those involved. This research involved two major groups: undergraduate interns and the schoolchildren with whom they worked. Data collection and initial analyses were undertaken by six of the undergraduate interns. They enrolled in a companion course for the internship, where, with the guidance of two graduate student instructors, they engaged in learning about and practicing participant observation, auto-ethnography, and interviewing. Interns also conducted oral histories with school staff engaged in the use of the garden for functional learning at each site and engaged in an experimental research project with the school children with whom they worked. The difficulties of collecting data that reflect children's understandings of the environment, including school gardens, are well documented (Wake 2008). We elected to follow Luttrell (2010; cf. Yates 2010) in attempting a photograph-based project. Schoolchildren were given cameras and asked to take pictures of things in the garden that were important to them. They were then asked to use these to tell a story to a friend or relative about what the garden space meant to them. The children wrote these narratives or told them to the intern-researcher (depending on the ages and preferences of the children).

The Sites

Interns from the UACSGP assigned to the Davis Bi-Lingual Elementary Magnet School garden in 2010 helped one of the teachers and a highly engaged parent revitalize the garden space, which had been dormant for over a decade. With frequent workdays at the school, parents and other family members, as well as the University of Arizona interns, are collectively maintaining and expanding the Davis school garden. In fall 2013, Davis was told that its eligibility as a magnet school was being evaluated and that it would lose that status, and associated grant monies, if it did not more fully integrate (Wallace 2013). Although Davis's award-winning mariachi and folklore programs are hallmarks of the school, its highly successful garden is increasingly being seen as an opportunity for recruiting new non-Hispanic students.

Like the Davis garden, the Manzo Elementary School ecology program including a school garden has had its fits and starts. When the school counselor arrived in 2006, the space had been neglected for ten years and was overrun with weeds and trash. Since then, and with initial help from a University of Arizona ecology faculty member, the counselor has directed and supported the installation of native habitat sites, gardens, and an orchard; installed multiple water harvesting cisterns and an aquaponic system; established a chicken coop and composting program, including a vermiculture bin; and built and installed a greenhouse. He has resourced these projects through grant writing, donations, and volunteer labor, including that of family members of the school's children, as well as through the cooperation of the Food Bank and other local institutions. The ecology program at Manzo (elected Best Green School of 2012 by the U.S. National Green Building Council) was originally oriented around counseling for students experiencing behavioral issues. Although it continues to be used for this purpose, student behavioral problems have decreased dramatically since the ecology program was started. The counselor believes that this is due to the effects of direct student involvement in building and maintaining the program's various components.

Manzo was threatened with closure in December 2012. With an enrollment of around 280 students and low scores in math, the district publicly proposed to transfer Manzo students to a larger nearby campus to generate cost savings and help address a $17 million district-wide budget deficit (Echavarri 2012). Manzo staff, family, University of Arizona interns, and friends of the school successfully protested the closure. In May 2013, its official Arizona's Instrument to Measure Standards math scores were reported as significantly improved and another sixty students were added to its roster in Fall 2013.

The Socioecological Value of School Gardens

In this section, we discuss two examples where garden project participants demonstrate socioecological imaginaries emerging from affective and playful labor. First, we describe the composting project at Manzo and then we turn to other affective labors in the Davis garden. In both, we highlight how traditional ideas of learning and individual achievement can be altered as groups undertake garden tasks through mutually negotiated organization. These examples, although not exhaustive, do provide insight into how changes in routine practices hold the possibility for fostering relationships of solidarity and belonging between people, as well as between people and nonhuman others.

Composting at the Manzo Garden

There are many tasks associated with maintaining a school garden. At Manzo Elementary, one daily activity in which children participate is composting. As Intern I describes, this task is accomplished in a regimented way: "A checklist is used to make sure the kids did all of their duties for compost including putting tools away, cleaning up around the compost area, washing out buckets, and feeding the chickens and worm bin" (Intern I, field notes, 9 November 2012).

Students are thus expected to be task-oriented and perform the necessary labor to produce compost. Adherence to schedules and rules is critical to a functional composting system. This process, though, also differs in important ways from more regimented classroom learning. It is premised on mutual learning that emphasizes relationality through affective and playful labor, while producing the material landscape of the garden. The children aid one another in each task and those with more experience help those with less. As one intern explained, "Two kids from the younger grades and two from fifth grade are assigned to be on compost duty for a day where they help kids done with their lunch to throw away, compost, and recycle their lunch. One kid assigned to the younger compost group wasn't sure about what could be composted" (Intern I, field notes, 9 November 2012). The less experienced composter was then shown how to do the task by more experienced children. In this case, children are unevenly created as "composters" and new composters are able to join the process through observation, experimentation, and imitation. In this case, an incremental change happens to a previously oppositional composter:

> The child, E, a third-grader who was slightly short and chubby with short black hair, who earlier in the semester dumped his entire tray of food into the trash and walked off back to his table, and I had to scold to get

him to come back and compost his food, came again to the compost line. This time he just stood there holding his tray and posing as if he were just going to put all of it in the trash. The assigned female composter [another student] put her hand out to stop him from putting it all into the trash. She pointed to his recyclables and asked, "Where do you put this?" And he motioned towards the recycling, she took it from his tray and put the milk carton and plastic into the recycling, and did the same for everything else on his tray. Although he didn't put the items in himself, he knew where some of them went and let her help him actually do the proper composting. Although he still seems to disagree with the process he is now more reluctant to throw his whole tray into the trash and leave because he knows it won't work. (Intern I, field notes, 21 November 2012)

Although such reluctant participation in composting could be viewed as an effect of neoliberal environmental governance, it is also an opportunity for students to educate one another and for some students to occupy leadership roles around environmental stewardship. It is further a task where playful and affective labor can be enjoyed. As Intern I observed of another child:

> Generally students will ask me if they can help with compost. It was an interesting contrast that he [V] just walked up and started doing it without asking. After the two composters assigned to compost that day came he stayed and helped along with them. The three of us leaned against the white cafeteria wall and watched as the boy [V] marched quickly along the trash bins and sang "I'm on patrol, I'm on patrol" . . . I laughed and the kids mentioned how weird they thought he is, and that he's always weird like that. I didn't stop him at all, the kids suggested that we didn't need his help, and I said, "No way, he's having too much fun." My thoughts were that it's great that he's so excited about helping out with composting. Not all kids can enjoy themselves while helping throw away kids' half eaten food. (Intern I, field notes, 21 November 2012)

Here, the student V joins in composting, even though he is not assigned as a monitor for that day. He performs the task, helping to reproduce the material landscape of the garden where the compost is used, while exuding enjoyment in becoming an environmental steward. These examples from the interns' participant observations of school children working in the Manzo garden are complemented by narratives of the children who were asked to take pictures and tell stories explaining what garden meant to them. Several of them emphasized the importance of the composting program in making their experience at Manzo special or unique. One first-grader, for example, took pictures of flowers and seeds in the greenhouse and explained to her cousin to whom her story was addressed that she "helped the school counselor with composting" so that the plants "can grow bigger like you see in the garden at Manzo." She goes on to describe another picture she took of the chickens that "eat cafeteria food that the students give to them." The child's knowledge of the importance of the compost for sustaining the other life of the garden highlights an understanding of ecological relationality shaped through her and others' material, affective, and playful labor in the space.

Other Forms of Affective and Playful Labor in the Davis Garden

In this section, we focus on one undergraduate intern who, following her graduation, became an employee of the Davis school garden. In these passages, students of different ages, along with nonhuman actors, contributed to a transformation in the intern's understanding of learning and education.

> At the same time as I have been trying to figure out respect in the garden, I have been trying to overcome the things that I have normalized through my seventeen years of schooling, such as rules and order. I have to balance my own internalized notions of education structure with the realities and constrictions of the school, and with the opportunity and freedom the garden offers. (Intern A, field notes, 15 December 2012)

Here, Intern A links "respect" to the different type of learning the garden provokes, equating what she at first thought was a lack of respect in the space with the hierarchical structures she had "normalized" during her own education. She concludes that the garden offers something different both from her personal experience and from the other spaces of the school. In this garden space with chickens, harvested rainwater, pollinators, and altars, a different learning community constituting alternative subjectivities emerges.

In this second passage, Intern A again reconfigures her previously unexamined epistemological position toward respect and proper behavior in the Davis school garden:

> Maybe climbing on top of that altar/statue isn't disrespect. Sure it shows not everyone knows why it's there. But M. does and she still climbs on it. She also cleans it

out, puts new flowers in the altar everyday. "Do we have a broom?" She's paying homage through that and what kid wouldn't want to be remembered through something that brings fun, joy, momentary thrill. So maybe I've got it wrong—climbing on the altar isn't necessarily disrespect, but a more honest and pure respect, an unconscientious [sic], unknowing respect. Paying homage through climbing on top of it. (Intern A, field notes, 24 September 2012)

The altar, a tribute to a school child that died, is a commemorative, sacred object that the intern originally views differently than the schoolchildren. The affective labor of this young girl in maintaining the space, changing the flowers daily, makes Intern A realize that it is acceptable to reappropriate this sacred object because she is witnessing "a more honest and pure respect." She recognizes that, as schoolchildren, interns, and other staff create a new set of rules or customs within their emerging community, homage and respect are embodied acts, which she describes as unconscious or "unknowing." Although her field notes capture a sustained grappling with her own authoritarian impulses to reproduce the same hierarchies that frustrated her during her own education, we also see in Intern A's words the emerging importance of affective and playful labor as reconstituting her own subjectivity and socioecological imaginary. The importance of these labors is similarly highlighted by a second-grader, M, who works in the Davis garden. Having taken and selected three pictures of the garden space (a photo of Intern A and three children hoeing, one of another child pushing a wheelbarrow full of dirt, and one of a chicken in a coop) he was unsure of how to proceed with his story. Intern A, having learned from previous attempts to explain the project to other children, reflects:

The way I explained it to M was essentially the same, but I said: "You're basically writing your story, because remember when you were taking your photos you're thinking of each photo as the story you're telling? And so now you're going to put words to your photos; it doesn't necessarily have to be a description of what's in the photos, it's more kind of how the photos and what you took in the photos relate to your garden experience. . . . So it's up to you, be creative, it's your choice how you want to write it." . . . I think the difference was here, when he looked at his old titles and asked "Do I have to think of a title for them?" and I said "Don't think about the titles too much, just focus on how it connects to your experience in the garden." (Intern A, field notes, 24 September 2012)

Given the freedom to decide if he wants to write about each photo individually or about the three as a group, M decides to write about them together (spelling and grammar as in the original handwritten document):

The story my photos tell is that the garden is awesome for everybody. It shows that people work hard and work as a team. My pictures show that we have friends in the garden with us. These pictures make me feel happy and exited to be a Davis kid and to be at this comunity garden. I care about this because without this we would not have garden workdays and things like that. This reminds me of a football game or a basketball game because we work as a team. Without everybody that helped this garden, it would not be anything. This influences on me alot because it was really fun. So come see are community garden. (Davis Garden Research Stories, Group 1, 3 March 2013)

M links his garden experience to playful and affective labor, associating it with team sports. At the same time, he draws attention to the way the school garden is a site of interaction for the neighborhood by referring to it repeatedly as a "community garden." This resonates with Intern A's experience as she credits her involvement in the UASCGP and the Davis garden for motivating her to continue her college education by embedding her in a Tucson community from which she had originally felt alienated.

Conclusion

In this article, we have emphasized how school gardens can work as sites of emerging transspecies relations through affective and playful labor. Thus, as standardized educational discourses and practices flood the schools with universal equivalencies and individuating expectations, other discourses and practices are also present. Although neoliberal disciplining ideologies and regimes might be in force in schools across the country, alternative futures can be fostered in the midst of school gardens, providing leverage for different ways of being and doing. As Haraway (2008) and others remind us, relationships with lively beings such as plants and compost, as well as other humans can affect new ethics and enable noncapitalist logics to take hold.

By viewing school gardens in underresourced schools as sites of potential socioecological transformation, we are not dismissing the very real political and economic challenges faced by public schools in resource-strapped urban school districts. Nor are we arguing that such sites are necessarily progressive or emancipatory. Rather, we drew on our experiences to demonstrate that, through consistent practice, centered on affective and playful labor, the diverse ecologies of many school gardens present opportunities to emphasize relationality and solidarity in school space. Through the creation and expansion of a service-learning course that places undergraduate interns in school gardens and other urban agriculture sites, we have been able to bridge university–public school–community divides and find moments of collective knowledge production and learning that can reshape the socioecological imaginaries of participants.

Acknowledgments

We wish to express our gratitude to the students and staff at Manzo, Davis, and Borton Elementary Schools as well as to our intern-researchers, Ylenia Aguilar, André Luiz Domingues, Iris Gishkin, Amy Mellor, Emily Mormino, and Yaqi Rong, for their participation in the research reported on here.

Funding

This research was funded by a grant from the University of Arizona Green Fund.

References

Alkon, A. H., and T. Mares. 2012. Food sovereignty in US food movements: Radical visions and neoliberal constraints. *Agriculture and Human Values* 29 (3): 347–59.

Allen, P., and J. Guthman. 2006. From “old school” to “farm-to-school”: Neoliberalization from the ground up. *Agriculture and Human Values* 23 (4): 401–15.

Bhatti, M., A. Church, A. Claremont, and P. Stenner. 2009. ‘I love being in the garden’: Enchanting encounters in everyday life. *Social & Cultural Geography* 10 (1): 61–76.

Clapp, H. L. 1901. School gardens. *Education* 21 (10): 611–17.

Colls, R., and B. Evans. 2009. Introduction: Questioning obesity politics. *Antipode* 41 (5): 1011–20.

———. 2010. Re-thinking ‘the obesity problem.’ *Geography* 95:99–105.

Crosby, C., L. Duggan, R. Ferguson, K. Floyd, M. Joseph, H. Love, R. McRuer, et al. 2012. Queer studies, materialism, and crisis: A roundtable discussion. *GLQ: A Journal of Lesbian and Gay Studies* 18 (1): 127–47.

De Lissovoy, N., and P. McLaren. 2003. Educational ‘accountability’ and the violence of capital: A Marxian reading. *Journal of Education Policy* 18 (2): 131–43.

Echavarri, F. 2012. Closer look at TUSD school closures. *Arizona Public Media* 4 December. https://www.azpm.org/p/top-news/2012/12/4/19756-a-closer-look-at-tusd-school-closures/ (last accessed 21 July 2014).

Eizenberg, E. 2012a. The changing meaning of community space: Two models of NGO management of community gardens in New York City. *International Journal of Urban and Regional Research* 36:106–20.

———. 2012b. Actually existing commons: Three moments of space of community gardens in New York City. *Antipode* 44 (3): 764–82.

Galt, R. E., L. C. Gray, and P. Hurkey. 2014. Subversive and interstitial food spaces. *Local Environment* 19 (2): 133–240.

Graeber, D. 2009. *Direct action: An ethnography*. Oakland, CA: AK Press.

Greene, M. L. 1910. *Among school gardens*. New York: Charities Publication Committee.

Guthman, J. 2008a. Bringing good food to others: Investigating the subjects of alternative food practice. *Cultural Geographies* 15 (4): 431–47.

———. 2008b. “If they only knew”: Color blindness and universalism in California alternative food institutions. *The Professional Geographer* 60 (3): 387–97.

———. 2008c. Neoliberalism and the making of food politics in California. *Geoforum* 39 (3): 1171–83.

———. 2009. Teaching the politics of obesity: Insights into neoliberal embodiment and contemporary biopolitics. *Antipode* 41 (5): 1110–33.

———. 2011. *Weighing in: Obesity, food justice, and the limits of capitalism*. Berkeley, CA: University of California Press.

Guthman, J., and M. DuPuis. 2006. Embodying neoliberalism: Economy, culture, and the politics of fat. *Environment and Planning D* 24 (3): 427–48.

Haraway, D. 2008. *When species meet*. Minneapolis: University of Minnesota Press.

Hardt, M. 1999. Affective labor. *Boundary 2* 26 (2): 89–100.

Hardt, M., and A. Negri. 2004. *Multitude: War and democracy in the age of empire*. New York: Penguin.

Hayes-Conroy, A., and J. Hayes-Conroy. 2008. Taking back taste: Feminism, food and visceral politics. *Gender, Place and Culture* 15 (5): 461–73.

———. 2010. Visceral difference: Variations in feeling (slow) food. *Environment and Planning A* 42 (12): 2956–71.

Hayes-Conroy, J. 2011. School gardens and “actually existing” neoliberalism. *Humboldt Journal of Social Relations* 33 (1–2): 64–96.

Jewell, J. R. 1906. The place of nature study, school gardens, and agriculture in our school system. *Pedagogical Seminary* 13 (3): 273–92.

Katz, C. 2011. Accumulation, excess, childhood: Toward a counter-topography of risk and waste *Documents D'Anàlisi Geogràfica* 57 (1): 47–60.

Kozol, J. 2005. *The shame of the nation: The restoration of apartheid schooling in America*. New York: Crown.

Lipman, P. 2007. Education and the spatialization of urban inequality: A case study of Chicago's Renaissance

2010. *Spatial Theories of Education: Policy and Geography Matters* 9:155–74.
———. 2009. The cultural politics of mixed-income schools and housing: A racialized discourse of displacement, exclusion, and control. *Anthropology Education Quarterly* 40 (3): 215–36.
———. 2011a. Contesting the city: Neoliberal urbanism and the cultural politics of education reform in Chicago. *Discourse: Studies in the Cultural Politics of Education* 32 (2): 217–34.
———. 2011b. *New political economy of urban education: Neoliberalism, race, and the right to the city*. London and New York: Routledge.
Lizotte, C. 2013. The moral imperatives of geographies of school failure: Mobilizing market-based reform coalitions. *The Canadian Geographer* 57 (3): 289–95.
Luttrell, W. (2010). "A camera is a big responsibility": A lens for analysing children's visual voices. *Visual Studies* 25 (3): 224–37.
McClintock, N. 2014. Radical, reformist, and garden-variety neoliberal: Coming to terms with urban agriculture's contradictions. *Local Environment* 19 (2): 147–71.
Miller, L. K. 1908. School gardens. *Elementary School Teacher* 8 (10): 576–80.
Miller, P. K. 1904. School gardens in their relations to the three R's. *Education* 25 (2): 531–42.
Niesz, T. 2010. "That school had become all about dhow": Image making and the ironies of constructing a good urban school. *Urban Education* 45 (3): 371–93.
Orfield, G., J. Kucsera, and G. Siegel-Hawley. 2012. E pluribus separation: Deepening double segregation for more students. http://civilrightsproject.ucla.edu/research/k-12-education/integration-and-diversity/mlk-national/e-pluribus...separation-deepening-double-segregation-for-more-students/orfield_epluribus_revised_omplete_2012.pdf (last accessed 21 July 2014).
Pudup, M. B. 2008. It takes a garden: Cultivating citizen-subjects in organized garden projects. *Geoforum* 39 (3): 1228–40.
Robinson-O'Brien, R., M. Story, and S. Heim. 2009. Impact of garden-based youth nutrition intervention programs: A review. *Journal of the American Dietetic Association* 109 (2): 273–80.
Ruddick, S. 2010. The politics of affect: Spinoza in the work of Negri and Deleuze. *Theory Culture & Society* 27 (4): 21–45.
Singh, N. M. 2013. The affective labor of growing forests and the becoming of environmental subjects: Rethinking environmentality in Odisha, India. *Geoforum* 47:189–98.
Wake, S. (2008). "In the best interests of the child": Juggling the geography of children's gardens (between adult agendas and children's needs). *Children's Geographies* 6 (4): 423–35.
Wallace, J. D. 2013. Parents and teachers defend TUSD magnet schools. *Tucson News Now* 25 September. http://www.tucsonnewsnow.com/story/23521847/parents-and-teachers-defend-tusd-magnet-schools (last accessed 29 November 2013).
Weeks, K. 2007. Life within and against work: Affective labor, feminist critique, and post-Fordist politics. *Ephemera: Theory and Politics in Organization* 7 (1): 233–49.
Yates, L. (2010). The story they want to tell, and the visual story as evidence: Young people, research authority and research purposes in the education and health domains. *Visual Studies* 25 (3): 280–91.

From Incremental Change to Radical Disjuncture: Rethinking Everyday Household Sustainability Practices as Survival Skills

Chris Gibson, Lesley Head, and Chantel Carr

Australian Centre for Cultural Environmental Research, University of Wollongong, Wollongong, Australia

Households within affluent countries are increasingly prominent in climate change adaptation research; meanwhile, social and cultural research has sought to render more complex the dynamics of domesticity and home spaces. Both bodies of work are nevertheless framed within a view of the future that is recognizable from the present, a future reached via socioecological change that is gradual rather than transformative or catastrophic. In this article, we acknowledge the agency of extreme biophysical forces and ask what everyday household life might be like in an unstable future significantly different from the present. We revisit our own longitudinal empirical research examining household sustainability and reinterpret key results in a more volatile frame influenced by political ecological work on disasters. We seek to move beyond incremental to transformative conceptions of change and invert vulnerability as capacity. Vulnerability and capacity are contingent temporally and spatially and experienced intersubjectively. The resources for survival are ultimately social and therefore compel closer scrutiny of, among other things, household life.

富裕国家中的家户，在气候变迁的调适研究中日益重要；同时，社会及文化研究，已开始展现居家性与家庭空间之间更为复杂的动态关係。上述两类研究，仍然是在一个对当下而言可指认的未来之观点中形成，而这样的未来是透过渐进式的社会生态改变、而非透过革命性或激烈的变迁达到。我们在本文中，承认极端生物物理趋势的作用，并质问在显着不同于当下的不稳定未来之中，家户的每日生活将会是如何。我们重探自身检视家户可持续性的长程经验研究，并在一个受到政治生态学的灾难研究所影响的更为多变的架构中，重新诠释主要的研究发现。我们超越渐进式的变迁概念，寻求革命性的变迁概念，并将脆弱性的概念转化为能力。脆弱性与能力，在时间及空间上具有偶然性，并且在相互主体性的层级中经验之。生存的资源，最终是社会的，因此必须对家户生活及其他事物进行更仔细的检视。

Los hogares de los países ricos son cada vez más importantes en la investigación sobre adaptación al cambio climático; entretanto, la investigación social y cultural ha buscado hacer más compleja la dinámica de la domesticidad y los espacios hogareños. No obstante, ambos cuerpos de trabajo se enmarcan dentro de una visión del futuro reconocida a partir del presente, un futuro alcanzado a través del cambio socioecológico, más gradual que transformador o catastrófico. En este artículo, damos crédito a la agencia de fuerzas biofísicas extremas y nos preguntamos cómo podría lucir la vida cotidiana del hogar en un futuro inestable significativamente diferente del presente. Volvemos sobre nuestra propia investigación empírica longitudinal examinando la sustentabilidad hogareña, y reinterpretamos resultados claves en un esquema mucho más volátil influido por el trabajo ecológico político sobre desastres. Buscamos ir más allá de las concepciones del incremento gradual del cambio a las concepciones transformadoras e invertimos la visión de la vulnerabilidad como capacidad. La vulnerabilidad y la capacidad son temporal y espacialmente contingentes y se ganan como experiencia de modo intersubjetivo. En últimas, los recursos de la supervivencia son sociales y por lo tanto demandan un examen más minucioso de la vida hogareña, entre otras cosas.

Households in the affluent West are increasingly present in sustainability discourse, in hazards management, and in visions of how to mitigate and adapt to climate change (Reid, Sutton, and Hunter 2009; Waitt et al. 2012; Eriksen 2013). At the same time, social and cultural research has sought to trouble the household as a bounded or stable entity—revealing complexities of family life and intersections between people, nature, objects, work, and practices within and beyond the home (Blunt and Dowling 2006; Gregson 2007; Pink 2012; Cox 2013). The microscopic details of household life matter enormously: They can be conceived as the cellular activity fueling climate change and shaping preparedness, whereas questions of family, relationships, and livelihood are what most people care and worry about.

Meanwhile, in research on preparedness for hazards and emergencies, the household is one scale at which to observe vulnerability and resilience. Capacities to respond to severe disruptions are shaped by underlying inequalities and social differences (Sevoyan et al. 2013) and complex place- and path-dependent factors including social cohesion and community networks (Prior and Eriksen 2013). A related, critical literature also considers the security and risk management apparatuses increasingly governing citizens amidst anticipated volatility (B. Anderson 2010; Methmann and Rothe 2012). Recent efforts have sought to link some of these otherwise disconnected threads (Shove and Walker 2010; Morrice 2013; Brown 2014) but, by and large, the two broad bodies of work—political ecologies of hazards and cultural geographies of home and household sustainability—exist in separate orbits. This article seeks to bring them together based on the premise that in the affluent West, where per capita carbon emissions are highest, the household is a key scale to envisage future transformations in socioecological relations.

Our contention is that much of this work, including our own, has nevertheless been overtaken by escalating events and is framed in ways that consequently now appear limited. Here, we seek to consciously unsettle normative framings of such established precepts as resilience and sustainability, in light of anticipated extremity and the likelihood of forced and unstable future socioecological relations rather than gradually planned transitions. We seek to bring cultural geographical work on households into conversation with some of the scenarios invoked by climate change science and with critical political ecologies of disasters.

Research on adaptation, vulnerability, preparedness, and resilience has emphasized urban and community scales, especially in the developing world, where populations are recognized to be particularly vulnerable to climate change (Adger et al. 2003; Bulkeley and Tuts 2013). Notwithstanding Fordham's (1998) call nearly two decades ago for disaster research to consider the private/household domain in the developed world as "a legitimate object of research" (126), analyses are still less advanced in the developed than in the developing world (Head et al. 2011), and the internal complexities of home life, families, gender, and emotion are only beginning to be appreciated (Proudley 2008; Caruana 2010; Whittaker, Handmer, and Mercer 2012; Eriksen 2013; Morrice 2013). This article accordingly focuses on developed world households as one important, but neglected, site of analysis.

Meanwhile, the cultural literature on households has other conceptual limitations. Nuanced ethnographic accounts are the norm (e.g., Gregson 2007; Pink 2012) but of the relativities and contingencies of the present—a framing that makes sense in an anthropological mode of inquiry but that inadvertently eschews abilities to conceive future climatic extremity. What if we acknowledge the agentic capacities of nonhuman others—including the climate assemblage itself—to produce more catastrophic results? The household literature might not have acknowledged such agency, but the scientific literature on climate change has (although not in exactly these words). What might everyday household life be like living on a 4°C hotter earth? What will happen to the notion of the everyday when haunted by the specter of biophysical catastrophe and associated economic hardship or when it routinely incorporates extreme weather events?

With climate change up until now viewed as somewhere over the near horizon, the emphasis has been on, inter alia, identifying policies for mitigation and adaptation, predicting unequal impacts on sections of society, identifying vulnerable populations, and determining policies to increase potential resilience (Brown 2014). All such actions foresee futures in specific ways that shape the style of investigations and policies—anticipating extremity but rarely catalyzing "exceptional extraordinary measures" (Methmann and Rothe 2012, 323). Application of concepts including resilience, vulnerability, and adaptation has proceeded without due acknowledgment of disciplinary and epistemological traditions that shape interpretations (Miller et al. 2010), instead advancing a "governmental scheme of risk management" (Methmann and Rothe 2012, 323) that has the effect of maintaining the status quo. The same could be said of "sustainability" in the household realm (Davidson 2010). Lurking are desires that seem increasingly impossible: to return back to a stable, Holocene condition (Hulme 2010), to somehow adapt gradually to changes in ways that obviate discomfort.

In what follows we seek to rethink climate change adaptation and sustainability agendas at the household scale, and through a more catastrophic lens. We first move from incremental to transformative conceptions of change and, second, invert concepts of vulnerability and capacity. Transformative change acknowledges that things cannot proceed as present and that a more disruptive framing enables scope for conceptions of uncertain futures. Likewise, vulnerabilities inverted open up possibilities to identify creative abilities and capacities.

The household is, we emphasize at the outset, not a neatly bounded or even a preeminent scale at which to pursue this agenda. The household is, however, a critical scale at which people encounter disruption and adjust amidst the dilemmas of everyday life (Eriksen 2013). In this important space, we argue, it is possible to observe practices that have been, within incrementalist logic, viewed as proenvironmental or sustainable, but that within a more volatile framing can be more meaningfully interpreted as survival capacities.

We then illustrate by briefly revisiting our own empirical findings on households—already communicated previously within a discourse of sustainability and climate change (Gibson et al. 2013)—to excavate insights for survival. We discuss what households worry about the future and what kinds of creativity might be needed for volatile futures. Although the household is the focus, we hope that, in the context of this special issue, the rethinking of vulnerability and capacity will have relevance to wider debates.

From Incremental to Transformative Change

Many of the biophysical processes generating a changed climate are already locked in (Solomon et al. 2009). We are looking at forced change and the need for responses to unpredictable events, rather than a period of controlled, staged adjustment enabling continued growth and prosperity (O'Neill and Handmer 2012). There are profound challenges here for the whole of society. The ground is shifting rapidly—it is not long since adaptation was considered taboo because it implied defeatism against the possibility of mitigation (Pielke et al. 2007).

From within the adaptation community come amplified warnings about the speed and magnitude of change and of uncertainty escalating in ways that make any sense of gradualist change impossible (Adger and Barnett 2009; Kates, Travis, and Wilbanks 2012). Incrementalist approaches limit capacity to acknowledge the power of individual climatic events to throw things into disarray (O'Neill and Handmer 2012). Transformational rather than incremental change appears essential (Park et al. 2012), referring not only to the possibility of a 4°C warmer world (Stafford Smith et al. 2011) but also to the increased level of surprise associated with rapid change in complex systems.

There are, nevertheless, tensions. The scientific community increasingly recognizes the need for transformative change. Australia's federally funded Commonwealth Scientific and Industrial Research Organization (CSIRO) recently argued that alternative governance arrangements might become necessary, in ways that are clearly challenging for politicians to hear (Head et al. 2014). Decision makers will have to "hedge bets," knowing that some will turn out to be wrong (Stafford Smith et al. 2011).

Yet, transformation will need more than accurate decision making or modeling and predicting of uncertainty. Significant transformations require multiscaled governmental responses, accommodative social contexts, and deep challenges to the economic status quo (Bulkeley and Moser 2007; Agrawal 2010; Kates, Travis, and Wilbanks 2012). And yet there are also societal limits to adaptation prospects shaped by values, ethics, knowledge, and culture (Adger et al. 2009). Vulnerability is multivalent, path dependent, spatially heterogeneous, and more difficult to model forward than backward (Adger 2006; Preston 2013). Preparedness is refracted through complex filters of social difference, including gender in the context of domestic spaces (Proudley 2008; Eriksen 2013) and within migrant, indigenous, and socioeconomically disadvantaged communities (Sevoyan et al. 2013). Application of concepts such as vulnerability and resilience has been criticized for downplaying capitalist social relations that produce uneven geographies of exposure to risk (MacKinnon and Derickson 2013) and for neglecting the emotional work of trauma recovery (Whittle et al. 2012). Failing to engage with advances in social and cultural research results in unwitting essentializing of axes of inequality and difference and neglecting how race, class, and gender intersect (Cupples 2007; Eriksen 2013). Many such experiences are galvanized within household life.

Moreover, governed populations react unpredictably to collapsing economic and social systems. Decision-making frameworks necessarily need both long and short lead times while acknowledging uncertain lag effects between decisions and consequences (Stafford Smith et al. 2011), all of which risks social unrest and unsettles attempts to quantify future national, community, or household vulnerabilities.

Recent experiences after the financial crisis show that dramatic political and economic shocks are already disrupting emotional ties and relationships in unpredictable ways (Gorman-Murray 2011). Upheaval and dissent is discomforting but might also be more truly transformative (Pickerill and Chatterton 2006). The professionalization of adaptation as a new form of management similarly risks overestimating human control and privileges elite experts and state agencies in the defining and documenting of vulnerability and resilience (MacKinnon and Derickson 2013)—again isolating households.

On the flip side, starting with a more urgent, catastrophist framing challenges much of the work on sustainability initiatives and policies for households—where the aim has been to overcome barriers of habitual everyday practice and transition incrementally to a low-carbon but still steady-state economy and society. Instead of contemplating what kind of technologies, policies, or behavioral changes are needed to reduce energy or water use with minimal impact on living standards, we ask how well households are positioned to respond to as-yet-unknown external forces.

Indeed, in much of the cultural and ethnographic work on households, remarkably little is said of uncontrollable external climate processes, of biospheres as dynamic living systems, or of the prospects of volatile events forcing change suddenly. Although there is increasing recognition of more-than-human agency—some of it unwelcome—in households (e.g., Kaika 2005; Power 2005), it is seldom about volatile or calamitous forces. Although this work provides much needed texture to understanding everyday practice within households, there is nevertheless a sense that time rolls on, rather than households being jolted around by forces beyond human control.

Inverting Vulnerability and Capacity

A heightened sense of urgency around potential catastrophe invites productive inversions: of vulnerability to risk and capacity to adapt. In question is what kind of agency is ascribed to households (and other everyday actors) in our framings (Pickerill and Chatterton 2006; Bulkeley and Newell 2010). Research is increasingly patterned into established ways of thinking and modeling vulnerabilities, based on antecedent traditions (Adger 2006; Miller et al. 2010). Income, education, social class, and geographical location (latitudinal position, coastal exposure, remoteness) have become common proxies for vulnerability (Beer et al. 2012). Yet, actions and people who seem resilient or adaptive under a gradualist framing of change might be stranded under conditions of catastrophe—different skills and capacities could come to the fore.

Ethnographic studies of diverse populations across rural–urban and developed–developing contexts have further demonstrated that beyond social class and cultural difference, capacities to cope and adjust are linked to vernacular and tacit knowledges, social cohesion, sense of place, memories, and cultural inheritances (Ford et al. 2008; Strengers and Maller 2012; Eriksen 2013). Developed world populations determined as vulnerable using quantitative demographic data are being shown through subsequent qualitative methods to have strong social bonds, from prior experiences of rallying together in response to extreme external forces such as droughts, wildfires, and floods (Beer et al. 2012). As Deb Anderson (2008) argued, seemingly vulnerable low-income, rural households apprehend climate change through shared discourses of endurance, uncertainty, advocacy, and local resolve.

Likewise, indigenous peoples globally (including within developed countries) have been identified as the most vulnerable to climate change impacts, and yet household-scale research is showing strong potential to cope with climate variability (Altman and Jordan 2008; Ford et al. 2008; Head et al. 2014). Maintenance of extended kinship networks can exacerbate residential overcrowding but also offsets decreased access to appropriate food and other resources (Altman and Jordan 2008). High degrees of mobility increase possibilities for relocation during or after extreme events, and being unfairly peripheral to government and civil support paradoxically fuels self-sufficiency (Howitt, Havnen, and Veland 2012). The ethics and sustainability of hunting look different if viewed through the lens of maintaining food supply to urban populations (Adams 2013). Inverting vulnerability and capacity is thus both an issue of conceptual framing and a function of the scales, sites, and modes of subsequent empirical research.

The Australian Households Study

With these thoughts in mind, we briefly revisit our own group's work on household sustainability. Viewed through the lens of rupture and potential chaos, it becomes less about generating sociocultural change and more like how to respond to forced circumstances. A list of what we thought of as sustainability practices becomes, in a more catastrophic headspace, a catalog of resources for survival. Our empirical work, in an Australian industrial region with strong working-class history and identification, combined a survey of 1,465 households and more than 200 ethnographic interviews, including a longitudinal study of a core group of sixteen households (Gibson et al. 2013).[1]

A first phase of statistical work from the survey suggested that, contrary to common assumptions, the households doing more of the "sustainability work" in our sample were low-income rather than affluent (Waitt et al. 2012). They were living in detached homes rather than apartments and were families headed up by women. Households participating in subsequent ethnographic interviews reflected this mix. They were green-leaning in environmental views, although diverse in typology, including single-person, extended family, same-sex couples, and nonfamilial households. Many were low-income families, including retired couples surviving on minimum government pensions. A vast majority brought with them a concern for unknown ecological futures, with only one household, a retired couple, explicit in their skepticism of climate change science. Revisiting our findings, it becomes clear that our understandings of household capacities and assemblages of wider forces must be made more complex.

Transformation

Our households have acute insights into how the future might look different but cannot see how to get there, except that it might need drastic change. Many expressed frustration with present political inaction; they implicitly critiqued risk management approaches that did little to shift the status quo. Hence, for Jen, a retired nurse who lives alone,

> The crux of the whole matter is that as long as we can make a quid out of what we do, if the world does crumble in sixty years' time we'll fix it when it's really bloody urgent. That's how I feel the world is handling it at the moment . . . nobody wants to do anything in case the guy in the next backyard doesn't . . . it's so pathetic.

Such sentiments were echoed by Bob, a permaculture enthusiast with a strong desire for a "relaxed, simple, family life," who lives with his partner and toddler daughter in a small cottage in a low-income industrial neighborhood:

> I think that households need to get more political. . . . But those big changes have to happen to rein in industry . . . it seems like the same shit has been talked in circles for years, just the same shit over and over.

Households are contemplating extremity, want to see transformational change, and are frustrated by the ineffectual nature of incremental approaches.

Households revealed complex understandings of socioecological relations that appear receptive to more dynamic possibilities. Janette is an enthusiastic gardener and a teacher in another low-income suburb, living with her partner and teenage children in a detached house with extensive food gardens. Her intimate understandings of growing cycles and a nuanced conception of seasonal rhythms suggest survival capacity, rather than socioeconomic vulnerability.

Many households related to cycles of abundance and scarcity across a broad spectrum of practices but particularly around practices of water use, a heightened but illustrative example from our ethnographic sample, given that interviews were undertaken amidst the worst drought in living memory in Australia. Households demonstrated unexpectedly high degrees of willingness to endure rationing (experienced in the 2000s via compulsory water restrictions) and used creative practices to capture, save, and reuse water outside formal regulation. Tom and Joan, for example, are a retired couple who have recently downsized their home. Their skepticism toward climate change sits in stark contrast with their enthusiasm for water-saving practices, including coordinating their toilet visits to minimize flushing and capturing gray water from the shower.

Meanwhile, reduced water use practices did not appear permanently locked in from experience of drought (as the incrementalist behavior change view might hope). But rather than seeing this as a sustainability failure, it alternatively suggests that households are prepared to "go with the flow" and enjoy or even indulge in abundant water in times of excess. The latter seemingly unsustainable practice viewed through an alternative framing shows a stronger sense of connectedness with the resource's materiality and cycles of abundance and scarcity. Consciousness of cycles of abundance and scarcity was also evident across a spectrum of practices, including gifting and everyday weather experiences. Households maintain continuities with past ways of interacting with non-human nature, without appearing backward looking, signaling everyday capacities to deal with increasingly likely sorts of flux.

Agency and Capacity

People feel complex things about their own agency in this process. One assumption is that household contents are cherished and that future policies should secure lifestyle maintenance around the quantity and newness of material things. The idea of accumulating possessions being a proxy for quality of life is seemingly sacrosanct, yet at the household scale there was real angst about it as well. Paul, for instance, who lives alone in a small public housing unit, was not alone in describing his desires to live more simply, to "be happier with less."

We find broad evidence of evolving and shifting attachments to material possessions. People might not have the agency to prevent catastrophic forces but enact a kind of agency in reevaluating the significance of material possessions in everyday life. Certainly in Australia, where climate change risk (and associated extremities of flood, fire, and drought) is seeping more deeply into everyday life, and where insurance costs are increasingly prohibitive, people are actively imagining losing everything—and many are not so disturbed by that prospect. In a sign of comprehending forced change and extremity in the present, households that acknowledged their lack of power over fire or flood were reevaluating which material possessions matter most—becoming accordingly willing to forsake home contents insurance or give things away to others more needy (cf. Martin 2002; Eriksen 2013).

Meanwhile, debate erupted about the loss of other kinds of vital skills and capacities—those that enable frugal existence in times of constraint and scarcity. Jen is an avid sewer and talks at length about its usefulness. Peter laments the loss of a disposition to rummage resources together from necessity and tinkering skills needed to deal with unforeseen change:

> It used to be "she'll be right and we'll fix it up with safety pins and four by two," or something like that, but that's gone. All of that's long gone.

Janette ruefully describes conspicuous consumption in a new residential estate rapidly developing around her century-old house, disparaging a perceived lack of ingenuity to make do:

> They can't even put a hem up. They can't sew. They can't design. They can't even borrow a book in the library, a lot of them.

In our ethnography, households told abundant stories of making do and marshaling resources with limited financial security. Sustainability thinking or concern over climate change was rarely present. They were instead frugal practices valued as necessary modes of living—old-fashioned values from working-class and rural upbringings that might be revived as an economic ethics for uncertain futures (Gibson-Graham and Roelvink 2010). Maria, for example, foresees an end to wasteful use of domestic electricity, driven by financial pressures, and Amy, a nurse who shares the spare rooms in her house with international students, describes her experience growing up in the remote Northern Territory, where

> you learned to make do with what you had. We had electric lights only when the visitors came.

People viewed the stuff within homes differently if forced to imagine it had to last for the rest of their lives. Peter believes this kind of thinking is innate for people of his generation:

> I'm a child of the depression of the 1930s . . . that was what I picked up and my siblings picked up as values. I have a feeling that you use things up 'til they haven't got much more value and only then you throw them out.

Another potential inversion relates to the accumulation of things within the home, encompassing collecting and storing (Gregson 2007). Large suburban homes that mainstream sustainability discourse would view as energy-inefficient were expedient in the context of survival and the need to store rather than divest useful things (cf. Dowling and Power 2011). A subset of our ethnographic participants was extended family households, many of migrant backgrounds, frequently formed from the fallout of broken marriages or out of necessity to care for sick, but poor, elderly parents. Extended family households needed larger homes to house larger numbers of people and things—precisely the kind of living arrangements more likely under extreme scenarios (Klocker, Gibson, and Borger 2012). Within extended family households, older family members valued accumulative practices that might be viewed as "hoarding" or as sustainability failures, as necessary for stewardship of still-useful material possessions. Garages were converted into storage zones for things that retained use value, so that those things could be subsequently redistributed among family members. Not only do sustainability practices become survival practices but also those putatively unsustainable practices that amidst volatility appear to engender both social bonds and responsible stewardship of materials. Such practices, which we were previously drawn to interpret within frames of household sustainability, now appear within a more transformative, catastrophic framing as both portentous and propitious.

Conclusions

We have argued here that the household is a key scale at which to envisage future transformations in socioecological relations in the developed world and that climate change adaptation and sustainability agendas need to be rethought through a more catastrophic lens. Although households are now largely aware of climate change and its consequences and are contemplating major upheavals, governments and climate change adaptation policies remain incremental and welded to a static view of the present. Yet householders might be readier to face transformative futures than is often imagined, and different patterns of vulnerability might appear. As Eakin (2005, 1936) has argued, more "grounded, locally relevant research" is needed to expose experiences of vulnerability "from the perspective of the vulnerable." This is especially relevant for households. Catastrophe is already with us and becoming embedded in everyday Western lives (B. Anderson 2010). We have long been aware that catastrophe is an everyday experience for many people in the world (such as we have seen recently with the Philippine typhoon). We are not so used to experiencing it as part of everyday affluent lives—or, indeed, researching it within that context.

Where households have figured prominently in political ecological research on disaster vulnerability and preparedness, such research has overwhelmingly (and not without good reason) focused on the developing world (Head et al. 2011) and has only recently begun to intersect with cultural and ethnographic work on everyday life, the home, gender, and emotions (Eakin 2006; Eriksen and Gill 2010; Eriksen 2013; Morrice 2013). We have sought to show what kinds of conceptual inversions are possible through bringing political, ecological, and cultural households research into closer dialogue.

Revisiting our own and others' empirical findings shows that the details of everyday life provide considerable resources to help engage with the challenges of climate change. In particular, this work suggests ways in which ideas of vulnerability and resilience need to be reframed as contingent temporally and spatially and experienced intersubjectively. The household scale of analysis shows some inversions from patterns analyzed at the macroscale. Vulnerability and resilience are cut across by underlying class, intergenerational, and cultural differences but also emergent in the moments and spaces of everyday life, including beyond the porous boundaries of the home itself.

There are immense opportunities to revisit the past and question what kinds of dormant practices could be retrieved anew (Shove 2012). People make sense of dramatic changes and crises and rekindle capacities to cope, through recalling past experiences (D. Anderson 2008; Head et al. 2011). Aspects of daily life that make use of diverse "analog," preelectrical, and nonmonetary techniques and technologies (line-drying, cycling, walking, sharing, water harvesting, foraging for food) both revisit past experiences and reconnect with biophysical rhythms of abundance and scarcity (cf. McLain et al. 2014). Given that conditions are becoming more extreme, modernist expectations of continuities of supply—in food and water, for instance—will presumably give way to seasonal availability and uncertainty. If we let go of the expectation of retaining existing comforts and ubiquitous abundance, then what we think of as failures might be reframed as productive disruptions (Adey and Anderson 2012) or vulnerabilities as capacities. Trauma stimulates resourcefulness and new relationships are built (Eriksen 2013). Meanwhile, from the perspective of those households measured as statistically vulnerable, resources for survival, including past values, memories, and shared experiences of disruption, are evident among hardships. Seemingly vulnerable working-class, migrant, indigenous, and rural populations have profound capacities to reengage with such resources or are already doing it out of necessity.

If our brief household examples seem a touch too small scale and not attentive enough to questions of big industry, government, and infrastructure, they do nevertheless provide one picture of what survival might look like when the big systems fail. In scenarios where governments or the private sector are unable to meet basic conditions for life, social and community capacity will be critical to survival. And, as the preceding examples attest, vulnerability might look quite different than it does now. This is not a rationale for absolving governments or industry of the responsibility to act. Rather, we suggest that amidst government and corporate intransigence, the resources for survival are ultimately social. Thus, when contemplating unthinkable uncertainty and possible socioecological disruptions, we must look deeper, and with less loyalty toward normative framings, to identify contradictions and unheralded survival capacities within households.

Acknowledgments

We thank Gordon Waitt, Nick Gill, and Carol Farbotko for collaboration on the households research. Our perspectives benefited from ongoing discussions with Christine Eriksen, Michael Adams, Natascha Klocker, and Stephanie Toole.

Funding

This research was funded by the Australian Research Council (DP0986041, FT0991193, and FL0992397).

Note

1. For further detail on method, sampling strategy, and representativeness, see Waitt et al. (2012).

References

Adams, M. 2013. "Redneck, barbaric, cashed up bogan? I don't think so": Hunting and nature in Australia. *Environmental Humanities* 2:43–56.

Adey, P., and B. Anderson. 2012. Anticipating emergencies: Technologies of preparedness and the matter of security. *Security Dialogue* 43 (2): 99–117.

Adger, W. N. 2006. Vulnerability. *Global Environmental Change* 16:268–81.

Adger, W. N., and J. Barnett. 2009. Four reasons for concern about adaptation to climate change. *Environment and Planning A* 41:2800–05.

Adger, W. N., S. Dessai, M. Goulden, M. Hulme, I. Lorenzoni, D. R. Nelson, L. O. Naess, J. Wolf, and A. Wreford. 2009. Are there social limits to adaptation to climate change? *Climatic Change* 93:335–54.

Adger, W. N., S. Huq, K. Brown, D. Conway, and M. Hulme. 2003. Adaptation to climate change in the developing world. *Progress in Development Studies* 3:179–95.

Agrawal, A. 2010. Local institutions and adaptation to climate change. In *Social dimensions of climate change*, ed. R. Mearns and A. Norton, 173–98. Washington, DC: World Bank.

Altman, J., and K. Jordan. 2008. *Impact of climate change on Indigenous Australians*. Canberra, Australia: Centre for Aboriginal Economic Policy Research.

Anderson, B. 2010. Preemption, precaution, preparedness: Anticipatory action and future geographies. *Progress in Human Geography* 34 (6): 777–98.

Anderson, D. 2008. Drought, endurance and "the way things were": The lived experience of climate and climate change in the Mallee. *Australian Humanities Review* 45. http://www.australianhumanitiesreview.org/archive/Issue-November-2008/anderson.html (last accessed 27 November 2013).c

Beer, A., S. Tually, M. Kroehn, and J. Law. 2012. *Australia's country towns 2050: What will a climate adapted settlement pattern look like?* Gold Coast, Australia: NCCARF.

Blunt, A., and R. Dowling. 2006. *Home*. London and New York: Routledge.

Brown, K. 2014. Global environmental change I: A social turn for resilience? *Progress in Human Geography* 38 (1): 107–17.

Bulkeley, H., and S. Moser. 2007. Responding to climate change: Governance and social action beyond Kyoto. *Global Environmental Politics* 7 (2): 1–10.

Bulkeley, H., and P. Newell. 2010. *Governing climate change*. London and New York: Routledge.

Bulkeley, H., and R. Tuts. 2013. Understanding urban vulnerability, adaptation and resilience of climate change. *Local Environment* 18:646–62.

Caruana, C. 2010. Picking up the pieces: Family functioning in the aftermath of natural disaster. *Family Matters* 84:79–88.

Cox, R. 2013. House/work: Home as a space of work and consumption. *Geography Compass* 7 (12): 821–31.

Cupples, J. 2007. Gender and Hurricane Mitch: Reconstructing subjectivities after disaster. *Disasters* 31 (2): 155–75.

Davidson, M. 2010. Sustainability as ideological praxis: The acting out of planning's master signifier. *City* 14:390–405.

Dowling, R., and E. Power. 2011. Beyond McMansions and green homes. In *Material geographies of household sustainability*, ed. R. Lane and A. Gorman-Murray, 75–88. Aldershot, UK: Ashgate.

Eakin, H. 2005. Institutional change, climate risk, and rural vulnerability: Cases from Central Mexico. *World Development* 33:1923–38.

———. 2006. *Weathering risk in rural Mexico: Climatic, institutional, and economic change*. Tucson: University of Arizona Press.

Eriksen, C. 2013. *Gender and wildfire: Landscapes of uncertainty*. London and New York: Routledge.

Eriksen, C., and N. Gill. 2010. Bushfire and everyday life: Examining the awareness–action "gap" in changing rural landscapes. *Geoforum* 41 (5): 814–25.

Ford, J. D., B. Smit, J. Wandel, M. Allurut, K. Shappa, H. Ittusarjuat, and K. Qrunnut. 2008. Climate change in the Arctic: Current and future vulnerability in two Inuit communities in Canada. *Geographical Journal* 174:45–62.

Fordham, M. H. 1998. Making women visible in disasters: Problematising the private domain. *Disasters* 22 (2): 126–43.

Gibson, C., C. Farbotko, N. Gill, L. Head, and G. Waitt. 2013. *Household sustainability: Challenges and dilemmas in everyday life*. Cheltenham, UK: Edward Elgar.

Gibson-Graham, J.-K., and G. Roelvink. 2010. An economic ethics for the Anthropocene. *Antipode* 41:320–46.

Gorman-Murray, A. 2011. Economic crises and emotional fallout: Work, home and men's sense of belonging in post-GFC Sydney. *Emotion, Space and Society* 4 (4): 211–20.

Gregson, N. 2007. *Living with things: Ridding, accommodation, dwelling*. Wantage, UK: Sean Kingston.

Head, L., M. Adams, H. McGregor, and S. Toole. 2014. Climate change and Australia. *WIREs: Climate Change* 5 (2): 175–97.

Head, L., J. Atchison, A. Gates, and P. Muir. 2011. A fine-grained study of the experience of drought risk and climate change among Australian wheat farming households. *Annals of the Association of American Geographers* 101 (5): 1089–1108.

Howitt, R., O. Havnen, and S. Veland. 2012. Natural and unnatural disasters: Responding with respect for Indigenous rights and knowledges. *Geographical Research* 50:47–59.

Hulme, M. 2010. Learning to live with recreated climates. *Nature and Culture* 5:117–22.

Kaika, M. 2005. *City of flows: Modernity, nature, and the city*. London and New York: Routledge.

Kates, R. W., W. R. Travis, and T. J. Wilbanks. 2012. Transformational adaptation when incremental adaptations to climate change are insufficient. *Proceedings of the National Academy of Sciences* 109:7156–61.

Klocker, N., C. Gibson, and E. Borger. 2012. Living together, but apart: Material geographies of everyday sustainability in extended family households. *Environment and Planning A* 44:2240–59.

MacKinnon, D., and K. D. Derickson. 2013. From resilience to resourcefulness: A critique of resilience policy and activism. *Progress in Human Geography* 37 (2): 253–70.

Martin, R. 2002. *Financialization of daily life*. Philadelphia, PA: Temple University Press.

McLain, R. J., P. T. Hurley, M. R. Emery, and M. R. Poe. 2014. Gathering "wild" food in the city: Rethinking the role of foraging in urban ecosystem planning and management. *Local Environment* 19 (2): 220–40.

Methmann, C., and D. Rothe. 2012. Politics for the day after tomorrow: The logic of apocalypse in global climate politics. *Security Dialogue* 43 (4): 323–44.

Miller, F., H. Osbahr, E. Boyd, F. Thomalla, S. Bharwani, G. Ziervogel, B. Walker, et al. 2010. Resilience and vulnerability: Complementary or conflicting concepts? *Ecology and Society* 15 (3): 11.

Morrice, S. 2013. Heartache and Hurricane Katrina: Recognizing the influence of emotion in post-disaster return decisions. *Area* 45 (1): 33–39.

O'Neill, S. J., and J. Handmer. 2012. Responding to bushfire risk: The need for transformative adaptation. *Environmental Research Letters* 7 (1): 1–7.

Park, S. E., N. Marshall, E. Jakku, A. Dowd, S. Howden, E. Mendham, and A. Fleming. 2012. Informing adaptation responses to climate change through theories of transformation. *Global Environmental Change* 22:115–26.

Pickerill, J., and P. Chatterton. 2006. Notes towards autonomous geographies: Creation, resistance and self-management as survival tactics. *Progress in Human Geography* 30 (6): 730–46.

Pielke, R., Jr., G. Prins, S. Rayner, and D. Sarewitz. 2007. Climate change 2007: Lifting the taboo on adaptation. *Nature* 445:597–98.

Pink, S. 2012. *Situating everyday life: Practices and places*. London: Sage.

Power, E. R. 2005. Human–nature relations in suburban gardens. *Australian Geographer* 36:39–53.

Preston, B. L. 2013. Local path dependence of U.S. socioeconomic exposure to climate extremes and the vulnerability commitment. *Global Environmental Change* 23 (4): 719–32.

Prior, T., and C. Eriksen. 2013. Wildfire preparedness, community cohesion and socio-ecological systems. *Global Environmental Change* 23 (6): 1575–86.

Proudley, M. 2008. Fire, families and decisions. *Australian Journal of Emergency Management* 23 (1): 37–43.

Reid, L., P. Sutton, and C. Hunter. 2009. Theorizing the meso-level: The household as a crucible of pro-environmental behavior. *Progress in Human Geography* 34:309–27.

Sevoyan, A., G. Hugo, H. Feist, G. Tan, K. McDougall, Y. Tan, and J. Spoehr. 2013. *Impact of climate change on disadvantaged groups*. Gold Coast, Australia: NCCARF.

Shove, E. 2012. The shadowy side of innovation: Unmaking and sustainability. *Technology Analysis & Strategic Management* 24:363–75.

Shove, E., and G. Walker. 2010. Governing transitions in everyday life. *Research Policy* 39 (4): 471–76.

Solomon, S., G.-K. Plattner, R. Knutti, and P. Friedlingstein. 2009. Irreversible climate change due to carbon dioxide emissions. *Proceedings of the National Academy of Sciences* 106:1704–09.

Stafford Smith, M., L. Horrocks, A. Harvey, and C. Hamilton. 2011. Rethinking adaptation for a 4°C world. *Philosophical Transactions of the Royal Society A* 369:196–21.

Strengers, Y., and C. Maller. 2012. Materialising energy and water resources in everyday practices: Insights for securing supply systems. *Global Environmental Change* 22 (3): 754–63.

Waitt, G., P. Caputi, C. Gibson, C. Farbotko, L. Head, N. Gill, and E. Stanes. 2012. Sustainable household capability: Which households are doing the work of environmental sustainability? *Australian Geographer* 43:51–74.

Whittaker, J., J. Handmer, and D. Mercer. 2012. Vulnerability to bushfires in rural Australia. *Journal of Rural Studies* 28 (2): 161–73.

Whittle, R., M. Walker, W. Medd, and M. Mort. 2012. Flood of emotions: Emotional work and long-term disaster recovery. *Emotion, Space and Society* 5 (1): 60–69.

Transforming Household Consumption: From Backcasting to HomeLabs Experiments

Anna R. Davies and Ruth Doyle

Department of Geography, Trinity College Dublin

Following the rhetoric of an impending "perfect storm" of increasing demand for energy, water, and food, it is recognized that ensuring sustainability will require significant shifts in both production and consumption patterns. This recognition has stimulated a plethora of future-oriented studies often using scenario, visioning, and transition planning techniques. These approaches have produced a multitude of plans for future development, but many valorize technological fixes and give limited attention to the governance and practice of everyday consumption. In contrast, this article presents empirical findings from a practice-oriented participatory (POP) backcasting process focused on home heating, personal washing, and eating. This process provided spaces for collaborative learning, creative innovation, and interdisciplinary interaction as well as producing a suite of ideas around promising practices for more sustainable household consumption. Further action is required, however, to explore how such ideas might be translated into action. The article concludes by outlining how collaborative experiments among public, private, civil society, and citizen-consumers, or HomeLabs, provide a means to test and evaluate the promising practices developed through POP backcasting.

随着对能源、水与粮食的需求逐渐增加的"超完美风暴"将至之修辞而来的是，人们已认知到，若要确保可持续性，那麽生产与消费模式皆必须有显着的转变。此一认知，已激发众多以未来为导向、并且运用剧本、想像力及变迁规划技术的研究。这些方法，为未来的发展生产出诸多计画，但其中许多仅稳定技术修补，并对每日消费的治理与实践，投以相当有限的关注。反之，本文呈现出聚焦家户暖气、个人洗涤和饮食、并以实践为导向的参与式（POP）回溯过程之经验发现。此一过程，提供了空间给合作式学习、创造性发明与跨领域互动，并生产一系列有关更具可持续性的家户消费的有为实践之概念。但仍需要进一步的行动，探讨这些概念如何能够被转译成行动。本文于结论中概述，公共、私人、公民社会与公民—消费者之间的合作式实验，抑或称之为"家庭实验室"，如何提供方法，测试、评价随着POP回溯方法建立的有为实践。

Siguiendo la retórica de una inminente "tormenta perfecta" relacionada con la creciente demanda de energía, agua y alimentos, se reconoce que para asegurar la sostenibilidad al respecto se requerirán cambios significativos tanto en los patrones de producción como en los de consumo. Este tipo de reconocimiento ha estimulado una plétora de estudios de dimensión futurista en los que se usan técnicas de planificación sobre escenario, visión y transición. De tales enfoques resultan infinidad de planes de desarrollo futuro, pero muchos de ellos dan prioridad a recetas tecnológicas, prestando atención muy limitada a la gobernanza y la práctica del consumo cotidiano. Por contraste, el presente artículo exhibe los hallazgos empíricos de un proceso participativo de prospección inversa orientado a la práctica (POP), focalizado en la calefacción de la casa, la lavandería personal y la comida. Este proceso generó espacios para aprendizaje colaborativo, innovación creadora e interacción interdisciplinaria, al tiempo que produjo una suite de ideas sobre prácticas prometedoras para un consumo hogareño más sostenible. No obstante, se requiere de mayor trabajo para explorar la manera como tales ideas podrían convertirse en acción. El artículo concluye delineando el modo como los experimentos de colaboración entre consumidores públicos, privados, sociedad civil y ciudadanía en general, o LabHogares, proveen unos medios para ensayar y evaluar las prácticas prometedoras desarrolladas a través de la prospección inversa POP.

Combining observations from the fifth Assessment Report of the United Nations (UN) Intergovernmental Panel on Climate Change (IPCC) with socioeconomic and demographic trends, predictions have been made for a "perfect storm" of food, water, and energy shortages by 2030 unless immediate and dramatic socioecological transformations are initiated. Explicit within all IPCC reports has been the need to change the way humans interact with their wider environment, essentially the need to transform socioecological systems of production and consumption (UN 2013). Meanwhile, the inherently normative, highly political, and ultimately fundamental questions regarding what kind of transformed socioecological system people might wish to inhabit in the future and, importantly, how such transformations should occur, remain marginal to IPCC outputs to date. Nonetheless, having some kind of navigational compass calibrated toward a vision of a desirable future is widely seen as an important component of creating impetus for transformation (De Gues 2002). It is such thinking that has stimulated a burgeoning array of futures-based techniques that seek to anticipate and shape the way we might live in times to come.

Beyond the modeling of climate change so influential in IPCC reporting, there are other anticipatory approaches emerging based on risks related to socioecological transformation in spheres such as transspecies epidemics, transboundary food risks, and even terrorism (e.g., de Goede and Randalls 2009; Anderson 2010). Within these spheres of anticipation, preemption, and precaution, action is taken in the present on the basis of what has not yet happened and might never come to pass at all, whether that be through commodification and trade on futures markets, the drafting of contracts such as household mortgages, or the forward planning of neighborhoods, cities, regions, and even nations. Inevitably, such actions taken in the present as a means to prevent, mitigate, or adapt to perceived risks are imbued with particular assumptions about "the future" that are frequently characterized by conditions of uncertainty and indeterminacy but are also infused with normative judgments of what is good or bad and ultimately what is valued (Anderson 2010).

In addition to attempts to extrapolate probable futures, there are manifold examples where creative and imaginative skills have been developed as a way of thinking through what the future might hold. Most visible in design and literary communities, and often characterized as part of a utopian–dystopian tradition, the goal of imagining futures is not prediction per se but rather a means to stimulate thinking about possible alternative futures. The future is seen as a safe space to envision disruptive innovations, to elaborate weak signals in social trends or roll-out of prototypical technologies. As a result, studies explicitly focused on imagining alternative socioecological futures are expanding through a range of foresight methods including visioning, backcasting, and transition planning, methods that are orchestrated by a variety of actors and organizations across scales and sectors (Tukker and Fedrigo 2009; Davies, Doyle, and Pape 2012). As critiqued by those emphasizing the importance of social practices, however, these techniques often focus on elite perspectives, valorize technological fixes, and give limited attention to the governance and practice of everyday consumption, thus limiting their capacity to achieve substantive societal transformations (Shove and Walker 2008).

In light of these critiques, this article reflects on the experience of developing and conducting an experimental practice-oriented, participatory (POP) backcasting process and its contribution to imagining socioecological transformations aimed at promoting sustainable consumption. Initially, core concepts of socioecological systems and transformations are briefly outlined, with attention paid to scholarly and policy developments informed by key insights from transitions management and social practice approaches. This is followed by an overview of the POP backcasting process, which adapted participatory backcasting techniques by taking social practices as the fundamental units for problem framing, solving, and innovation. The process then sought to counterbalance visioning and backcasting exercises that prioritize ecological modernization strategies, instead exploring alternative possibilities that place sufficiency, well-being, and sociocultural change as central strategies for transitions in everyday practices (Jackson 2009). Reflecting on this experiment, it is concluded that the POP backcasting procedure created unique and valued moments for civic engagement, collaborative learning, and transdisciplinary interaction among participants, particularly around the dynamics of consumption practices. As such, POP backcasting and similar endeavors could offer creative spaces (both material and virtual) where attention to the transformation of everyday practices can be collectively imagined and critically debated. The article closes with a brief discussion of ongoing HomeLabs research that is interrogating the conversion of promising practice ideas derived from POP backcasting into practical household interventions.

Imagining Futures: Socioecological Systems and Practicing Transformations

As detailed elsewhere in this special issue, connecting the terms *social* and *ecological* has become associated with emergent discourses of resilience and sustainability transformation (Fischer-Kowalski and Rotmans 2009). In particular, transitions theorists and transition management advocates have examined how large-scale transformations in sociotechnical regimes, such as shifts from horse-drawn to motor vehicles, or from water wells to water main systems, occurred (see Rip and Kemp 1998; Geels 2010). This research led to the emergence of the multilevel perspective (MLP) for understanding how innovations (mainly technological) come to be adopted, upscaled, and ultimately mainstreamed through system transitions. Drawing on the MLP, and with a view to designing deliberate interventions within socioecological systems, transition management techniques have been applied within policy settings as tools of governance and innovation. These frequently encompass exploratory or predictive scenario-building processes in collaboration with stakeholders ultimately leading to the formation of long-term policy and technology innovation plans (Meadowcroft 2009). Although the forward-looking and collaborative mechanisms of transitions management are often lauded (Tukker and Fedrigo 2009), their conception of sustainability problems as issues of resource management, to be addressed through hierarchical transformations (from niche to mainstream) in systems of provision, primarily supported by innovative technological fixes, has been subject to critique.

It has been argued that there is insufficient consideration of power and politics, the nature of human behavior–technology interactions, and lifestyle change within much transitions research (Bailey and Wilson 2009; Brown, Vergragt, and Cohen 2013). In addition, the importance of horizontal transfers of technologies, norms, and expectations across cultures—such as the spread of washing machines redefining the meaning of cleanliness and the skill of clothes washing or trends for air conditioning leading to expectations of standardized indoor temperatures irrespective of climate—that can affect the configuration of everyday practices remains weakly articulated (Shove and Walker 2008). As a result, researchers have begun to look toward a well-established, albeit highly diverse body of work on social practices that identify the complex of social and material elements that shape the way people live, including stuff (for personal washing, stuff would include showers and taps); skills (practical know-how on how and when to wash); understandings (social expectations of cleanliness; Warde 2005; Shove et al. 2008); and, adopting a macro focus, broader rules (that often relate to systems of provision and regulation; Spaargaren 2003). Stability in social practices arises when stuff, skills, understandings, and rules are integrated and reproduced, and practices are transformed as the links between elements are broken or new elements added (Pantzar and Shove 2010). Conceptualized in this way, any deliberate attempts to transform specific practices require coordinated and complementary actions across these social and material elements by a range of actors including citizens as the ultimate performers of practices. Yet, this presents challenges for contemporary governance of household consumption that has tended to favor voluntary agreements, information campaigns, and fiscal measures as isolated interventions (Davies et al. 2010). It also raises highly contentious specters of social engineering for sustainability (Shove and Walker 2007).

Although social practice approaches provide for a more nuanced understanding of the complexity of these diverse factors and actors, they have also been critiqued for their tendency to focus on configurations of past or current social practices, rather than contributing to debates about how life might be lived more sustainably in the future (Kuijer and De Jong 2012). Equally, practice-oriented empirical studies are frequently design led, exhibiting an overriding emphasis on the stuff of practices (Scott, Bakker, and Quist 2012), with less attention to broader questions of governance and how practices can or should be shaped through nonmaterial interventions. Although recent social practice research has begun to engage explicitly with the diffusion of resource-intensive technologies from a complex system perspective, it remains the case that consideration of how technologies might lead to more rather than less sustainable practices is marginal to current activity (Brown, Vergragt, and Cohen 2013).

Clearly, social practice and transition approaches exhibit different vocabularies and emphases. Nevertheless, both are bound by a common recognition of the need to move beyond product innovation, eco-efficiency, or redesign strategies, to consider integrated sociocultural, technological, and organizational changes to achieve transitions toward sustainable ways of life. It is with these commonalities in mind that the following sections describe and reflect on the explicitly POP backcasting experiment. POP backcasting aimed to build on the strengths of backcasting (with its normative future focus and practical participatory governing techniques) while recasting the unit of analysis from environmental resources or technologies to broader cross-cutting social practices.

POP Backcasting: Combining Practice Thinking and Transition Approaches

Backcasting describes an overarching, multiphase process involving the cocreation of desirable future vision(s), followed by working back (or backcasting) from that future alternative to the present to design sequential steps for its achievement. The normative dimension of backcasting, focusing on desirable visions, contrasts with explorative scenario traditions, exemplified in Shell's energy scenario research and commonly applied in transitions studies where likely futures are extrapolated from technological, political, or socioeconomic trends and are said to be limited in their ability to achieve trend-breaking solutions (Vergragt and Quist 2011). Backcasting studies vary greatly in their units of analysis, ranging from urban visioning projects (Eames and Egmose 2011), to sustainable residential energy (Svenfelt, Engström, and Svane 2011), to water governance (Kok et al. 2011). Others have adopted a lifestyle approach, such as the European Union project SPREAD (2013) and SusHous (Quist et al. 2001). Interestingly, however, a review of the literature found no backcasting study with an explicit focus on everyday social practices as units of problem definition, problem solving, and innovation (Doyle and Davies 2013). Equally, although emergent projects like SPREAD are producing a plethora of recommendations for social innovation, such as community gardens or shared tool facilities, the implications of such innovations on everyday practices (their skills, stuff, and understandings) remains underevaluated.

Set within a wider project examining consumption, environment, and sustainability (see Davies, Fahy, and Rau 2014), the POP backcasting procedure was developed to explore the possibilities and challenges created by adopting a future focus and participatory ethos while also adopting social practices as the unit of analysis. As such, the research was experimental and examined both the process enacted and its emergent products with respect to practices of personal washing, home heating, and eating. It is this explicit and foundational focus on practices and practice transformation (rather than technology diffusion or social acceptance) that differentiates it from many other backcasting procedures. Essentially, our intent was to explore what it means and what might be gained by putting practices at the center of analysis of everyday household consumption. The multiphase POP backcasting process (Figure 1) included the following essential phases, many of which are common to participatory backcasting experiments: problem definition, a stakeholder visioning workshop, elaboration of scenarios, scenario sustainability evaluation, citizen-

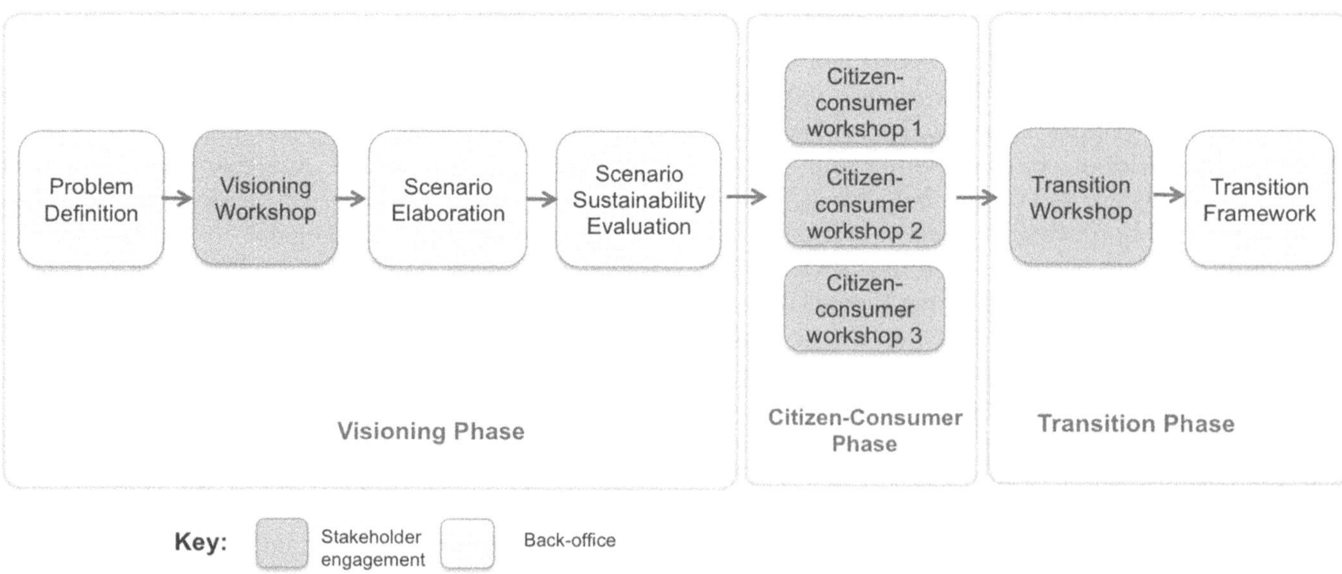

Figure 1. Practice-oriented participatory (POP) backcasting. (Color figure available online.)

consumer workshops, and a final stakeholder transition workshop. This process was applied in parallel for each of the three practices studies: heating, washing, and eating.

Having established the main sustainability challenges for each practice, participation was secured from a range of stakeholders from public, private, and civil society to take part in the first visioning workshop phase of the backcasting process. This diversity of interests and expertise sought to encourage a fusion of "knowledge across disciplines, sectors and institutions" seen as integral to creative problem solving and overcoming institutional and political impasses to making transitions to more sustainable living (Jansen 2003, 237). More than eighty stakeholders ranging from designers and communications experts to policymakers and commercial operators attended visioning workshops for the heating, washing, and eating studies. As elaborated in Doyle and Davies (2013), structured workshop sessions were orchestrated where participants were encouraged to imagine alternative sustainable practices in the year 2050. The year 2050 was invoked to allow participants to perceive the possibility of radically altered ways of living, while also being close enough to the present to connect to lived experiences and system interventions in the short, medium, and long term. Brainstorming took place in small multidisciplinary groups, each with a facilitator to stimulate discussion, record ideas, and ensure that participants focused on all practice elements. Participants were prompted to brainstorm ideas for innovations and interventions in four key dimensions of social practices, namely, skills, stuff, understandings, and rules (including regulations and infrastructures of provision) that together could promote sustainability transformations in the everyday practices under consideration.

The visioning brainstorms thus drew on microconceptualizations of social practice theory while also considering the sociotechnical regimes (energy, water, and food) within which these practices were situated and that are areas of typical concern in transitions visioning studies. Attention was paid to the role of commercial and government forces in shaping market conditions, infrastructures, and the rules of access to resources that simultaneously influence socially constructed and contextually dependent expectations, norms, and needs of home energy, water, and food consumption. A key focus was identifying or creating new ways to achieve the desired needs and end results of the practices being considered, noting that these needs might evolve through time. For example, in the case of washing, key needs identified included cleanliness, refreshment, and hygiene; heating needs included comfort and warmth; and eating needs were grounded in ideas of health, sustenance, and relationships. This practice perspective elevates emotional and cultural needs, rather than having a primary focus on functional needs as, for example, applied in transition management studies such as the Dutch Sustainable Technology Programme (Quist, Thissen, and Vergragt 2011). It was thus intended that the resultant outcomes would include not only new technology ideas but also social innovations and interventions designed to challenge unsustainable expectations and norms. This framing contrasts with the "productized," corporate visions of the future as seen in Philips's Design Probes, IBM's Smarter Planet, or Sony's FutureScapes, in which new technologies are often envisaged as being transplanted into typically Western social contexts where values and norms of the present day pervade.

The visioning brainstorms each generated more than 130 raw ideas for innovation in practice elements—from the linking of new material developments for odor-eating clothing to enhanced understanding of what is required for effective bodily hygiene to new norms for community-based food sharing facilitated by social networking and smart technologies. During the closing phase of the workshops, these ideas were clustered and ranked by participants. Following further participant deliberation and voting on these proposals through an online portal, three distinct scenarios were formulated by the research team (elaborated in Doyle and Davies 2013), each representing a new configuration of daily washing, eating, and heating practices. Each scenario implied varying degrees of sociocultural, technological, and organizational change and was represented visually and in "day in the life" narratives explaining how a person would carry out that practice. This involved articulating the underlying cues, motivations, and drivers along with the tools and institutional settings that might be involved in living in this imagined future. Citizen-consumer workshops provided further participatory engagement with the POP backcasting process. Three citizen-consumer workshops were conducted for each practice (washing, heating, and eating) across the Republic of Ireland and Northern Ireland. Participants were drawn from already formed groupings such as community clubs or residents' associations and the workshops were primarily discursive in nature but also included preference indication through voting on preferred scenario proposals. Initial

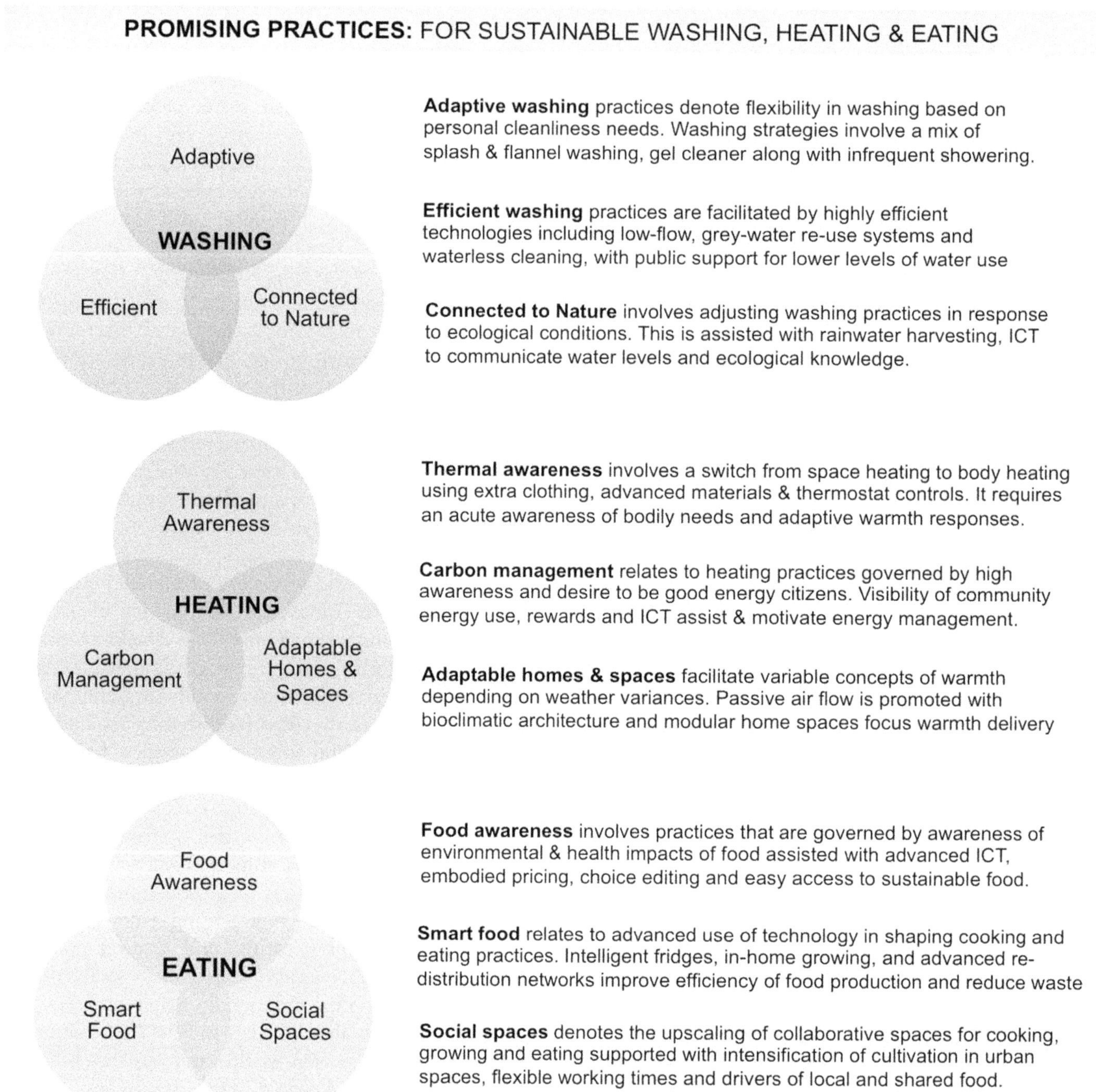

Figure 2. Promising practices for heating, washing, and eating.

personal reactions were recorded individually and then shared among the group to generate discussion. In-depth analyses of these debates and participatory procedures are detailed extensively elsewhere (see Davies [2014] and Davies, Fahy, and Rau [2014]) and were found to reveal important insights on the degrees of willingness to accept reconfigured allocations of responsibility and competency (Jelsma 2003) among people, technology, and government in the performance of the future practices. It was on the basis of the stakeholder

TRANSITION FRAMEWORK EXTRACT
'CONNECTING WITH NATURE' PROMISING PRACTICE

Short-term (2012 - 2020)

- Ⓟ Pilot retrofit for RWH & GWH - link with energy retrofits
- Ⓟ Build skills & accreditation for water retrofit programmes
- Ⓡ R&D for rainwater monitor & rainwater filters
- Ⓔ Myth busting on greywater & rainwater & health risks
- Ⓡ Investment in 'Hydro-nation' economy

Medium-term (2020 - 2035)

- Ⓟ Nationwide retrofit GWH, RWH & rainwater monitors
- Ⓔ Retrofitters provide education on water efficiency
- Ⓟ Building regulations for RWH & GWH systems
- Ⓡ R&D dual water systems to match water quality with use
- Ⓡ R&D for 'Smart water grid'

Long-term (2035 - 2050)

- Ⓟ RWH systems & monitors are mainstreamed
- Ⓟ Dual water systems & GWH mainstreamed
- Ⓟ 'Smart water grid' implemented
- Ⓔ Water use matches local supply availability
- Ⓔ Splash washing, lower cleanliness expectations

LEGEND Ⓟ Policy
Ⓔ Education & Community
Ⓡ Research & Business

Figure 3. Illustrative transition frameworks.

visioning process and citizen-consumer findings that promising practices were identified for each area of study (summarized in Figure 2). These represent alternative strategies to be explored and contain combinations of complementary tools, skills, norms, regulations, and systems of provision.

Each promising practice was distinguished by its promotion of a particular guiding principle or "how-to" rule for that practice; for example, "adaptability," where, in the case of washing practices, people adjust their responses in accordance with natural fluctuations in water availability. This draws on work by Akrich and Latour (1992), whereby everyday objects are considered to reflect and reinforce, or script certain values (e.g., sustainability) and rules of practice (e.g., product design guiding room temperature settings or signaling appropriate frequencies of usage). In this case, however, it was considered that systems of provision and related regulations should also be considered for their capacity to script particular principles; for example, rainwater harvesting systems to build understanding of fluctuations in water supply or variable water charges based on water availability, both of which would encourage and necessitate adaptive washing practices.

Although transitions studies do consider cultural drivers, these are typically ascribed to the landscape level, often considering polarities in cultural trends like localism versus globalism, and cultural conditions affecting acceptance of technology that are typically considered as external forces beyond the remit of influence (Bailey and Wilson 2009). This limited understanding of how cultural norms and motivations are cultivated by everyday systems, services, and technologies was then addressed in POP backcasting through its scripting approach. In the promising practices that emerged in relation to heating, eating, and washing, commonalities can be seen in the scripting of principles of adaptiveness (challenging standardized, unreflective practices), efficiency (rules to curtail resource use and technologies to enable efficient consumption), personal awareness (e.g., of personal bodily warmth, cleanliness, and nutrition needs), ecological connectedness (practices that are responsive to natural limits, seasonal or daily resource fluctuations), and sociality (e.g., where collaborative consumption promotes sustainability). In the final transition phase of the research, stakeholders were invited to brainstorm interventions to build toward the future promising practices that had been identified. During this second phase of brainstorming they were encouraged to think of complementary policy interventions (e.g., economic tools, voluntary codes of practice, or design and building regulations), education and engagement activities, and even new business models (e.g., interactive experiential learning programs, ICT-enabled peer–peer sharing initiatives), as well as research and development strategies nested along a short-, medium-, and long-term timeline. These ideas were discussed among the participants and formulated into transition frameworks for each practice. An illustrative promising practice from one transition framework is provided in Figure 3.

As outlined earlier, although critical analyses of the relative strengths and limitations of both transitions and social practice research processes are relatively well developed, constructive dialogue between the two perspectives remains comparatively rare. Instead, interaction has predominantly been manifest through critique, defense, and clarification of established positions (Rotmans and Kemp 2008; Shove and Walker 2008; Geels 2011). Even where attempts have been made to foster dialogue, the work has either remained at the conceptual level (McMeekin and Southerton 2012) or has been a retrospective analysis of empirical studies focused on current practices rather than on how practices might evolve more sustainably in the future (Hargreaves, Longhurst, and Seyfang 2013). Nonetheless, there appears to be a growing appetite to "forge intellectual bridges ... and [develop] ... a shared language of discourse" (Brown, Vergragt, and Cohen 2013, 3). As such, the POP backcasting experiment presents an explicit attempt to engage with perspectives proposed by both approaches through empirical research in a way that involves creatively and collaboratively imagining how socioecological transformations of core household practices might unfold in the future in a more sustainable fashion. So what lessons can be drawn from this experiment in putting practices at the center of participatory backcasting techniques when the goal is socioecological transformation toward more sustainable consumption practices in the home?

Conclusion: Creating Space for Imagining SocioEcological Transformations

Qualitative surveys were disseminated to more than eighty stakeholders who participated in the backcasting study. These were explicitly designed to evaluate learning processes and stakeholder perceptions of POP backcasting approaches. The responses from the survey indicate that many found the experiment provided them with a new space for interaction, collaboration, and nexus thinking (Davies, Doyle, and Pape 2012). The process brought together stakeholders from different arenas, many of whom are either frequently in conflict (e.g., environmental groups and industry) or operating in dislocated spheres of activity (e.g., community activists and product developers). Although long-established rivalries or poor interpersonal relationships are hard to overcome, the future perspective and the focus on solutions rather than critique within POP backcasting was seen as assisting in the development of a generally cooperative ethos among participants. As found in the work of Quist, Thissen, and Vergragt (2011), such benefits may be found in many participatory backcasting processes, but did the practice orientation bring any additional benefits to the interactions?

Certainly, participants mentioned that reorienting discussions away from end-of-pipe environmental problems (e.g., water shortages or climate change) toward questions of the needs that practices fulfill enabled them to drill down to the underlying drivers of unsustainable consumption in ways that took account of the manifold interactions of actors, agencies, and technologies. Rather than simply extrapolating current behaviors (and hence resource consumption) into a technologically enhanced "tomorrow's world," the approach permitted attention to how more sustainable practices (new and familiar) might be developed to meet or indeed disrupt current needs and desires. Such reframing also, at least to some extent, diffused potential tensions between stakeholders precisely because it removed the focus from particular industries, products, or services and concentrated instead on how needs might be met differently in the future through the combined influences of new technologies, policies, or social change. This solutions-oriented perspective encouraged participants to think about how practices might be recrafted such that their resource-intensive elements might be addressed; for example, how less sustainable elements embedded within practices might be substituted by more sustainable ones and how practices might interlock, as is the case between personal washing and home heating (Spurling et al. 2013). Essentially, beyond the interpersonal learning, capacity-building, and networking that occurred both during and beyond the workshops, taking practices as the fundamental unit of analysis created space for participants to rethink the ways in which they engaged with heating, washing, and eating. Such problem and solutions redefinition suggested higher order learning (Davies, Doyle, and Paper 2012), said to allow space for the emergence of behavioral or procedural alternatives (Quist, Thissen, and Vergragt 2011) and potentially increasing the likelihood of adoption of the backcasting proposals.

Yet, as with all participatory processes, POP backcasting experiments struggle to ensure complete inclusivity or representativeness. Nor do the resulting outputs necessarily have the authoritative and affective power to transform socioecological systems. For example, the formation of the new centralized state body to manage water provision in Ireland (Irish Water) by its very creation allocates significant power to actors within the organization (some of whom were involved in the POP backcasting process) to form new directions for the collection, treatment, and provision of water to householders. Decisions and actions by Irish Water will inevitably contribute to the shaping of household washing practices, particularly through proposed mechanisms for water charging. It would be simplistic, however, to assume that the activities of Irish Water alone shape how and why people wash. The horizontal circulation of multiple messages about health and hygiene, along with promises of new bathing sensations, are constantly relayed through advertising by geographically dispersed commercial actors seeking to compete for enhanced market share. These deliberate and explicit interventions are [re]interpreted through peer networks (online and offline) and familial socialization processes, which in turn could feed into future market research and development for washing products and services. All of these actors, including citizen-consumers, as well as the devices and regulatory frameworks they formulate and implement, contribute to the way practices endure or change over time. Not only do they affect the way that practices are performed, they could also—through combination, tension, and sometimes contradiction—create more widely experienced shifts in the practices of washing, heating, or eating.

The POP backcasting experiment detailed in this article explicitly created spaces for the delineation, discussion, and debate of coordinated interventions expressly for the purpose of reducing the material intensity of everyday practices that involve the use of constrained resources but also seeking to build social capacity and economic security. It was ultimately an experiment in collaboration for future-oriented governance of everyday household activities with an explicit focus on sustainability innovation, which adopts social practices as the basic unit of enquiry. How much further POP backcasting gets us along the road to actualizing more sustainable consumption depends on the collective imagination and actions of multiple actors (in organized and informal processes) to think about and enact new modes of interaction and hence social relations across spheres of society (public, private, and civil society); to untangle the complex and often global forces lying behind production and consumption patterns; and to face head-on the ways in which sociotechnical and socioecological interrelations play out in quotidian household practices.

As it is changes in practices rather than the articulation of visions and plans that will ultimately lead to socioecological transformations, additional research is necessary to experiment with implementing the promising practices identified and to evaluate the resulting outcomes. Building on the POP backcasting visions and transition frameworks, the HomeLabs project (see http://www.consensus.ie/homelab) responds to this need by drawing inspiration from collaborative "Living Lab" strategies taking place between industry and research institutes (Green 2007), as well as academic studies that are actively prompting altered practice performances (e.g., Higginson, Thomson, and Bhamra 2014). The HomeLabs are constituted by combined information, technology, and regulatory interventions that are introduced into households and then tested and evaluated by householders in conjunction with the research team for their capacity to support practice transformation. For example, during the washing HomeLabs, households are being provided with information designed to promote understanding of the origins of their water supply (e.g., location information for reservoirs or aquifers, photographs, data on capacity), targets or rules for reducing water use for washing (based on international guidelines for sustainable water use), enabling tools to enhance visibility and consciousness of water use in real time (smart water meters, timers), and products to assist in meeting the needs associated with personal washing in ways that are not so consumptive of water (from spot cleaning to dry shampoos). Within the HomeLabs experiment, the researcher acts as a filter, providing information on, for example, ways in which washing needs might be achieved with lower resource use and also how needs that are currently met through the use of water but that do not necessarily require water (e.g., to wake up in the morning or relax in the evening) could be satisfied by other nonconsumptive means (e.g., mindfulness or stretching). The researcher also acts as a facilitator, prompting discussion about the implications of alternative ways of meeting washing needs. Given the prototypical nature of many enabling devices for more sustainable consumption, the researcher plays a pivotal role as an interface between these technologies and the participants, particularly in the ICT-led smart water meter arena. Enacting such experimental HomeLabs affords another layer of meaning to better understand the relationship between sociotechnical change and its socioenvironmental consequences. It critiques the assumption that radical technical change takes place "in the context of relative social stasis, rather than technological and social change being interwoven through social practices" (Spurling et al. 2013, 7) and puts attention to the mechanics and cultures of everyday practice back on a par with novelty and innovation transfer (Hargreaves, Longhurst, and Seyfang 2013). Nevertheless, a challenge for these experiments, and indeed for all research projects focused on both understanding consumption practices and seeking to provide assessments of interventions to shift those consumption practices in more sustainable directions, is how positive outcomes identified in bounded experimental sites might be rolled out. The intensity of human resources involved in establishing, running, and evaluating HomeLabs-style initiatives means a simple scaling up nationwide is unrealistic. Equally, issues of power, knowledge, and politics course through the veins of everyday life, easily derailing any assumption that findings from one setting can be simply replicated in another. Rather than seeing HomeLabs themselves as outcomes to be rolled out in this way, we argue instead that they are more usefully seen as test beds for grounding and interrogating collaboratively designed scenarios for more sustainable household consumption futures. Further research and analysis is required before it is possible to say whether particular configurations of interventions (including the crucial roles of actors as filters, facilitators, and interlocutors) have wider potential to disrupt unsustainable household consumption practices in different settings. Nonetheless, the risks of untrammeled household consumption are so great, and gains made from governing approaches to date are so limited, that experimenting with different ways of approaching consumption governance is essential.

Acknowledgments

Thanks go to all of the research participants who contributed to the backcasting and HomeLabs experiments.

Funding

We would like to acknowledge funding from the Environmental Protection Agency of Ireland through the STRIVE Research Programme that supported the research on which this article is based: Project Number 2008-SD-LS-1-S1.

References

Akrich, M., and B. Latour. 1992. A summary of a convenient vocabulary for the semiotics of human and nonhuman assemblies. In *Shaping technology, building society: Studies in sociotechnical change*, ed. W. Bijker and J. Law, 259–64. Cambridge, MA: MIT Press.

Anderson, B. 2010. Preemption, precaution, preparedness: Anticipatory action and future geographies. *Progress in Human Geography* 34:777–98.

Bailey, I., and G. Wilson. 2009. Theorizing transitional pathways in response to climate change: Technocentrism, ecocentrism, and the carbon economy. *Environment and Planning* A 41:2324–41.

Brown, H., P. Vergragt, and M. Cohen. 2013. Societal innovation in a constrained world: Theoretical and empirical perspectives. In *Innovations in sustainable consumption: New economics, socio-technical transitions and social practices*, ed. M. Cohen, H. Brown, and P. Vergragt, 1–30. Cheltenham, UK: Edward Elgar.

Davies, A. R. 2014. Co-creating sustainable eating futures: Technology, ICT and citizen-consumer ambivalence. *Futures: The Journal of Policy, Planning and Futures Studies* 62:181–93.

Davies, A. R., R. Doyle, and J. Pape. 2012. Future visioning for sustainable household practices: Spaces for sustainability learning? *Area* 44 (1): 54–60.

Davies, A. R., F. Fahy, and H. Rau. 2014. *Challenging consumption: Pathways to a more sustainable future*. London and New York: Routledge.

Davies, A. R., F. Fahy, H. Rau, and J. Pape. 2010. Sustainable consumption and governance: Reflecting on a research agenda for Ireland. *Irish Geography* 43:59–79.

de Goede, M., and S. Randalls. 2009. Precaution, preemption: Arts and technologies of the actionable future. *Environment and Planning D: Society and Space* 27:859–78.

De Gues, M. 2002. Ecotopia, sustainability and vision. *Organization and Environment* 15:187–201.

Doyle, R., and A. R. Davies. 2013. Towards sustainable household consumption. *Journal of Cleaner Production* 48:260–71.

Eames, M., and J. Egmose. 2011. Community foresight for urban sustainability: Insights from the Citizens Science for Sustainability (Suscit) project. *Technological Forecasting and Social Change* 78 (5): 769–84.

Fischer-Kowalski, M., and J. Rotmans. 2009. Conceptualizing, observing, and influencing social–ecological transitions. *Ecology and Society* 14:3–20.

Geels, F. W. 2010. Ontologies, socio-technical transitions (to sustainability) and the multi-level perspective. *Research Policy* 39:495–510.

———. 2011. The multi-level perspective on sustainability transitions: Responses to seven criticisms. *Environmental Innovation and Societal Transitions* 1:24–40.

Green, J. 2007. *Democratizing the future: Towards a new era of creativity and growth*. Philips Electronics N.V.

Hargreaves, T., N. Longhurst, and G. Seyfang. 2013. Up, down, round and round: Connecting regimes and practices in innovation for sustainability. *Environment and Planning* A 45:402–20.

Higginson, S., M. Thomson, and T. Bhamra. 2014. "For the times they are a-changin": The impact of shifting energy-use practices in time and space. *Local Environment* 19:520–38.

Jansen, L. 2003. The challenge of sustainable development. *Journal of Cleaner Production* 11:231–45.

Jelsma, J. 2003. Innovating for sustainability: Involving users, politics and technology. *Innovation: The European Journal of Social Sciences* 16:103–16.

Kok, K., L. Van Vliet, M. Dubel, J. Sendzimir, and I. Bärlund. 2011. Combining participative backcasting and exploratory scenario development: Experiences from the SCENES project. *Technological Forecasting and Social Change* 78:835–51.

Kuijer, L., and A. De Jong. 2012. Identifying design opportunities for reduced household resource consumption: Exploring practices of thermal comfort. *Journal of Design Research* 10 (1–2): 67–85.

McMeekin, A., and D. Southerton. 2012. Sustainability transitions and final consumption: Practice and socio-technical systems. *Technology Analysis and Strategic Management* 24:345–61.

Meadowcroft, J. 2009. What about the politics? Sustainable development, transition management, and long term energy transitions. *Policy Sciences* 42 (4): 1–18.

Pantzar, M., and E. Shove. 2010. Understanding innovation in practice: A discussion of the production and re-production of Nordic walking. *Technological Analysis and Strategic Management* 22:447–61.

Quist, J., M. Knot, W. Young, K. Green, and P. Vergragt. 2001. Strategies towards sustainable households using stakeholder workshops and scenarios. *International Journal of Sustainable Development* 4 (1): 75–89.

Quist, J., W. Thissen, and P. Vergragt. 2011. The impact and spin-off of participatory backcasting: From vision to niche. *Technological Forecasting & Social Change* 78:883–97.

Rip, A., and R. Kemp. 1998. Technological change. In *Human choice and climate change: Resources and technology*, ed. S. Rayner and E. L. Malone, 327–99. Columbus, OH: Batelle Press.

Rotmans, J., and R. Kemp. 2008. Detour ahead: A response to Shove and Walker about the perilous road of transition management. *Environment and Planning* A 40:1006–12.

Scott, K., C. Bakker, and J. Quist. 2012. Designing change by living change. *Design Studies* 33 (3): 279–97.

Shove, E., and G. Walker. 2007. CAUTION! Transitions ahead: Politics, practice and sustainable transition management. *Environment and Planning A* 39:763–70.

———. 2008. Transition Management^{TM} and the politics of shape shifting. *Environment and Planning A* 40:1012–14.

———. 2010. Governing transitions in the sustainability of everyday life. *Research Policy* 39:471–76.

Shove, E., G. Walker, and S. Brown. 2014. Transnational transitions: The diffusion and integration of mechanical cooling. *Urban Studies* 51 (7): 1506–19.

Shove, E., M. Watson, J. Ingram, and M. Hand. 2008. *The design of everyday life*. New York: Berg.

Spaargaren, G. 2003. Sustainable consumption: A theoretical and environmental policy perspective. *Society and Natural Resources* 16 (8): 687–701.

SPREAD. 2013. *SPREAD sustainable lifestyles 2050*. Brussels, Belgium: European Commission FP7 Project. http://www.sustainable-lifestyles.eu/

Spurling, N., A. McMeekin, E. Shove, D. Southerton, and D. Welch. 2013. Interventions in practice: Re-framing policy approaches to consumer behaviour. *SPRG Report 2013*. http://www.sprg.ac.uk/ (last accessed 27 February 2014).

Svenfelt, A., R. Engström, and O. Svane. 2011. Decreasing energy use in buildings by 50% by 2050—A backcasting study using stakeholder groups. *Technological Forecasting and Social Change* 78:785–96.

Tukker, A., and D. Fedrigo. 2009. *Blueprint for European sustainable consumption: Finding the path of transition to a sustainable society*. Brussels, Belgium: EEB.

United Nations. 2013. Latest IPCC findings a clarion call for global community to accelerate efforts to combat climate change and steer humanity out of danger zone. United Nations Official Press Release on IPCC 5AR.

Warde, A. 2005. Consumption and theories of practice. *Journal of Consumer Culture* 5 (2): 131–53.

Index

Note: **Boldface** page numbers refer to figures and tables, Page numbers followed by "n" denotes notes.

For Product Safety Concerns and Information please contact our EU
representative GPSR@taylorandfrancis.com
Taylor & Francis Verlag GmbH, Kaufingerstraße 24, 80331 München, Germany

www.ingramcontent.com/pod-product-compliance
Ingram Content Group UK Ltd.
Pitfield, Milton Keynes, MK11 3LW, UK
UKHW052053230926
475702UK00025B/1176

* 9 7 8 1 1 3 8 6 6 8 6 9 0 *